THE POISON PARADOX

'. . . and she had never forgotten that, if you drink much from a bottle marked "poison", it is almost certain to disagree with you, sooner or later.'

From *Alice's Adventures in Wonderland*, by Lewis Carroll

THE POISON PARADOX

Chemicals as friends and foes

JOHN TIMBRELL

OXFORD
UNIVERSITY PRESS

Great Clarendon Street, Oxford OX2 6DP

Oxford University Press is a department of the University of Oxford.
It furthers the University's objective of excellence in research, scholarship,
and education by publishing worldwide in

Oxford New York

Auckland Cape Town Dar es Salaam Hong Kong Karachi
Kuala Lumpur Madrid Melbourne Mexico City Nairobi
New Delhi Shanghai Taipei Toronto

With offices in

Argentina Austria Brazil Chile Czech Republic France Greece
Guatemala Hungary Italy Japan Poland Portugal Singapore
South Korea Switzerland Thailand Turkey Ukraine Vietnam

Oxford is a registered trade mark of Oxford University Press
in the UK and in certain other countries

Published in the United States
by Oxford University Press Inc., New York

First published 2005

British Library Cataloguing in Publication Data
Data available

Library of Congress Cataloging in Publication Data
Data available

Typeset by RefineCatch Limited, Bungay, Suffolk
Printed in Great Britain
on acid-free paper by
Clays Ltd., St Ives plc

ISBN 0–19–280495–2 978–0–19–280495–2

1

For Anna, Becky and Cathy

Preface

To appreciate the dangers and the risks from chemicals of all kinds it is necessary to understand how, when and why they are toxic. This falls within the field of toxicology and is what this book is all about.

I was encouraged to write the book by Dr John Emsley, who is well known for his popular science books on the chemical sciences. He had reviewed my introductory textbook on toxicology and suggested it might be the basis for a popular science book written for the general public. That is what I have tried to produce.

I have approached the task of writing this book by using examples known to me through my teaching and research. The examples have been chosen to illustrate particular points and principles or because they are, I think, interesting stories.

However, the book is not one to consult for a review of the hazards of a particular drug or pesticide or industrial chemical. Such a book, covering all of the possible chemicals to which we might be exposed would be a reference text in many volumes and would be a lifetime's work to produce and yet probably out of date when it arrived!

This book is also not yet another attempt to worry people unnecessarily about the potential chemical dangers around them, but neither will it suggest that every chemical is perfectly safe.

I hope by the end of this book that you, the reader, will have a better understanding of these things and so can more easily appreciate the often conflicting and unsettling information with which we are all increasingly assailed. It is my belief that members of the public need objective, scientific information in order to make up their own minds and balance the risks with the benefits.

Using this book

Although I have tried to keep jargon and technical terms to a minimum, they are at times necessary for proper explanation. So throughout the book if not explained at the point of first use, such words will be highlighted in bold. This means the word will be explained in the glossary at the end. More detailed explanations of certain points are given in boxes in the text.

Finally I must acknowledge the help of a number of people. Firstly

David and Lorraine Matthews who kept the idea of the book alive for so many years and Andrew Clay for his crucial help in contacting the publisher. I have already mentioned the encouragement of John Emsley, who also offered constructive comments on outlines and a draft chapter. Special thanks to Professor Iain Purchase, who was kind enough to look at the whole manuscript and made many critical and informed comments.

Thanks also to Professors Andy Renwick and Juoko Tuomisto and Dr Sue Barlow for providing me with specific information at critical times. And lastly special thanks to Cathy for her unfailing support and Anna, Becky, Nick and Jon who have offered many useful suggestions and comments on sections they have taken the time to read.

John Timbrell
February 2005

Contents

List of Tables

List of Figures

1

Poisons

Old Art, New Science

TOXICOLOGY is the science of poisons but what is a poison? The word 'poison' conjures up many different ideas in people's minds and is often associated with the word 'chemical'. Consulting various dictionaries, I found several definitions which were basically similar: 'A substance that when introduced into or absorbed by living organisms causes death or injury—especially one that kills by rapid action even in small quantities'.[1] Another definition had an additional concept: 'Any substance which taken into or formed in the body which destroys life or impairs health'.[2] Both definitions tell us that a poison is a substance that is hazardous to life, possibly lethally so. The first indicates that a poison is also likely to be potent and the second that it might be naturally produced.

According to another source, in the United States a poison is defined legally as a substance which is lethal at a dose of 50 mg per kg of body weight or less.[3] This means that ¾ of a teaspoon of the substance could kill a normal man.

Dosages are often given as 'milligrams (mg) per kilogram (kg) body weight'. This is because doses of therapeutic drugs given to humans are often calculated on the basis of their body weight. Therefore, as an average man weighs 70 kg, a lethal dose of 50 mg/kg is equivalent to $50 \times 70 = 3{,}500$ mg.

So poisons are substances that can be lethal—but what are these substances? They are chemicals of all kinds, both synthetic and natural—but are all chemicals poisons? The answer is: it depends on a number of things, especially the amount of the substance given. This is best illustrated by a 400-year-old quotation:

All substances are poisons; there is none that is not a poison. The right dose differentiates a poison from a remedy.

Paracelsus (1493–1541)

This is an underlying theme of my book, and it is crucial to an understanding of chemicals, of their hazardous effects and their safe use.

The word 'toxicology' is derived from the Greek words *toxikon* (arrow poison) and *toxikos* (the bow). Poisons were known about and used in warfare from the earliest times. However, there were other reasons for interest in poisons, and study of them had begun by 1500 BC. The Ebers Papyrus, the earliest medical records, include references to and recipes for poisons. Many of these are recognizable, such as hemlock, the plant toxin that the Greeks used for the execution of Socrates in 399 BC. Other substances listed are aconite, which was used as an arrow poison by the Chinese, and poisonous metals such as **lead** and antimony. Opium, which was used for both poisonous and beneficial purposes, has been known for at least 5,000 years.

Many references to poisons are found in the literature and mythology of ancient Greece, where a more rational treatment for poisoning started to emerge. Poisoning was relatively common in ancient Greece and so the treatment of the effects and the use of **antidotes** became important. One of the first to apply rational principles to poisoning was Hippocrates who, around 400 BC, certainly understood the importance of reducing the absorption of a substance from the **gut** in the treatment or alleviation of poisoning.

Legend has it that King Mithridates VI of Pontus (132–63 BC) used criminals to search for antidotes to poisons and regularly protected himself with a mixture of many of these. Unfortunately, when he wanted to poison himself his antidotes were still active and so he had to fall on his sword instead! The term 'mithridatic', meaning antidote, is derived from his name. Another such person was the physician Nicander of Colophon (185–135 BC) who also experimented on criminals in order to discover antidotes. In his treatise on antidotes he mentioned a number of poisons, including ceruse (white lead), litharge (lead oxide), aconite (wolfsbane), cantharides, conium (hemlock), hyoscyamus (henbane), and opium.

In Roman times poisoning became, if anything, more common, when a conspiracy of women poisoners began practising the art for profitable purposes. In 82 BC a law against poisoning was passed by Sulla. This eventually became a protection against the careless dispensing of drugs. Both Agrippina, wife of Claudius, and Livia, wife of Augustus, used poisoning and employed belladonna (atropine), which was derived from deadly nightshade. Livia eventually killed Augustus by injecting the poison into figs on his personal tree.

While many of the poisons used at this time were of plant origin, **arsenic** compounds were also known and used. For example, Claudius and his natural son Britannicus were assassinated by Agrippina using arsenic. A devious plan was adopted after an earlier poisoning attempt had failed to kill Britannicus, but had made him ill instead. Suspicion having been aroused, a taster for the food was employed. In order to circumvent this, very hot soup (which had already been tested by the taster) was served to Britannicus while the arsenic was added to the water supplied to cool the soup.

An important milestone was the Materia Medica, a text produced by Dioscorides in AD 50 in which he classified poisons as animal, plant, or mineral, described them, and included drawings. This remained one of the major sources of information on poisons for sixteen centuries. Dioscorides also recognized the importance of **emetics**, which induce vomiting, in the treatment of poisoning.

It was the Italians who, in the Middle Ages, developed the art of poisoning for political purposes, financial gain, and marital reasons. The records of city councils, for example in Florence and Venice, testify to the use of poisoning in which victims were named and the prices for their removal given. Apart from the use of poisoning for political ends, there were those who engaged in the practice for other reasons, for example a lady by the name of Tofana, produced and sold 'Aqua Tofana' (which contained arsenic), with accompanying instructions. Another femme fatale, Hieronyma Spara, developed the practice even further, in particular for marital and financial purposes. A club of wealthy young women was formed, with the intention of removing unwanted husbands, a practice not unlike that in ancient Rome.

Members of the Borgia family, especially Cesare and Lucretia, were active poisoners in Rome in the Middle Ages and the papacy was probably an important financial beneficiary. A relative, Catherine de Medici, practised as a poisoner in France and was probably one of the first experimental toxicologists. Purporting to be engaged in charitable work, she experimented with her poisons on the poor and sick. She was able carefully to record important observations such as how rapidly the poison took effect, what parts of the body were affected, the symptoms in relation to the potency of the preparation of the toxic agent.

Another individual who contributed to the treatment of poisoning was Maimonides, a physician who lived in the twelfth century. His *Treatise on Poisons and their Antidotes* was particularly significant for the time, due to its recognition both of the effect of oily or fatty food in reducing the absorption of poisons from the stomach and of the use of a tourniquet on a limb to reduce the effect of an animal bite or sting.[4]

Apart from their documented use thoughout history, poisons were also mentioned in literature, for example in Shakespeare's *Macbeth*: 'Double, double toil and trouble; . . . Root of hemlock, digg'd i' th' dark . . .'. In Flaubert's *Madame Bovary*, the victim was famously poisoned by arsenic, and arsenic again featured in the poem *Under Milk Wood*, by Dylan Thomas, in which arsenic biscuits were mentioned.

The use of chemicals to dispatch enemies is not the sole prerogative of humans. Animals and plants have also adopted what is known as chemical warfare. Both animals and plants, as well as bacteria and fungi, can produce and contain some of the deadliest chemicals for the purpose of discouraging a predator or killing a potential meal. We come across such poisons in our everyday lives in the form of ant bites and wasp, bee, and nettle stings. In some countries the indigenous plants and animals may be especially hazardous, as we shall discover in Chapter 6.

All the substances that cause these unpleasant and possibly lethal effects are chemicals, albeit manufactured by a plant, micro-organism, or animal. They may be simple irritant chemicals such as the formic acid in ant bites (*formica* is the Latin for ant), or complex **protein** molecules such as is found in bee venom. Proteins are relatively large **molecules**, one of the main building blocks of the body and also the main component of **enzymes** (biological catalysts). The venom of animals such as snakes often contains enzymes which degrade flesh. Mushrooms and toadstools are another source of poisonous chemicals, for example the Death Cap mushroom found in Britain which can be lethal if eaten.

Thus humans have learnt to avoid eating plants that contain poisonous chemicals and to steer well clear of venomous animals. When so avoided, the chemicals in the plant or animal are no longer a significant **risk**.

The same approach can and should be applied to man-made chemicals. A healthy respect for and understanding of them allows them to be used safely. Man-made as well as naturally occurring chemicals impinge on almost every aspect of our lives, in most cases to our benefit. That is not to say there is no risk, but we simply have to minimize this and accept that there will always be some risk, however small.

Apart from the deliberate use of poisons for murder and assassination, man's use of chemicals was also associated with poisoning, for example, in the mining of metals such as **mercury** and lead and their subsequent smelting and working. Mercury, found both as the pure metal and as an ore, was mined in Idrija in Slovenia from the sixteenth century, and some of the toxic effects of the metal to the nervous system were probably known to the miners and their families (see pp. 110–118, 166–7). The occupational diseases of miners and the long-term effects of mercury were first described and documented by Paracelsus, who was perhaps the most important figure in the subject of toxicology. He appreciated the importance of experimentation, understood the significance of the dose and that the difference between the therapeutic and toxic effects of chemicals may be a difference of dose, and knew that chemicals could cause very specific effects. It has been only very recently, however, that the study of the toxic effects of chemicals has been truly scientific and concerned with more than descriptions of poisoning.

Humans have been aware of and have used potentially poisonous chemical substances for thousands of years. Most of them were derived from plants or occurred naturally in rocks. Only relatively recently have new substances been synthesized most of which are unknown in nature. In recent years we have become accustomed to headlines in our newspapers such as 'Chemical company poisons our water', 'The poisons in our food', 'Poison oil scandal', and so on. In the popular mind the words 'poison' and 'chemical' have become synonymous, which has led to what Alice Ottoboni calls 'poison-paronoia' or 'news media toxicology'. According to Edith Efron in her book *The Apocalyptics*, this unreasonable fear of chemicals probably started in 1976 when the Administrator of the Environmental Protection Agency in the United States told newsmen that 'Most Americans had no idea, until relatively recently, . . . they were often engaging in a grim game of chemical roulette whose result they would not know until many years later'.[5]

We shall revisit these concerns later, but first we need to find out what the word 'chemical' means. What are chemicals?

Chemicals 'R' us

A chemical can be natural or synthetic. There is nothing intrinsically different between a natural and a synthetic chemical and they can be equally hazardous. To a scientist a chemical is a collection of **atoms**, ranging from one or two to hundreds of thousands. Collections of atoms are called molecules, thus H_2O (water) is a molecule composed of two hydrogen atoms and one oxygen atom. Water is a chemical and a very unusual one at that. Without it life as we know it would be impossible.

Some chemicals are well known to us and vital, such as oxygen, salt, water, and sugar, although many people may not think of them as

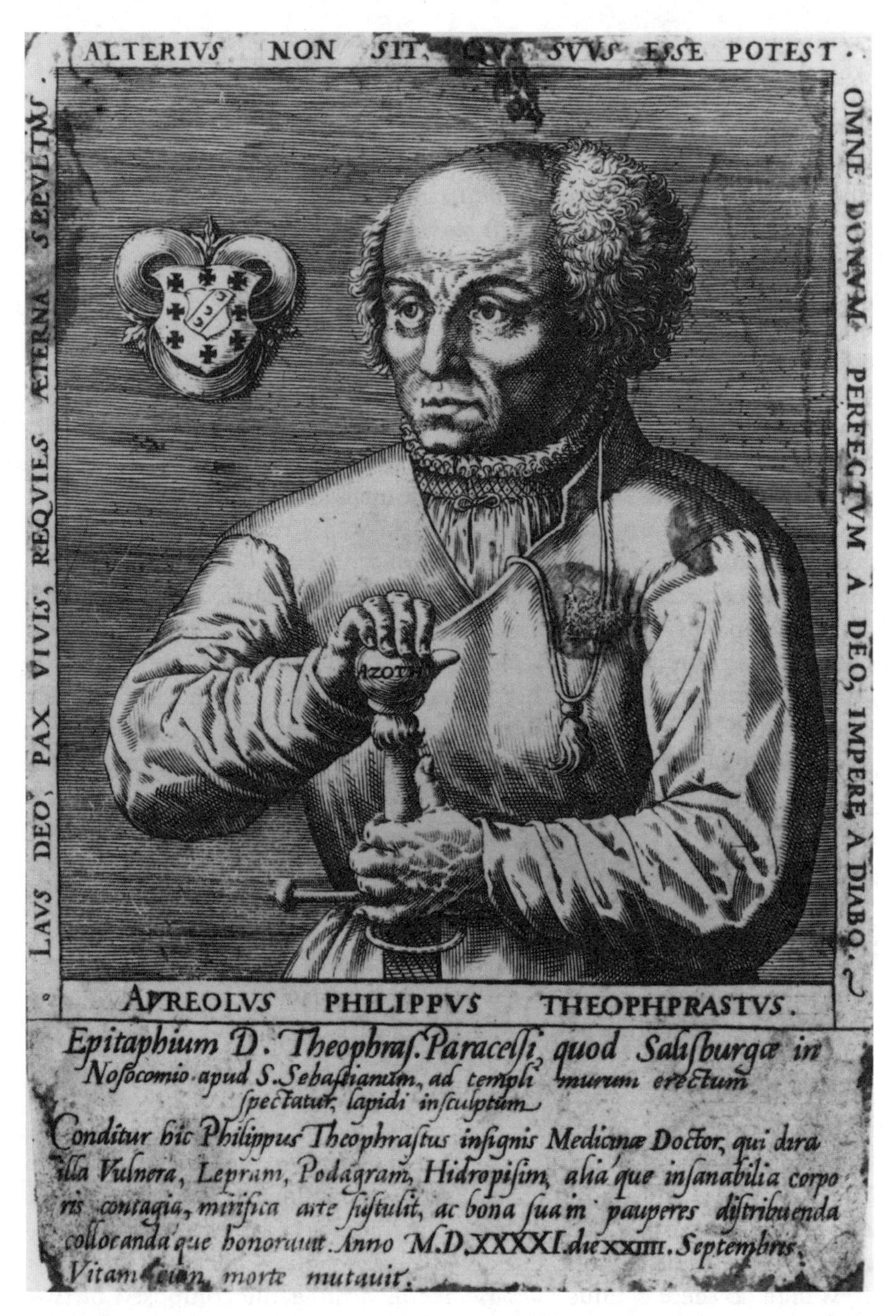

1. Paracelsus. This Swiss physician and scientist, born in 1493, questioned and rejected the irrational medicine of his time. He realized especially the crucial importance of dose in relation to both the adverse and the beneficial effects of chemicals.

chemicals. The world is made of chemicals, some simple such as water or salt, others very complex such as **DNA**, which is found in virtually every **cell** of every living organism on earth. Our bodies are composed of chemicals that range from simple to complex, and use chemicals as part of the living process. Chemicals are an integral part of us. Chemicals really 'R' us! Hence we should not be scared of chemicals.

Chemicals come in all shapes and sizes and derive from a variety of sources. While to many people the words 'chemical' and indeed 'poison' are synonymous with man-made things such as **dioxin**, organophosphates, pesticides, and nerve gas, there are many, many other chemicals that occur naturally. These may be produced by plants, for example belladonna in deadly nightshade, and ricin in the castor bean, or bacteria, for example botulinum toxin. Various animals such as snakes and spiders produce toxic venoms, and some poisons occur naturally in the ground, like **asbestos, cadmium**, and lead.

Just as chemicals come in all shapes, sizes and origins, so our exposure to them depends on circumstances. We may eat them, inhale them, drink them in our water at home, use them in the garden or at work. We may imbibe them intentionally, as with sugar in our coffee, salt in our food, and alcohol in our vodka and tonic. But surely some chemicals, such as those made by man, are harmful to us, you may argue, as we read in the newspapers? The answer is yes, in some cases they are, but it depends on the circumstances. A chemical is not harmful just because it is man-made nor is it harmless just because it is natural. As we shall see, some chemicals (both man-made and natural) are relatively harmless to some kinds of living organisms but harmful to others. We live in a world surrounded by chemicals and we ingest thousands of chemicals all the time, most of which are natural (99.9 per cent is one estimate); we do not live in a 'sea of pollution', as some headlines would have us believe. Some of these chemicals may be potentially hazardous or even poisonous but most are benign and some even beneficial.

What do the chemicals that are not part of our bodies do to us? Do they cause us harm? Where do they come from? Can we avoid them? The answer to these questions is again 'It depends': on where we live, what we eat, where we work, and who we are. It is one aim of this book to try to answer some of these questions.

Where do these chemicals come from? Most chemicals that are foreign to the body and potentially poisonous (known as xenobiotics, from *xeno*, foreign, *bios*, life) are found in the food we eat, whether we are vegetarians or meat-eaters. Many will be natural constituents of vegetables and fruits, while others may be naturally occurring contaminants or produced by cooking. Because humans and their ancestors have been exposed to many of these chemicals for millions of years, the body has developed ways of detoxifying them, generally ridding itself of them as quickly as possible. Humans have also learnt what plants are hazardous and so avoid exposure to some of the more poisonous chemicals.

There may, however, be additives in our food as well, depending on where we live and what we eat, and some of our food may also contain man-made contaminants such as pesticides or synthetic hormones. The water we drink may contain metals and other substances which occur naturally in the rock through which the water has flowed. Water may also contain pesticides and fertilizers from agricultural land, industrial chemicals from factory effluent and chemicals from sewage.

Another source of chemicals to which we can be exposed, apart from food, is our workplace. This could be a factory where **solvents** or metals are used or chemicals synthesized, a printing press, a photographic film developer or a scientific laboratory. Even workers in offices can be exposed to photocopier toners, for example. Then there are the chemicals we knowingly imbibe, such as drugs, stimulants like coffee, and of course alcoholic drinks.

Thus we are exposed to many chemicals on a daily basis, but what do they do to the body and do they cause us real harm? The purpose of this book is to explore these concerns and to try to answer the questions. Chemicals are therefore an important and integral part of our everyday lives. So why do they get such a bad press and why do they conjure up visions of poisoning, pollution, and **hazard** in the mind of the general public?

Chemicals are often portrayed by the media as dangerous and poisonous, and rarely are their beneficial effects reported. There are many reasons for this. It may be partly due to problems that have occurred but, as suggested by Edith Efron, it may also have resulted from certain members of regulatory bodies and scientists misinterpreting data or possibly deliberately exaggerating the dangers for their own ends. After all, disasters make much better news than stories about positive achievements. There have been films and documentaries depicting workers fighting chemical industries for compensation for illness and death from exposure to substances such as chromium, as featured in the film *Erin Brockovich*. While this was a true story, a film like this adds to the public prejudice about chemicals. It is to be hoped that another message also comes across: that the management of such companies not only has scant regard for their workers but also no respect for the chemicals they use, and that this is the real problem.

Chemicals can certainly be hazardous and some are extremely dangerous, but a balanced view is necessary. This can be achieved only by understanding how, why, and under what conditions chemicals are toxic, and the true nature of risk. One of the purposes of this book is to show what these conditions are, to explain how chemicals are toxic, and conversely how they can be used safely.

The principle of Paracelsus that the dose makes the poison is paramount. It means that all chemicals, from whatever source, human manufacture or natural, are potentially toxic at some dose. This is a fundamental concept which underlies toxicology and is crucial to the assessment of risk from chemicals and their safe use. The corollary to this principle is that all chemicals are potentially safe at some, perhaps extremely low, dose and therefore most can be used safely. The relationship between the dose and toxic effects of chemicals will be discussed later (see pp. 36–9, 298–301).

Chemicals are used, as drugs, to heal us when we are sick; as plastics, they are employed in a range of products or their components which we take for granted; as preservatives and sweeteners, they are added to our food; we use them to clean and sanitize our kitchens and as antiseptics; they are used to dye cloth and to brighten our lives, and we use them, as perfumes, to make us smell more appealing. In agriculture chemicals are used as fertilizers and pesticides to increase crop production and reduce starvation and to decrease disease borne by insects. They fuel our cars and aircraft, allowing us to travel.

The widespread use of chemicals in our everyday lives requires a chemical industry to manufacture the end product and other chemicals as intermediates and starting points. Some of these chemicals are potentially dangerous, such as corrosive acids and gases, and some of the by-products are poisonous, such as dioxin; but they and the end products can be used and handled safely. If the enclosed systems in which they are used are well maintained, the risks are minimized. If the employees using them are given protective clothing and factories are sited away from urban areas, the chances of adverse effects will be greatly reduced. The chemicals that are designed to be toxic, such as pesticides, must be stored in the right conditions and handled carefully, with protective clothing and in the correct amounts; then the risks can be low and acceptable. Chemicals should be used for the correct purpose, and not be assumed to be harmless.

As we shall see later in the book, chemical disasters have occurred as a result of unsafe practices or of misuse. This brings me back to an earlier point: chemicals must be used with respect.

Are we all at risk of developing incurable diseases such as cancer from living in a sea of dangerous chemicals, as some would have us believe? I do not think so but it may suit the purpose of some people to promote this idea. It also may be a reflection of our relatively safe and affluent lives. These ideas were discussed in a recent article in the *New York Times*:

> Spared from worry about whether they will have enough to eat today or a roof over their heads tomorrow, most Americans have the luxury of worrying about the hazards that may be lurking in the air, water and food as a result of all this progress and affluence. We are healthier, live longer, have more sources of pleasure and convenience and more regulation of industrial and agricultural production than ever.

This scenario now applies to many Western industrialized countries as well as the USA:

> Remember too that natural is not necessarily safe and just because something is manufactured does not make it a potential hazard. Nature is hardly benign.[6]

The conclusion of this article is that people do worry unnecessarily and fail to understand the risks of different activities, and that perhaps billions of dollars are spent on regulation or other measures which could be better spent elsewhere. This 'poison paranoia' means that the public is understandably confused, but the doomsayers have cried wolf so often that when a real toxic hazard is present it may be missed. The constant bombardment by the media of claims and counter-claims, raising concerns about contaminants and additives in food or pesticides in breast milk often serves only to confuse and cause either apathy or anxiety. Some of these issues will be visited again in the final chapter.

2

Bodily Functions

What Chemicals Do to Us and What We Do to Them

All of us are exposed to many potentially hazardous chemicals every day, as we saw in the last chapter. How are we exposed to these chemicals? How do they gain entry into our bodies? What do they do to our bodies once they are inside? Do they cause us real harm or should we not worry about them?

The answer to these questions is 'It depends': on where we live, what we eat, where we work, and who we are. In this chapter I shall try to answer these questions.

Exposure

First, how are we exposed to a chemical? There are normally three ways this can occur: by absorption through the skin, by breathing it in, and through the mouth, stomach, and intestines (the gut). Which route a chemical takes will depend on the nature of the chemical and the circumstances of exposure. We are most commonly exposed to chemicals through the gut when we eat and drink or take medicines.

Exposure to substances by breathing, through the lungs, will depend on where we are and live and what we are doing at the time. For instance, if we are painting the woodwork at home we will be exposed to the solvents in the paint as we breathe. Travelling to work in our cars in a big city, we will breathe in gases and perhaps particles of lead or a mixture of substances from car exhausts.

Exposure of the skin can occur when we use a solvent at work in a laboratory or in a garage to clean an engine, or at home to clean our paintbrushes or strip paint. Exposure can also occur from our clothes, for example from residues of washing powders and fabric conditioners.

The nature of a chemical will, obviously, affect its disposition and its effects on the body (the nature of a chemical can be described in terms of its so-called **physico-chemical characteristics**). These various characteristics will affect both the site of exposure and the consequences of the exposure. A chemical may be a solid, a liquid, or a gas. A solid may be in solution in water, for example sugar in a cup of tea, or in another solvent, for example alcohol, which is used to dissolve the fragrances in perfume. Liquids may be volatile such as petrol or white spirit. A solid may be in the form of lumps, crystals (for example, salt), or very small particles. Furthermore, the chemical could be irritant or corrosive, such as an acid like battery acid (hydrochloric acid) or kettle descaler (formic acid), or an alkali like caustic soda (sodium hydroxide), which is found in oven cleaners. The latter may not be well absorbed from any of the three sites of exposure but will still cause damage to the tissues with which they come into contact. Substances that are not at all soluble in fat will not be well absorbed, nor will substances that are very soluble in fat but not soluble in water. However, sufficient of the chemical may be absorbed for it to be toxic even if it is a very small amount. Substances that are soluble in fat will also be more readily distributed around the body and metabolized.

Another consideration in relation to exposure is whether it is a single occurrence or is continuous over days, weeks, or months (that is, whether it is acute or **chronic** exposure). This can make a very big difference to the effects a chemical can have. The size of the dose will also be crucial to the effect. These factors will be discussed later in the chapter.

Local damage

Exposure to a chemical can cause damage at the point or site of exposure if the substance is reactive, irritant, corrosive, or caustic. Thus substances such as kettle descaler (formic acid), battery acid (sulphuric acid), caustic soda (sodium hydroxide), and bleach (sodium hypochlorite/hypochlorous acid) can cause serious, and maybe permanent, damage to the skin, the eyes or the oesophagus, and stomach if they come into contact with these parts of the body. Unfortunately, people sometimes attempt suicide by swallowing such substances; for example, drinking kettle descaler or bleach causes serious damage to the lining of the gut (see case notes, p. 193).

Irritant gases such as ozone, sulphur dioxide, and nitrogen oxides can occur in the atmosphere. These may be products of industrial activity or be produced by car exhausts. They can damage the lungs, leading to bronchitis if exposure continues over a period of years.

Other chemicals may after repeated exposure cause more subtle effects on the skin such as allergic reactions. Skin sensitization can be caused by nickel in jewellery or the constituents of some washing powders. Sensitization of the skin leading to allergic, contact dermatitis can be very serious as well as disfiguring and is the most common industrial disease (see Chapter 7). Some natural toxicants, such as nettle sting (formic acid) and the very potent substances in the plant poison ivy, can be skin irritants.

Many chemicals however, will have no direct adverse effect on the tissues of the skin, lungs, or gut when these are exposed. In order to have any effect they must be absorbed into the body. There are some chemicals that are not absorbed and are not irritant or corrosive, which consequently will have no effect on the body.

Absorption into the body

So how are chemicals absorbed into the body and what happens after they have been absorbed? Let us suppose that an amount of a chemical, for example a drug like paracetamol, has just been swallowed by a person. If this chemical is slightly soluble in water (paracetamol is slightly soluble) it will easily come into contact with the cells lining the stomach and intestine. If it is soluble in fat (as paracetamol is) it will dissolve in the membrane of these cells (see Figure 2). If it is composed of large molecules such as a protein, or is not very soluble in fat (for example, the **herbicide** paraquat), then absorption is much less likely or only limited. However, sufficient **paraquat** is absorbed from the gut to cause toxicity. Because the amount of the substance on the outside of the cell is greater than the amount inside the cell, a process called diffusion will carry the molecules of the chemical (for example, paracetamol) into the cell. This cell is part of the lining of the intestine.

Blood circulates around the body in blood vessels. The larger vessels carrying blood away from the heart, and usually deep inside the body, are arteries; those carrying blood back to the heart are veins. The very smallest vessels are called **capillaries**. These may have walls only one or two cells thick and so substances can pass into and out of the blood in the capillary very easily. The inside lining of the intestine is folded, and on the folds are tiny protrusions called villi which offer a very large surface area for absorption. Inside the villi are blood capillaries into which substances in the intestine will be absorbed, as shown in Figure 3. Blood from the

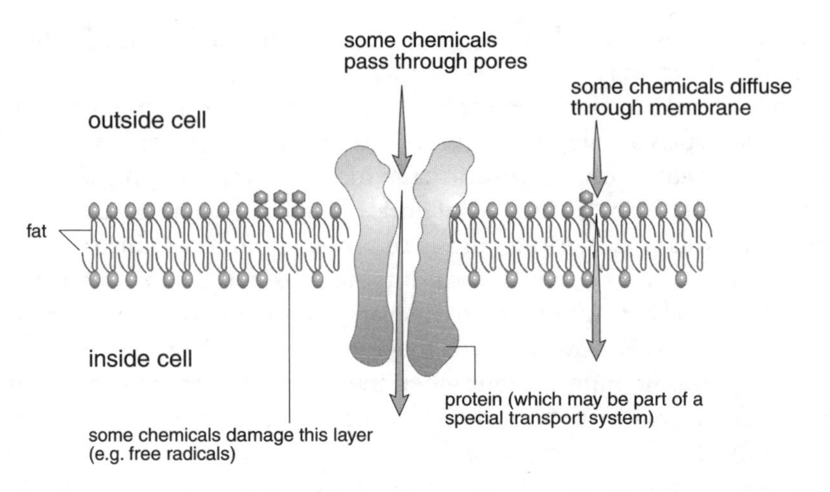

2. The basic structure of the cell membrane in animal cells. Chemicals can get into the cell by diffusing through the fat molecules, passing through pores, or being transported by special systems.

stomach and intestine is taken by blood vessels, which merge into the portal vein, which feeds into the liver. At first, there are none of the molecules of the chemical (for example, paracetamol) in the blood in the capillaries, but again the process of diffusion carries these molecules across the membranes of the cells and capillaries and into the blood. The intestine is structured to maximize absorption (of foodstuffs), and this characteristic means that chemicals such as drugs can also be rapidly absorbed by this route. The molecules of the chemical are now in the bloodstream and being carried rapidly away from the intestine. Where does the blood take them? From the intestine the blood goes straight to the liver (see Figure 3 and box, page 17).

Apart from absorption from the gut, chemicals can also be absorbed from the lungs and skin. The lungs have a huge surface area in the many air sacs and tiny tubes that make up the lung tissue. In the average human this area is equivalent to the area of a tennis court! The flow of blood is very rapid and the walls of the blood capillaries and the air sacs of the lungs are very thin. Consequently it is very easy for volatile chemicals like the solvents in adhesives (for example, toluene) and cleaning fluids or gases (like oxygen) to pass into the blood and be carried away. The breathing rate and rate of blood flow are important factors. The faster the rate of breathing the more rapidly a volatile chemical or gas will be absorbed. This is the basis of the use of birds (canaries) in mines to detect the accumulation of dangerous gas in the past. When gas was present the

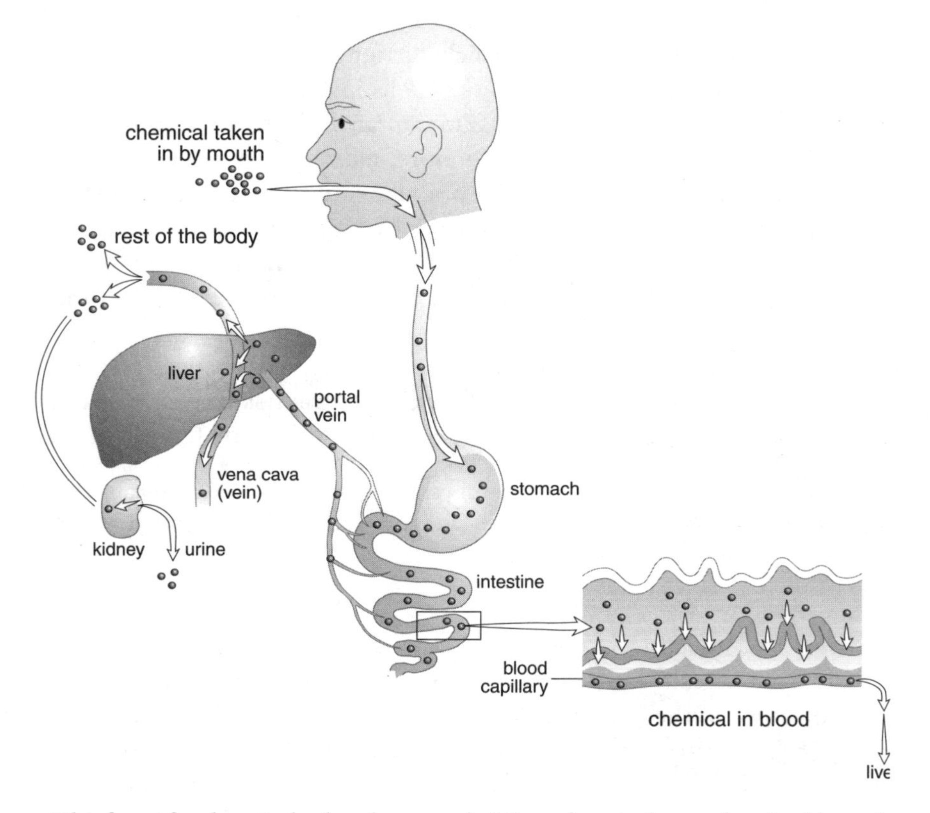

3. The fate of a chemical taken by mouth. Most chemicals are absorbed into the blood from the intestines, which are folded and so have a large surface area. Chemicals then pass directly to the liver via the blood.

canary, with its much faster breathing rate, would fall unconscious before the slower-breathing miners and so alert them to the danger.

Particles can also be absorbed from the lungs but their size is a crucial factor. Those that are too big will be removed in the uppermost parts of the respiratory system, while those that are too small will not settle. However, the particles that are deposited in the air sacs may be absorbed into the blood, for example lead from car exhausts. Asbestos fibres that are absorbed into the cells of the lungs are not transferred into the blood, simply staying *in situ*. Hence they can eventually cause asbestosis and lung cancer.

Once absorbed from the lungs, chemicals will travel very rapidly to the brain via the heart. Consequently the effects of volatile chemicals and gases, such as anaesthetics or solvents, on the brain can be very rapid.

In contrast, absorption through the skin is relatively slow and inefficient. This is because the skin is a thick barrier between the outside world and the blood. There are many layers of cells in the skin and some are not very permeable to chemicals. The skin contains few blood capillaries and the rate of blood flow is slower. Therefore, although some chemicals can pass through the skin and enter the bloodstream, this is generally slower than the other two routes of exposure. However, for chemicals like solvents and liquids that are soluble in fat (for example, some organophosphate **insecticides**) this can still be an important route of entry into the body.

Thus factors that can affect the amount and speed of absorption are the thickness of the tissue through which the chemical has to pass, the number of capillaries present in the tissue, and the flow of blood through those capillaries.

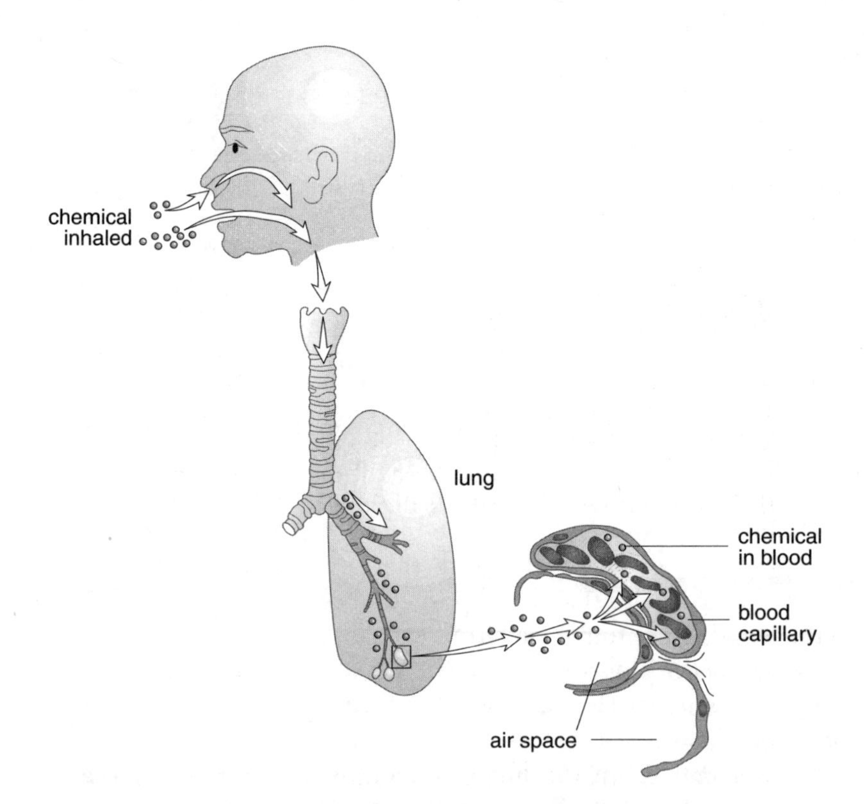

4. The fate of inhaled chemicals. The lungs have a very large surface area and a good supply of blood so volatile chemicals and gases are easily and rapidly absorbed into the body.

Fate of the chemical in the body

Absorption of a chemical from the stomach or intestine delivers that chemical directly to the liver. The liver plays a crucial role in the fate of chemicals (see box) as it is the organ that is the most metabolically active in the body: it transforms foodstuffs and other chemicals that are absorbed from the gut.

The liver

The liver is the largest gland in the body, weighing about 2.5 kg in the average man. It is responsible for most of the metabolic activity in the body, breaking down chemicals absorbed from food such as starch, fat, and protein into their component carbohydrates, **triglycerides**, and **amino acids**. These may then be further broken down to provide energy or synthesized into new proteins, large carbohydrates, or fats and either stored or transported out of the liver into the blood.

The liver is also the organ that metabolizes drugs and other chemicals to more water-soluble substances which can then be excreted into the urine.

The liver produces bile, a green liquid which is secreted into and stored in the gall bladder and then delivered to the small intestine after a meal. Bile contains detergents which disperse the fats in our food and allow them to be absorbed more readily from the intestine.

The liver is made up mostly of similar cells and, compared to many organs in the body, it is relatively homogeneous. Consequently, if damaged it will still function, even if only half of it is normal, but if a greater proportion of the liver is damaged or suffers dysfunction other parts of the body will suffer. One of the important functions of the liver is to remove ammonia which is toxic and results from the breakdown of proteins and their constituent amino acids. In liver failure this function is reduced, with the result that ammonia accumulates in the body and seriously affects the brain, causing convulsions and coma.

In the liver, a chemical absorbed from the intestines (for example, paracetamol) may have one of several fates. It will probably pass into the cells of the liver, where it may be excreted into bile, or (as happens with paracetamol) converted, or metabolized, into something else (see box). The chemical and its breakdown products then usually pass into the blood flowing out of the liver into the vena cava, the largest vein in the body. From there the molecules of chemical or the products of metabolism are carried around the rest of the body.

The chemical may interact with proteins in the blood (as happens with many drugs), or diffuse into the cells of other organs or tissue and remain there. Alternatively it may be filtered out of the blood by the kidneys and appear in the urine, and so be eliminated from the body (see below). If the chemical or its products passes into other tissues or organs it may remain there inert for some time, or it may cause some, possibly adverse, effect (for example, PBBs; see below). For example, the metal lead will eventually be deposited in bone where it becomes part of the structure of the bone. Here it can affect the growing ends of the bone in children. It also localizes in teeth, which affects their appearance: 'lead lines' can be seen (see pp. 136–142). Molecules that are not metabolized and are soluble in fat can find their way into areas with a high fat content (adipose tissue). Here they may remain for days or weeks or possibly even years. Substances such as the insecticide **DDT** (see pp. 90–98) and polybrominated and polychlorinated biphenyls (**PBBs and PCBs** respectively), which are used in fire retardants and electrical insulators see pp. 127–131, have this characteristic. These substances are very soluble in lipid but are hardly metabolized and so not readily excreted. They therefore localize in tissue with a high level of fat or in fat itself, where they remain for years. Their slow release into the blood and distribution to other organs and tissues can have adverse effects (see the Michigan farm disaster, p. 258).

How long a molecule remains in the body will depend on how easily it is excreted. Excretion is most commonly through the kidneys, with the chemical being passed out into the urine, but some chemicals may be eliminated into the expired air or in bile. Chemicals, for example volatile solvents (like white spirit) and anaesthetic drugs (like halothane) that are also volatile, are exhaled, but elimination from the body may take hours or even days depending on the amount absorbed. Perhaps you have noticed that you can detect the smell of a solvent on someone's breath some hours after they have been working with it in a laboratory or workshop, for example. Such chemicals are usually very soluble in fat and so will dissolve in the fatty tissue. Because the level of the volatile chemical in the blood drops as it is exhaled, the chemical, which is dissolved in the fatty tissue, diffuses into the blood in order to maintain an equilibrium. As the chemical is volatile it can then diffuse from the blood into the air spaces in the lungs and be eliminated by breathing.

Chemicals that are easily absorbed into the body are usually soluble in fat, which makes them difficult to eliminate from the body (for example, paracetamol). For them to be eliminated in the urine (which is basically water), the molecule has to be water-soluble, and chemicals are rarely soluble in both fat and water. Consequently the body has a mechanism for

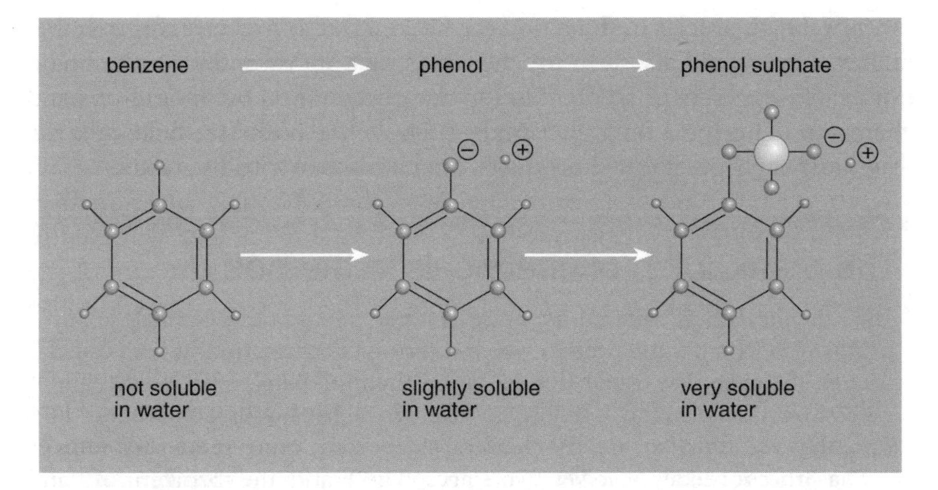

5. Using a simple chemical, benzene (a solvent and contaminant in petrol), for illustration, this figure shows the way chemicals are changed (metabolized) in the liver. The end product is usually more soluble in urine and so easily excreted.

dealing with this. It is called metabolism or biotransformation, which yields a product(s) that is more soluble in water (see box).

The more soluble in water a chemical or its products are, the more quickly and efficiently it will be eliminated in the urine. The function of metabolism is to convert a chemical from a fat-soluble substance, for example the benzene present in petrol (see Figure 5 and box), into a water-soluble one. If the process is successful, the substance is eliminated quickly into the urine within a few hours. The process of metabolism ensures that the body is protected from prolonged exposure to a potentially hazardous chemical. The process has evolved over millions of years to allow humans and other animals to eliminate chemicals absorbed from food.

Elimination of chemicals from the body

The process of excretion of chemicals into the urine (and bile) has developed so as to rid the body of noxious substances. Some of these are products of normal bodily activity such as the breakdown of proteins and removal of the ammonia produced by this process, which is converted into the substance urea in the liver and excreted into the urine.

The processes of excretion have therefore developed to protect the body. Urinary excretion largely involves filtration of the blood, in which most of the chemicals in the blood are filtered out as it passes through the kidney. This would also remove chemicals that are essential for the body (for example, glucose), so there follows a process of reabsorption for such chemicals. The remaining chemicals, such as the waste product urea or drugs and their breakdown products, are eliminated into the urine.

The breakdown of chemicals in the body

After chemicals have entered the body of a human and other animals, many are changed by a natural process which is termed **detoxication**. It is so called because the change(s) usually results in the chemical being more easily eliminated from the body and therefore less likely to be hazardous. Sometimes this is not the case, however, and the chemical can become much more poisonous.

The process usually involves two stages. The first is the alteration of the chemical sructure to provide a 'handle'. This handle is then used to add on another molecule, something normally found in the body, which makes the breakdown product more soluble in water. It can then be eliminated into the urine.

This process of detoxication, by metabolism to water-soluble products applies to all chemicals that are 'foreign' to the body, whether they are synthetic or natural. This process of change, or metabolism, occurs with the help of biological catalysts called enzymes. Enzymes are proteins, large molecules composed of amino acids (some of which we get from our diet) in chains and the whole molecule is folded into a particular shape. An enzyme is like a lock, and the chemical to be metabolized fits into the lock like a key. This allows the chemical to undergo a change, perhaps by reacting with another substance. There are many enzymes in the body, each doing a particular job, most carrying out the normal processes involved with digestion of food, or the manufacture of replacement parts for the body, or generating energy.

The more soluble the chemical is in fat, the more easily it will fit into the enzyme (the 'lock') and be metabolized into something more soluble in water. This is well illustrated by the chemical **benzene**, a volatile solvent and contaminant in petrol (gasoline) (see Figure 5). This process of metabolism allows the body to eliminate benzene effectively.

However some chemicals are not metabolized, such as the environmental contaminants dioxin and PCBs. Because these chemicals are soluble in fat, they can be absorbed into the body where they will remain for long periods of time.

Excretion is the process of removal of chemicals and their breakdown products by the kidneys. Most drugs and other chemicals are removed from the body by the kidneys, usually in a process called filtration. Chemicals and their breakdown products circulate in the blood throughout the body. A lot of blood flows through the kidneys (a quarter of the blood pumped out by the heart). When the blood flows through the kidneys, in which some of the blood capillaries have relatively large holes or pores, it is under some pressure, and the molecules of chemicals floating in the blood (for example the breakdown products of paracetamol) pass out through pores, rather like water, and small particles flow through a sieve. The kidney then reabsorbs some of the essential chemicals, such as glucose, along with some of the water but leaves the waste products behind. These waste products result from normal metabolic processes but need to be eliminated. This selective loss ensures that potentially noxious and unwanted chemical substances, including natural waste products, are eliminated relatively quickly and efficiently, while most of the essential substances are retained. Water is one essential substance that is partly lost which has to be replaced along with salts.

The function of the kidney decreases with age, and in the elderly the ability to excrete substances is reduced. Therefore drugs and their breakdown products can accumulate in the bodies of elderly patients if they take repeated doses of a drug (see box on Opren, page 29).

The majority of chemicals are removed from the body in this way. Those that are volatile, such as gases and volatile solvents, are exhaled through the lungs. Some chemicals, with a particular structure or characteristics, are excreted into the bile and passed out into the intestine and then the faeces.

As well as excretion into urine, bile, and expired air, chemicals, especially those that are very soluble in fat, can be eliminated into breast milk. It can be shown that DDT, for example, is sometimes present in samples of human breast milk. This is potentially very significant as it exposes the newborn baby to chemicals, some of which may be harmful and present in concentrations sufficient to cause effects.

The processes of metabolism and excretion try to protect the body against the possible adverse effects of chemicals. If so, why are chemicals sometimes toxic and specifically damaging to one organ? One reason is that the process of metabolism sometimes produces chemicals that are more poisonous than the original. Whether or not a chemical or its metabolic products causes any damage depends on its chemical properties, how much of it is present, and where it goes in the body. This will be discussed further in later chapters, but some illustrations here may show how the sequence of events is crucial in determining what toxic effects occur and where.

Paracetamol, as we have seen, is an example of a drug that is absorbed from the gut into the blood and then passes directly into the liver, where it is changed into mainly innocuous and water-soluble products which are easily passed out into the urine. At the same time, a more poisonous product is also made in small quantities in the liver, which is normally changed into another harmless product. Only when an overdose is taken does the system fail, and the liver is damaged (but usually no other organ) by the poisonous product of paracetamol that is manufactured in the liver itself (see further discussion on pp. 49–56).

The second example that illustrates the importance of the distribution of a chemical in the body is the herbicide paraquat (Weedol). Paraquat is poorly absorbed from the gut but sufficient enters the bloodstream to cause effects, sometimes with fatal consequences. It damages primarily the lungs, and does not affect all organs, because its structure causes it to be accumulated specifically in the lungs, with the result that the concentration reaches levels high enough to be toxic (see page 104–6). There are many other instances, some of which are mentioned later in the book, where one particular organ or system is affected by a chemical as a result of a flaw, peculiarity, or alteration in the sequence of processes through which chemicals pass, that is, absorption of the chemical, distribution of it in the body, its metabolic breakdown, and its final excretion and elimination.

Metabolism can give rise to various products, some of which are chemically reactive rather than stable and inert. This means that such products can react with parts of the cell and cause damage. Some of these reactive substances are known as **free radicals**, which can react with fats and cause chain reactions that lead to the destruction of cells and tissues. Other types of reactive chemicals may interact with protein or DNA molecules (see box).

The human body, indeed the body of any mammal, is immensely complicated and its functioning is a wonder to behold. Given this complexity, it is remarkably tolerant of the foreign chemicals to which it is exposed but, not surprisingly, at times these chemicals do interfere with the normal workings of the body. A chemical that is foreign to the body, such as a drug, can interact with the substances of which the cells are made, for example to mimic them, such as when a drug interacts with a receptor. The result is the alteration of a normal process: switching it on or off, or decreasing or increasing its effect. Alternatively a foreign chemical may react chemically and irreversibly with a vital structure, protein, or enzyme that is crucial for a particular process. This could then destroy, or lead to the destruction of, part of a cell or even all of it or stop the vital process from occurring. These are some of the ways in which chemicals can be toxic.

Thus if the chemical or its products can react chemically with other

Reactive metabolites

Metabolism, primarily in the liver, can sometimes yield a product that is chemically reactive. Although generally the liver can detoxify such metabolites, if the amount is excessive these detoxification processes can be overwhelmed. When this happens the chemically reactive metabolite can react with constituents of the cells in which they are produced. This can cause damage to important molecules such as DNA or structures inside the cell, which can lead to the death of the cell.

There are many different kinds of reactive metabolites which it is beyond the scope of this book to discuss. However, one type is particularly important. This is the free radical, and the important feature of these is that on reacting with another chemical another free radical is produced, a so-called chain reaction. This leads especially to the destruction of fats and the membranes of which they form part. In the presence of oxygen this causes a destructive process call lipid peroxidation. A similar process causes butter to go rancid.

Sometimes highly reactive and toxic forms of oxygen such as superoxide radical and hydroxyl radical, known as reactive oxygen species, are also produced.

---

molecules found naturally in the body such as DNA, proteins, or fats, this may cause damage to the structures in the cell formed by these molecules, to their function, or to the function of the whole body. The results of these interactions will be discussed below.

A chemical may be innocuous but become potentially dangerous after metabolism and activation in the liver. A single such molecule can be dealt with but if there are many such activated molecules, the ability of the body to remove them is exceeded and the result can be widespread damage to tissues and organs. Fortunately there are protective measures in the body, often inside cells, which can help to reduce the likelihood of damage at lower doses. The reactive chemicals are removed by **antioxidants** such as vitamin E or vitamin C and other substances in the body called **thiols** (see box).

These protective agents prevent damage to organs and tissues, but if there are too many toxic molecules they are soon overwhelmed, and damage and destruction can follow. As we shall see in the next chapter, this is what happens in paracetamol poisoning following an overdose.

On the whole the body can withstand insults from chemicals: the liver will detoxify and remove the potentially hazardous substances, provided the dose is not excessive. Protective agents will react with and remove

How the body protects itself

The liver and other organs, have evolved methods of protection against reactive, potentially dangerous chemicals. For example, there are antioxidants like vitamin C and vitamin E, which will remove chemicals such as the various forms of reactive oxygen and other free radicals. There are various kinds of thiols, molecules containing **sulphur**, which will also react with such reactive radicals, and the most important of these is **glutathione**. There are also specific enzymes that can remove certain types of reactive chemical such as superoxide. The liver especially contains the highest amounts of such protective agents, while the lungs contain systems for removing reactive oxygen. The amounts of these protective agents usually present in the body will generally be enough to remove noxious chemicals encountered naturally but may not be enough to cope with a large chemical insult.

Some of these protective substances are found in the diet, such as the vitamins, the sulphur-containing amino acid taurine, and methionine which is an important precursor of protective thiols. The sulphur amino acids are found especially in foodstuffs containing protein and, in the case of taurine, only in meat and fish. Vegans and vegetarians are therefore deficient in this important substance.

noxious chemicals, again provided the amount does not exhaust the available protective agent. Thus there is a very clear underlying principle as to why the dose of a chemical is so important. It underlies what Paracelsus concluded in the sixteenth century: 'All substances are poisons; there is none that is not a poison. The right dose differentiates a poison from a remedy.' The body can protect itself against small doses of a chemical, and these small doses may even be beneficial, but a large dose overwhelms the available natural resources and causes toxic, adverse effects. Therefore chemicals can be used safely if exposure is small enough. It also shows the falsity of the view that if a small amount is good for you then a larger amount must be better, as some people seem to believe about vitamins for example.

Exposure to drugs and other chemicals may be a single occurrence or may occur almost continuously. Continual doses of a drug or continuous exposure to a chemical can have very different effects to a single dose. This may be due to a change in the metabolism of the chemical which can occur with repeated exposure to some chemicals (see below), and which may increase the detoxication and elimination of the chemical, as happens with some barbiturate drugs. Indeed some chemicals stimulate protective mechanisms in the body, and so their toxicity decreases after several doses. Conversely, the result could be the production of more of a toxic product.

A theory that has been gaining acceptance recently is that many chemicals have beneficial effects at very low doses and may perhaps stimulate protective mechanisms, and that as the dose increases the effects become adverse. This is known as hormesis and may occur as a result of more than one exposure to a chemical (see box).[1]

Hormesis, or, a little bit of poison is good for you

There is an old saying about a peck of dirt being good for you. This probably refers to the effects of early exposure to bacteria and other foreign particles, which are all around us in the environment, on the immune system. It has been suggested that early exposure of the immune system to foreign agents such as bacteria helps to protect young children when they are older and are exposed to infectious diseases. There is growing evidence that this concept may have more widespread relevance in that low levels of chemicals may be not only harmless but positively beneficial. This means the shape of the dose–response curve shown in Figure 6 on page 37 and Figure 29 on page 300 may have to be redrawn as a J rather than an S. There is some scientific basis for this in that exposure to low levels of chemicals will stimulate the body and individual cells to increase protective measures such as the so-called stress proteins, proteins which help to protect the cell.

This means that extrapolating from a high dose of a chemical to a very low dose to determine a threshold and calculate risk may not be appropriate (see Chapter 12), especially if a *linear model* is used which implies there is no safe dose (for example, for a **carcinogen**). If true this has profound implications for risk assessment, suggesting that we may sometimes have been more cautious than necessary. Thus attempting to reduce exposure levels for chemicals excessively may be unnecessary and, worse, a waste of effort and money.

This effect seems to apply to many different types of chemicals as well as to radiation, and is associated with the phenomenon of **tolerance** (see below). Its implications for the assessment of risk from chemicals are profound (see Chapter 12).

Factors that make chemicals more or less toxic

Repeated exposure to a substance can stimulate or alert the immune system and result in an allergic reaction. This can occur, for example, with the exposure of some individuals to penicillin, with serious results (see pp. 66–8). Many years ago an enterprising American had the seemingly clever idea of putting penicillin in lipstick, so that the constant protection it afforded would combat sexually transmitted venereal diseases, which some women might contract. Unfortunately, the constant exposure resulted in severe allergic reactions in some women.

Constant or repeated exposure can also lead to irritation and cancers, and constant damage may eventually lead to the failure of an organ. This is what happens in chronic alcoholics whose livers finally succumb to hepatitis and/or cirrhosis after years of damage. Repeated doses of a drug can lead to accumulation in the body and eventually toxic effects. This happens with aspirin when it is taken repeatedly by patients with influenza or arthritis; it can lead to the build-up of toxic levels. Many of us are bitterly aware of the consequences of accumulation of alcohol: too many glasses of wine in too short a space of time (due to the desirable effects of alcohol) leads to drunkenness followed by a hangover (due to the unpleasant and undesirable breakdown products).

All of us are exposed to chemicals daily in our food, both natural constituents and contaminants. Some of the contaminants derive from cooking, while others are naturally occurring substances, such as those produced by fungi, and still others may be man-made environmental pollutants. All these chemicals may affect the disposition of other chemicals such as drugs.

Some of us intentionally expose ourselves to substances like alcohol and tobacco. Both of these are capable of changing the metabolism and toxic effects of other substances. For example, the drug paracetamol is more poisonous in individuals who are habitual and heavy drinkers, while tobacco smoke and charcoal-grilled meat contain substances called polycyclic aromatic hydrocarbons which can increase the metabolism of some drugs.

The exposure to chemicals in the environment is often a cause for concern, yet the levels are often extremely low. The effects of such low levels of chemicals can be difficult to measure or predict. If they are readily eliminated and before the next exposure, the exposure may well be inconsequential. If they are not eliminated, however, and accumulate in the body, it may be of more significance. As well as the accumulation of a chemical substance, the effect can also accumulate. This occurs with organophosphate insecticides, for example; therefore agricultural workers using these pesticides should be monitored regularly (see pp. 99–103).

Species

Other species as well as humans are exposed to chemicals, including drugs (for example, veterinary drugs) and environmental pollutants. Different species of animals sometimes metabolize a chemical differently. Small animals will generally metabolize and eliminate chemicals more quickly than larger ones. This may affect the subsequent toxic effects of the chemical, in that it may be less poisonous if the substance itself is toxic, but more poisonous if a metabolic product is more toxic.

A particular animal species may have a deficiency and be unable to metabolize and so eliminate a chemical. For example, the domestic cat has a particular deficiency which affects the handling of the drug paracetamol. If pet owners give doses of paracetamol to their cats, it may kill them. Understanding such differences is crucial for the use of veterinary drugs.

Wild animals may be particularly affected by a chemical. For example, DDT had drastic effects on the bird population by influencing the development of eggs (see pp. 92–7) and accumulating to toxic levels in predatory birds. Mammals, by contrast, are not especially affected by the pesticide. In some cases wild animals may be much less sensitive to a chemical for other reasons. For example, frogs are twenty-two times less sensitive to organophosphate insecticides than laboratory mice because the enzyme with which the insecticide reacts is different in the frog.

The safety of chemicals, like drugs for human use, relies on testing and evaluation in other animal species such as rats and mice. Therefore the differences between species in this context have to be recognized. For example, the disposition of a particular chemical in a rat or mouse may be different and so studies in such species do not always show how poisonous it will be in humans. The drug may be more or less poisonous to a rodent than to a human. For this reason human volunteers are exposed early on in the development process of drugs and human tissues are used ***in vitro***. Knowledge of the disposition of the drug in humans reveals whether the species initially used (for example, rats) are appropriate. If not, other species will be studied and used for the safety evaluation. These points will be discussed in more detail in Chapter 12.

Factors that affect the disposition of chemicals in humans

Within human beings many factors can affect the processes of absorption of chemicals, their distribution throughout the body, metabolism, and excretion. Factors such as the age of the person, their state of health, their genetic make-up, their diet, and what drugs they are taking can all influence the disposition of a drug.

Age

Age can affect the uptake of drugs from the gut, so very young children, who have low levels of acid in the stomach, absorb penicillin more readily than adults but paracetamol less so. Once absorbed into the body, chemicals are distributed around the body in the blood (as explained earlier). Chemicals commonly bind to proteins in blood which limits the level available to distribute into organs and tissues. This binding may be reduced to as low as a third in newborn babies, which means that the circulating free level of the substance is effectively higher than in an adult. This can result in an overdose in a newborn baby.

The ability of an individual to metabolize a chemical is also affected by age, being generally less in elderly individuals than in young adults. Newborn and very young children also generally have a reduced ability to metabolize chemicals, which may make them more or less susceptible than adults. Some drugs, for example, are less toxic in newborn animals.

The excretion of drugs from the body is often reduced in old age such that normal doses of some drugs may become dangerous as they accumulate (see box).

Diet

Although most people in the affluent West have an adequate, well-balanced diet, this is not the case in some parts of the world. Protein and vitamin deficiencies occur and can have effects on the toxicity of chemicals. Lack of protein in the diet reduces the level of many of the enzymes necessary for the metabolism of chemicals and reduces the level of proteins in the blood. If the chemicals are extensively bound to proteins in the blood, this can influence its disposition, as with less protein more of the drug will be freely available in the blood, which may lead to an increase in toxicity. However, less of the drug will be metabolized as the

Opren

Opren (Benoxaprofen) was a very effective drug developed for the treatment of arthritis. Repeated doses were given to elderly patients but, unfortunately, some of the patients suffered adverse effects to the skin and liver, and over seventy of them died. The elimination of the drug from the body into the urine was less efficient in these elderly patients, compared to healthy young adults, owing to reduced metabolism and kidney function. Consequently, after multiple doses were given to elderly patients, the drug accumulated in some of them until it reached toxic levels, which resulted in liver damage. This could have been avoided if the drug had been more carefully evaluated in elderly patients before the drug was marketed. Had the level of the drug in the blood been monitored in the patients after repeated dosing, it would have become apparent that the **half-life** was longer in some elderly patients than in the healthy young volunteers in which it was tested, and that this half-life increased as more doses were given.

necessary enzymes will be depleted. The effect of dietary deficiency on the toxicity of **cyanide** is an example discussed in Chapter 10 (see pp. 255–6). Thus dietary deficiencies can make individuals more (though sometimes less) susceptible to the effects of chemicals.

State of health

Similarly the state of health of a person can affect the way they handle a drug or other chemical substance by altering, for example, their metabolic capability or kidney function. Serious diseases affecting the function of a particular organ will have major effects if that organ is involved in the metabolism or excretion of the drug.

Genetic factors

Genetic differences are among the most important reasons for variability in the response to chemicals between individuals. These differences often affect the way a chemical is metabolized but can also affect the way the body responds to a chemical. Many of us have heard someone claim, 'My grandad lived to be 95, drank like a fish and smoked like a chimney—and it never hurt him', suggesting that smoking and excessive drinking are harmless. We know this is not really the case and that, while there may be some individuals who seem more resistant, such a lifestyle leads to serious illness and premature death. Of course we are all different, and we respond slightly differently to chemicals. In many cases these differences may not be critical, but sometimes they can be a matter of life and death.

Why do individuals vary in the way they respond to drugs and other chemicals? It is now known that there are many genetic factors that can influence a chemical's effects on the body. These factors influence for the most part the metabolism of the substance but sometimes the way the person responds (see box on page 32 on Favism). These genetic differences are often due to variations (polymorphisms) in the enzymes involved with metabolism. These variations result from **mutations** (damage or changes) in the section of DNA that contains the code for the particular enzyme. If enzymes responsible for detoxification are missing or don't function properly, the unfortunate person may be poisoned with a normal dose of a drug or show an unusual adverse reaction to it. For example, before a patient has an operation in a hospital operating theatre, they are often given a drug that relaxes muscles: suxamethonium (alternative name succinylcholine, trade name Anectine). The drug acts quickly but lasts only for a few minutes, because it is broken down very quickly in the body and inactivated. However, in some individuals (about 3 per cent of the population) the effects on muscles can be prolonged and excessive, and they may even stop breathing and require artificial respiration. This is because in these affected individuals the enzyme responsible for the detoxication is deficient and therefore the dose given is effectively excessive, leading to prolonged effects on the muscles.

Other mutations in the DNA could lead to lack of enzymes required for the production of protective substances or repair of damage. In the next chapter there are several examples of drugs that show variable effects in different humans as a result of this type of variation in disposition.

As we have seen, the breakdown or detoxication (metabolism) of a chemical is carried out by enzymes. These biological catalysts are proteins that are produced with information encoded in DNA. If there is an error (known as a mutation) in the DNA code that holds the information, the enzyme (protein) that is produced may be faulty. Mutations occur naturally and are passed on from one generation to the next. Some are benign, others potentially lethal. An example of this is the mutation that causes the disease haemophilia, which was carried by Queen Victoria and which afflicted various male members of the royal families of Europe who were descended from her.

If a mutation results in the production by the body of a faulty enzyme, then the metabolic processes catalysed by that enzyme will either not occur or occur only slowly. The consequences of this will depend on the particular chemical and the products of the metabolism by the enzyme in question. For example, if a person is exposed to a chemical that is itself toxic, then breakdown (by metabolism) is detoxication. A reduction in this process will mean that the individual will be more susceptible to the chemical. Conversely, if the chemical is not toxic but the breakdown products are poisonous, then the individual with the deficiency will be less susceptible to exposure to the chemical.

The breakdown of alcohol in the body, for example, is catalysed by two enzymes, both of which can show variation between individuals (genetic variability / genetic polymorphisms). The first enzyme converts the alcohol into a product called acetaldehyde. Because the well-known effects of alcohol are due to the chemical itself, a deficiency or fault with this enzyme in an individual will mean that the alcohol remains unchanged in that individual's body for longer. This happens more often in some racial groups such as Native Americans, the Inuit, and Orientals such as the Japanese. Individuals with this deficient or faulty enzyme would get drunk on smaller amounts than those without the mutation. The second enzyme converts the acetaldehyde into acetic acid. There are also individuals with a deficiency or fault in this enzyme. They experience unpleasant symptoms when they drink alcohol such as becoming flushed and feeling sick, because the acetaldehyde accumulates in their bodies. This genetic predisposition also occurs more frequently in people of Japanese origin.

Further examples of the effect of genetic make-up on the handling and response to chemicals will be discussed in more detail in the next chapter.

Other genetic variations may alter the way in which the body reacts to a chemical rather than what the body does to the chemical (see box overleaf). Some of these genetic deficiencies or polymorphisms may be common, occurring in perhaps 50 per cent of the population, or they may be much rarer, occurring in perhaps only 1 or 2 per cent of the population.

Lifestyle

An important factor that can determine the disposition of a chemical, whether it is more or perhaps less toxic than expected, is exposure to other chemicals. For example, we often take drugs in combination but do we know what effects they have on one another? With common drug combinations the interactions should be known by the pharmacist, but with unexpected mixtures, which may be unintentional, dangerous interactions can occur. For example, both alcohol and smoking can affect the toxicity of the drugs that people take and the chemicals they may be

Favism

An example of greater susceptibility to chemicals resulting from a genetic deficiency is shown by the case of Favism. This can be quite common in men in some regions of the eastern Mediterranean, reaching an incidence of 50 per cent in certain communities. The trait occurs only in men and is therefore 'sex linked'. It results in serious damage to the red cells in the blood when men with the affliction eat Fava beans or use certain drugs such as sulphonamides (antibiotics). The susceptible men lack an enzyme called glucose 6 phosphate dehydrogenase, which is essential for an important process which metabolizes sugars and maintains the level of an important substance (the coenzyme NADPH). In turn NADPH is needed to maintain the levels of a protective substance (a thiol called glutathione) in the red cells in the blood. In the individuals lacking the enzyme, the level of glutathione is too low and hence their red cells are not protected against metabolites of drugs and substances found in food such as Fava beans. The red cells are therefore vulnerable to damage. If these men eat Fava beans or take sulphonamides the red cells in their blood are destroyed and they suffer **anaemia**.

exposed to. Food contains many chemicals which may have significant interactions with drugs, and some people are also exposed to chemicals in their workplace. Even more uncertainty surrounds the possible effects of environmental chemicals on one another or on the effects of drugs.

Two ways in which one chemical can change the effects of another are by inhibiting an enzyme or by inducing an enzyme.

Induction and inhibition of enzymes

Enzymes may be inhibited in many ways, which are not within the scope of this book to discuss in detail. However, it is an important reason for variation in the disposition and toxicity of chemicals. If we think of an enzyme as a lock and the chemical it metabolizes as a key, then an inhibitor is like a bent key which becomes stuck in the lock. Alternatively it might destroy the lock (enzyme) completely. The result is that the enzyme does not work properly or at all, and the drug or chemical is not metabolized. By decreasing the detoxication and excretion this can increase the toxicity of the drug.

Induction describes the phenomenon where the amount of an enzyme is increased following exposure to a chemical. This can occur as a result of the substance, for example dioxin, interacting with DNA. The interaction switches on the process that makes enzymes and so more enzyme is produced in the liver for example.

Polycyclic aromatic hydrocarbons such as benzo(a)pyrene (which is present in cigarette smoke and meat cooked over charcoal), dioxin, many drugs, and substances found in plants like St John's wort all **induce** the enzymes involved in the detoxication of drugs and other chemicals. The effect of this induction can be seen as an increase in the metabolism of a drug, which therefore increases the elimination. However, it can also increase the toxicity of a drug. This is the case with paracetamol, whose toxicity is increased in chronic alcoholics and in people taking certain other drugs (see pp. 54–5).

If a chemical inhibits the enzyme that catalyses the metabolism of a particular drug, the disposition and hence toxicity of that drug can be affected. Stopping the metabolism of the drug can increase or decrease its toxic effects, depending on whether or not it is detoxified by metabolism. Because of the concentrations necessary, inhibition is most likely to be caused by drugs although some naturally occurring substances in foods,[2] industrial exposure to solvents, and exposure of agricultural workers to pesticides have been shown to be at high enough levels to inhibit enzymes (see case notes).

A similar example which has been known about for a long time is the so-called cheese reaction which can occur in patients taking

CASE NOTES

Fatal juice

A young man, just 29 years old, had been taking the antihistamine drug terfenadine twice daily for two years because he suffered from hay fever. One day he took his normal dose, drank two glasses of grapefruit juice and went into his garden to mow the lawn. He suddenly became ill, stumbled back into his house, collapsed, and died. The coroner reported that there were increased levels of the drug in his blood and concluded that he had died from the adverse effects of terfenadine.

The man had been using the drug for two years and had taken the recommended dose as before, but the fatal overdose he suffered apparently happened because he had drunk grapefruit juice. This contains various substances, one of which (possibly a flavonoid, which is often found in plants) is a potent inhibitor of the enzyme most commonly involved in the metabolism of drugs. The reason the young man died of an apparent overdose is that the metabolism of the drug was inhibited and so the level of the unchanged drug in the body remained too high, causing a fatal alteration in the rhythm of the heart (arrhythmia).

antidepressant drugs. Some kinds of cheeses (Gruyere for example) and certain other foodstuffs, such as yeast extract and Chianti wine, contain a substance called tyramine. Tyramine is metabolized in the body by an enzyme that is inhibited by certain antidepressant drugs. Therefore, in patients taking these drugs who have eaten foods high in tyramine, the substance will reach toxic levels. This causes high blood pressure which can rise high enough to be fatal.

In contrast to the inhibition of an enzyme, which usually only needs a single dose, induction requires repeated exposure to a chemical but often only small amounts are needed. As well as some drugs, many other chemicals can cause this effect including hydrocarbons found in smoke and substances found in vegetables. This effect, unlike inhibition, can occur as a result of exposure to environmental chemicals and things in food which often are present at lower levels in our bodies than the levels of drugs. The phenomenon of induction was first discovered to be caused by a barbiturate drug, which was used to put patients to sleep but was found to become less and less effective after repeated doses. It was eventually found that this tolerance was due to induction of the enzyme that is responsible for metabolizing the drug itself. Therefore as the patient continued to take the drug its metabolism increased and so its effect on the patient decreased. Hence higher and higher doses of the drug became necessary. This phenomenon was then found to be caused by many other chemical substances including natural chemicals. For example, Brussels sprouts and other common vegetables contain substances known to induce enzymes involved in detoxication. A popular herbal remedy, St John's wort, has recently been found to be a potent inducer of these enzymes and to cause changes in the metabolism and therefore effect of other drugs taken at the same time. The induction process involves increasing the production of the enzyme (see p. 85).

Alternatively, chemicals can interact to increase or decrease the effects of one another by directly altering the effect itself. For example, two chemicals may act in a similar way and when someone is exposed to the two together they may be much more potent. This is known as synergy (see below). There is concern about this phenomenon among some scientists because of the potentially large numbers of chemicals to which humans are exposed.

Tolerance, synergy, additivity, potentiation, antagonism

Tolerance is the situation where the body (or a single cell) becomes tolerant after repeated exposure to a chemical. This means the effect of the chemical is diminished after several exposures. Tolerance can occur because of a change in metabolism, because of an increase in the production of protective substances, or because of a change in a receptor.

Synergy applies to a situation involving exposure to two or more toxic chemicals. If this occurs the result is that the combination of chemicals is much more toxic than expected (that is, it is greater than the sum of the parts). In contrast, *additivity* is simply the sum of the two effects of the individual chemicals, that is the expected effect.

Potentiation is similar to synergy in that the combination of two chemicals has much greater toxicity than expected. However in this case only one of the chemicals is toxic; the non-toxic chemical simply increases the effects of the other. This means that although exposure to an individual chemical is at a safe level, exposure to several chemicals, each at a safe level, may, because of synergy or potentiation, become a risk. This has certainly been suggested for chemicals known as **endocrine disruptors** (see pp. 131–6).

Antagonism is the reverse of synergy, and is where two chemicals together are less toxic because one antagonizes the effects of the other.

All these factors can have an impact on the toxicity of a chemical and indeed may make one individual human being more or less susceptible than another to a particular chemical substance. They may need to be taken into account when setting limits of exposure for a particular chemical in a particular population or environment. Many of these factors will be explored again with examples in later chapters.

Effects of chemicals on the body

Having considered how a chemical gets into the body and what the body can do to the chemical, let us look at what the chemical or its breakdown products can do to the body. One of the most important things that determines toxicity is the amount of the chemical—the dose or exposure level.

The dose–response relationship

We have seen that, in order to be poisonous and before it can cause harm, a chemical first has to get inside the body. The question is how much of it gets into the body. What is the real dose? The internal or real dose of a chemical may be very small even if the amount swallowed or inhaled is large. This is because some chemicals (including some drugs) are very poorly absorbed. However, this is not the only reason that some chemicals are of low toxicity. The internal dose of the chemical may be detoxified very quickly by metabolism, be absorbed into fat, be bound to some other molecule, or just be excreted rapidly. But as the external dose rises so does the internal dose and a point may be reached where the detoxification or excretion processes are overwhelmed. Then the chemical will become toxic. This is one reason why we can see a relationship between the dose of a chemical and the response (effect). The chemical may cause an effect or even damage at low doses but this may not be detectable until some function is compromised.

The dose of a chemical is crucial to the toxicity because all chemicals are potentially toxic. It is simply a matter of the dose—the Paracelsus Principle. As the dose level rises, the cells or systems in the body (human or other animal) show an increasing level of dysfunction and damage. For example, if different doses of a drug that lowers the blood pressure are given to a patient and the effects on blood pressure are measured, as the dose is increased the decrease in blood pressure will increase. At very low doses there would be no change in the blood pressure at all, whereas at high doses there would be a maximal effect, which could be lethal. If this is plotted as a graph, a curve results, similar to that shown in Figure 6.

Another way in which the dose–response relationship can be represented is if the effect is measured as simply present or absent. For example, a toxic effect either occurs or does not occur; the number of people showing an effect after different doses is used to plot the graph. This relationship between the dose of a chemical and the effect can be appreciated from the graph using a simple example like alcohol. At low doses (1 glass of wine) there is no effect on a person's ability to walk, but after a large amount (10 glasses) any person would be affected to some extent (that is, 100 per cent of a group would show some effect on their ability to walk). This kind of curve can be generated for any effect of a chemical, whether it is toxic or beneficial (see Figure 6).

So why do small doses of a drug or other chemical have no effect on us, while large doses do? For most effects in the human body caused

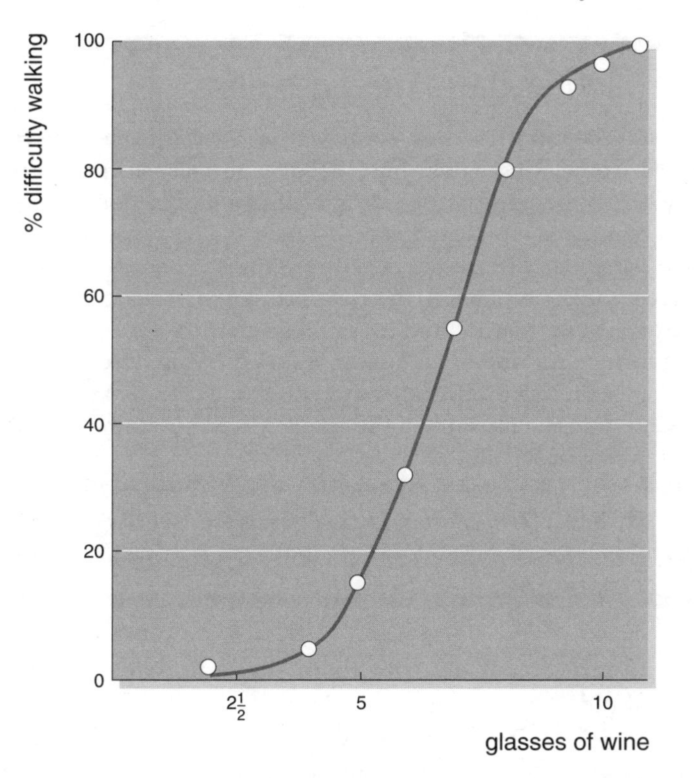

6. The typical relationship between the dose of a chemical and its effect. This is a visual representation of the Paracelsus Principle. The chemical used in this illustration is alcohol and the effect measured is ability to walk. As the dose of alcohol increased the inability would increase until the maximum effect was attained (the subject would be unable to walk!)

by chemicals, interaction of some kind with a molecule is necessary. This may be a receptor or an enzyme. In order for there to be an effect, sufficient receptors must be occupied, and this depends on the concentration of the chemical in relation to the number of receptors. Therefore at very low doses of a drug or other chemical the level of the chemical is insufficient to occupy enough receptors (see box p. 39).

There is therefore a threshold, a dose below which there will be no effects, either desirable or poisonous. This is a very important concept which is used in the assessment of risk from chemicals (see Chapter 12). For some chemicals, in particular those that cause cancer, there is uncertainty over whether there is a threshold. Consideration of the factors that restrict and impede entry of chemicals into the body, impede

Paul Erhlich and the magic bullet

The idea of **receptors** was first conceived by Paul Erhlich (1854–1915) who noticed that some dyes would bind strongly to cells such as bacteria and animal cells. He thought that if a drug could be designed to contain a toxic metal that bound only to the target cells—such as a bacteria—it could be used to treat a particular disease selectively. He investigated a number of compounds (606 in fact) containing the toxic element arsenic, one of which was arsphenamine. It did not kill the isolated bacteria (that is, it did not work *in vitro*), but fortunately one of his colleagues tried giving doses to rabbits. The animals changed it to a product that *was* effective. Humans were able to carry out the same transformation and the result was Salvarsan, a life-saving drug for the treatment of syphilis.

The ideas of Erhlich were continued by John Langley who proposed that animal cells had receptive substances on the surface, which became known as receptors.

access to the target, and detoxify them suggests that all chemicals will show a threshold.

A knowledge of the relation between the dose of a chemical and the response it produces enables its safety to be assessed and predictions to be made. By knowing the relationship and the threshold it is possible to estimate a safe dose of a chemical (see Chapter 12).

In general we can expect that very potent, poisonous chemicals are those that are easily absorbed but not removed or detoxified easily, and that have a specific target which is crucial to the function of the body. Natural poisons such as ricin and botulinum toxin, for example, fall into this category (see pp. 150–2, 249–51). The consequences of the interaction between a chemical and cells in the body or with a molecule such as DNA, a protein, or fat inside or outside cells are variable and depend on many factors. In the case of damage to DNA this can lead to a mutation or even cancer, though not necessarily. Where proteins are damaged or are otherwise the targets, the consequences will depend on the function of the protein. If it is an enzyme this may fail to catalyse a reaction that could be crucial to every cell in the body. This is what happens in cyanide poisoning, as will be discussed in Chapter 9 (see p. 220). Alternatively the result may be the build-up of excessive amounts of a natural substance which has adverse effects when the level reaches a particular point, as happens in organophosphate pesticide poisoning (see pp. 100–2). However, the interactions could be of little or only minor consequence.

Receptors

A receptor is a part of a large molecule (often a protein) or structure which binds another, usually smaller, molecule. This 'binding site' will be specific for a particular chemical or group of chemicals, normally hormones or transmitters, and the binding may be very tight. The receptor is like a lock into which the chemical substance (the ligand) fits as a key. So receptors are similar to enzymes.

The result of this interaction is that some change occurs in the cell or in the body. For example, if the chemical substance is a hormone, such as **adrenaline (epinephrine)**, which interacts with a receptor the whole of the body will undergo changes. The heart starts to pound (the heart rate increases), the rate of breathing increases, the muscles start to twitch, and the face goes pale as the blood is diverted to the muscles. These are all part of the 'fight and flight' reaction, as it is called. Adrenaline is meant to prepare us either to run away from danger or to fight an enemy.

Receptors are found in the membranes of cells in the muscles, for example. Transmitters are substances produced by nerves in organs or muscles. For every hormone or transmitter there will be a specific receptor located in the membrane of particular cells. For some hormones there may be more than one receptor.

Receptors are crucial to the action of many drugs, which are often designed to act on a particular receptor or process.

Although receptors are generally sites for binding chemicals found naturally in the body such as hormones, other chemicals can also bind to them. This may be desirable for a drug but if other chemicals bind to them the result could be a toxic reaction.

If the chemical or one of its products destroys tissues that are part of an organ, that organ may fail which will in turn have other effects and consequences, depending on the organ and its functions. The effects of these interactions depend on many factors, and it is not within the scope of this book to discuss them in detail.

There are a limited number of types of adverse or toxic effects that chemicals may cause. These are described below.

Physiological effects

These include changes in blood pressure or heart rate. This type of effect is commonly caused by drugs, and so are also known as pharmacological effects. They are often due to the drug or chemical interacting with a particular receptor (see above). This effect may be the desirable action of a drug, for example salbutamol. This drug is known as a β-agonist and is prescribed for the treatment of asthma. The drug acts by dilating the airways (bronchioles) in the lungs and it does this by interacting with β-receptors present in the cells in these airways. These interactions do not require the drug molecule to be reactive, do not damage the protein, and are reversible. A reversible and normally desirable effect, in the case of a drug, is caused by this interaction with a receptor. But if the effect is exaggerated as a result of the dose being too high, or if the effect is a side effect of a drug or an effect of another chemical, it may be a toxic effect. For example, exposure to organophosphate insecticides causes an increase in the level of a substance called **acetylcholine** which occurs naturally in the body and is involved with the function of certain nerves (it is a neurotransmitter; see Figure 12, p. 101). Acetylcholine affects receptors in other tissues such as the lungs, as well as the nervous system. As will be explained in Chapter 4, toxic doses of organophosphates cause unnaturally large amounts of acetylcholine to accumulate, causing the airways to contract, hence people exposed to it will experience difficulty in breathing (and a variety of other effects).

Damage to tissues

This kind of effect can occur, where part or all of a tissue or an organ is damaged by the chemical and then may undergo a process called **necrosis** where the tissue degrades and disappears (see box). This may be due to a reactive chemical such as a free radical interacting with proteins or fats in cells and destroying them. The consequence can be that the internal structures and those that hold the cell together are damaged beyond repair and the cell breaks open. Alternatively or additionally the chemical or its products may interfere with vital metabolic processes in the cell by stopping the action of enzymes, for example. This is the kind of effect caused when people take overdoses of the drug paracetamol or drink solutions of the weedkiller paraquat (see pp. 51–4, 104–6).

Effects on metabolic processes

This type of effect can occur when a chemical interferes with a particular metabolic pathway, such as one which supplies energy, and individual cells or even a whole organ stops functioning. When this happens cells eventually stop functioning or slow down sufficiently so that the organ of which

Necrosis and apoptosis

Necrosis is the process of degeneration of tissue which can be caused by a chemical. The cells in an organ or tissue disintegrate, with the cell membrane breaking open and releasing the contents. Inside the cell there is a structure (called an organelle) known as the lysosome, essentially a bag containing enzymes, which break down proteins and other constituents of the body. If this bag bursts, as can happen during necrosis, the cell and those surrounding it are damaged and destroyed. Therefore necrosis spreads damage to the surrounding area, which becomes inflamed.

In contrast, **apoptosis**, which may also be caused by chemicals, is a more precise and controlled type of cell death. It is called 'programmed cell death' and is also a natural process involved in processes such as the development of the embryo, the removal of cancerous and damaged cells, and the shedding of tails by tadpoles. The name comes from the Greek word for the dropping of leaves in autumn. When cells die by apoptosis neighbouring cells are not involved; the affected cell simply fragments and disappears. Hence single cells that could be dangerous for the organism if allowed to remain can be removed.

they are part also stops functioning. If this is the heart or brain, death is the result. Cyanide and the natural pesticide fluoroacetate cause this kind of effect, which is often lethal (see pp. 152, 220). Aspirin will also cause this type of effect after overdoses that may be lethal (see pp. 61–2).

Effects on the embryo and foetus (teratogenesis)

This is an effect specific to the unborn baby. The kind of effects caused are early abortions, malformations, reduced growth, and low birth weight, and more subtle effects such as on behaviour or development after birth. **Teratogens** are particularly insidious because they are generally not harmful to the mother and affect only the developing and growing baby. An example of a teratogen is thalidomide, the sedative drug that caused serious malformations in babies when their mothers took it at a particular time during pregnancy (see pp. 56–9). As part of the process of development in a newborn baby, as well as of growth and repair, cells must divide in order to multiply. This process involves dividing up the DNA molecule into equal halves so that each new daughter cell has the same information. In the developing baby this process follows a defined sequence and after the initial stages, when all the cells are similar, cells take up particular positions and eventually become part of a structure, a leg for example.

There can be many different causes of adverse effects on the developing embryo but a chemical that interferes with the processes of cell division and cell growth is likely to result in this type of toxicity (see Figure 7). The chemical could interact with a molecule such as DNA or a protein. If the sequence of division and growth of cells is interfered with in any way, the development of the particular limb or organ is arrested or slowed. Such an interference can occur as a result of exposure to drugs, for example thalidomide, and to other chemicals. If the chemical is very toxic, usually the cells that form part of the developing embryo will be destroyed and the embryo will be aborted at an early stage. At a later stage this interference can lead to severe and lethal malformations. More subtle effects can give rise to malformations that are not life-threatening or slowed growth resulting in babies of low birth weight. As well as causing damage to the cells, interference that reduces the supply of energy, or the level of a vitamin or other important intermediate substance such as a sugar or protein, is often sufficient to cause a malformation. A drug or other chemical that has a very specific effect, such as inhibiting a process peculiar to the embryo, would be a very potent teratogen. Thalidomide is one well-known example (see pp. 56–9). Thus a chemical need not be reactive or kill cells to be dangerous to the developing embryo, for any interference with normal processes or conditions is potentially disruptive to the process of development.

Effects involving the immune system

The immune system is designed to protect us against foreign invaders, usually bacteria, viruses, or other organisms recognized as foreign. Our immune system recognizes any foreign bodies provided they are of a sufficient size, such as bacteria and even pollen and animal proteins. Exposure to such things does not always stimulate the immune system but some people are particularly sensitive to them. Many of the chemicals we are exposed to are usually much too small to be recognized as foreign by the immune system, but chemicals or drugs that are reactive or whose metabolic products are reactive can react with proteins in the body and alter them so that they are then recognized by the body as foreign. This response by the immune system can set up a reaction such as inflammation or asthma. The drug penicillin can cause several different types of immune reaction in some people by these types of mechanism (see pp. 66–8).

An alternative way in which chemicals can affect the immune system is by depressing the system so that it does not function properly. Dioxin is a

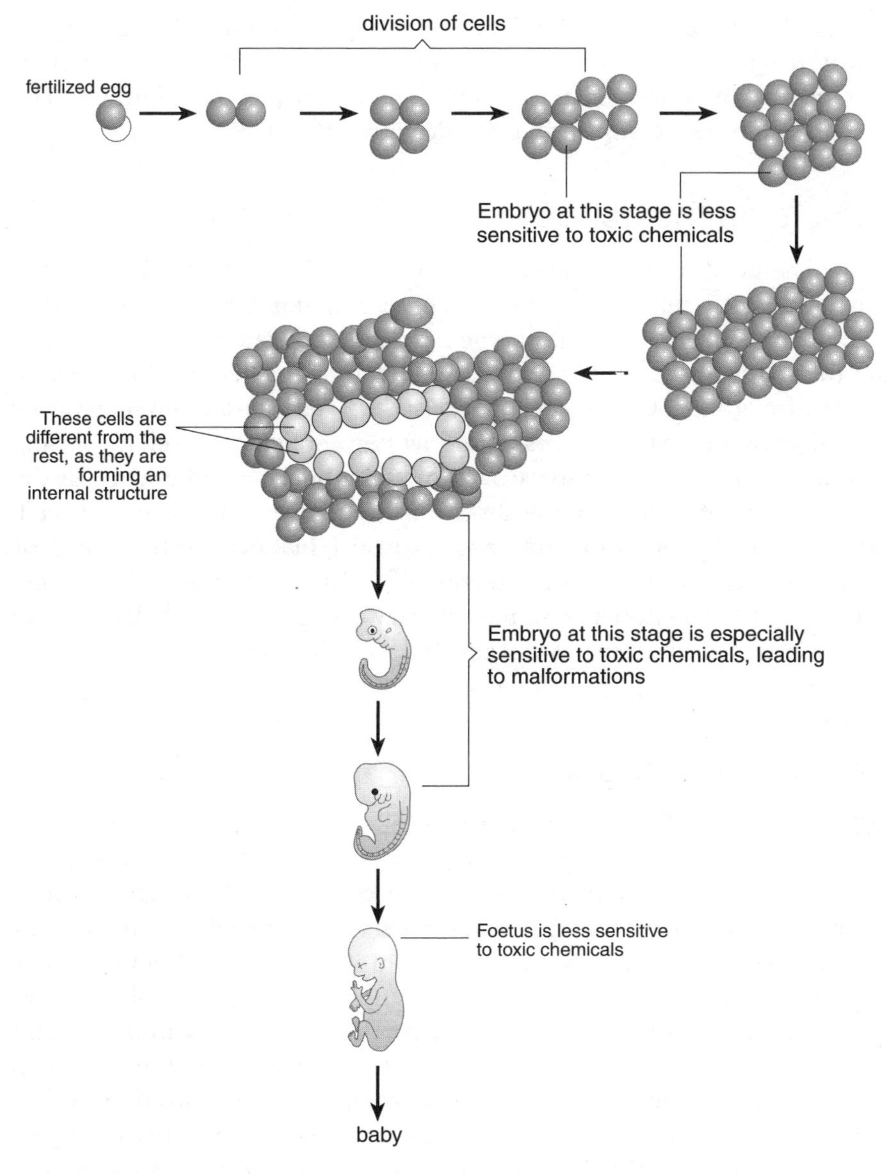

7. Schematic diagram of the development of a newborn animal. The process can be disrupted by chemicals at various stages so that the embryo develops with malformations or other abnormalities. The stages when the limbs and organs are developing are the most sensitive to disruptions that cause malformations. Disruption of early and late stages is more likely to result in smaller babies or abortions.

chemical that causes this kind of effect by interfering with the organs involved with and essential to the immune system (the thymus, which produces white blood cells). A person or animal with a depressed immune system will be more susceptible to infection.

Effects on the genetic material (mutagenesis)

Some chemicals can interact directly with the DNA or chromosomes in the nucleus of a cell. If this causes changes or major damage it can lead to a mutation. This means that the genetic code is altered so that cells that are derived from the cell(s) with altered DNA are different. If this occurs in germ cells such as male sperm or female eggs and is passed on to a son or daughter, an inherited disease such as haemophilia can result.

Alternatively the mutation can lead to major dysfunction such as the development of cancer or the death of the cell or of the daughter cells after it divides. Not all mutations are harmful, however. Some may have no effect at all on the cell if the section of DNA is not important or is not used at all. Sometimes a mutation may be positive and lead to an improvement in the function of the cell or animal or plant (the basis of evolutionary change).

Cancer (carcinogenesis)

When chemicals damage the DNA of a cell they can cause the appearance of a cancer or tumour at a later stage. If damage and a mutation occurs in a crucial region of the DNA molecule, or particular genes such as **oncogenes** are affected, the control of cell division may be affected so that the cell divides repeatedly and without control and a cancerous growth appears.

The process involves a number of steps: the interaction with or damage to DNA in a cell will lead to a tumour only if the cell divides and that damage is expressed. Only then does the uncontrolled division of the cell result in cancer. There are, however, protective mechanisms which can come into play. The DNA damage may be repaired, or if it is too extensive the cell may undergo a process called apoptosis or programmed cell death (see box above). This is a way in which the body removes single cells that have been damaged. Only if these processes are ineffective or overloaded will a tumour result.

There are other means by which chemicals can cause cancer that do not directly involve DNA damage. For example, some chemicals stimulate cells to grow and divide, some are irritants, and some inhibit the processes

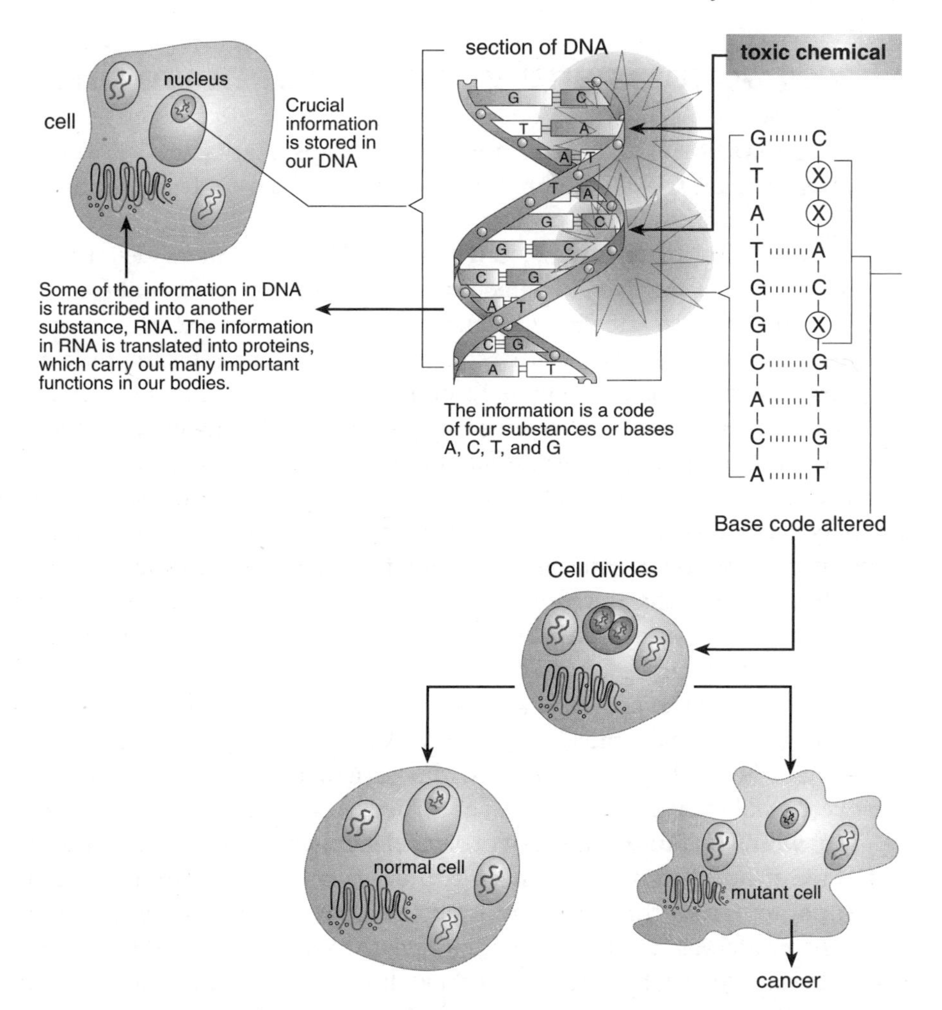

8. The process by which the information in DNA in cells is utilized by translation into proteins. This can be disrupted by chemicals if they are reactive and damage the substances (bases) A, C, T, or G which make up the DNA code, leading to mutations and possibly cancer.

of DNA repair (DNA is constantly being damaged in our cells and so must be repaired). Some forms of asbestos are carcinogens but they do not directly damage DNA. After inhalation, the fibres remain in the cells of the lung for many years and cause constant irritation which eventually leads to tumours.

The diagnosis and treatment of poisoning

Poisoning may take many forms, resulting perhaps from acute overdoses of a drug or from repeated (chronic) low-level exposure, say for a period of weeks or months, to a chemical. Both scenarios can lead to toxic effects. Some of these poisonings can be treated but not others, depending on the substance involved. For acute poisoning with many substances there are treatments available and standard treatments used in Accident and Emergency departments at hospitals. These will be revisited in later chapters, but the principles will be given here.

What happens if a doctor is called by distraught parents because their child has swallowed something that might be poisonous, such as paracetamol tablets or weedkiller? If the doctor is a general practitioner he/she may not know specifically what to do but can telephone the nearest Poisons Information Service in the United Kingdom or Poison Control Centers in the United States, who may be able to give advice on treatment. Often the patient will be taken to an Accident and Emergency department at the nearest hospital.

If the nature of the poison is known, treatment can begin immediately. If the drug or other chemical has been taken by mouth, the stomach can be washed out with water, provided the patient is conscious. This may remove the substance before it is all absorbed. Alternatively an emetic can be given which will make the patient sick, ejecting the remaining chemical out of the stomach. This won't be done, however, if the patient has taken something corrosive or a solvent, or if the patient is unconscious, as it may end up damaging the oesophagus or it may get into the lungs and cause damage or asphyxiation. Another alternative treatment is to give activated charcoal which is a very good adsorbent of chemicals, absorbing many times its own weight. An alternative adsorbent is fuller's earth.

If it is known which chemical the patient has taken, an antidote can be given if one is available, as for example with paracetamol (see pp. 53–4). Whether an antidote is used will depend on whether the overdose is at a dangerous level. Antidotes have been used (not always successfully) for centuries and one of the earliest mentions is in Homer's *Odyssey*, where it is suggested that Odysseus take 'moly' (possibly snowdrop) to protect himself against the poison of Circe. Apparently, in ancient China around 2000 BC antidotes were tried experimentally.

If there is no antidote and the identity and amount of the chemical is known, other measures can be taken to reduce the likelihood of poisoning, for example increasing the elimination by increasing the amount of urine produced by giving fluid or by changing the acidity or alkalinity of the urine. This can be done by administering bicarbonate of soda, which makes the urine more alkaline, or a salt called ammonium chloride, which makes the urine more acid. This increases the excretion of acids and alkalis respectively. If these treatments are not appropriate or the amount taken is very large, other techniques such as haemodialysis or haemoperfusion might be used. These techniques pass the blood of the patient through a machine which removes the chemical either by dialysis through a special membrane or by allowing it to bind to a special resin. For some chemicals, for example the weedkiller paraquat, there is no antidote or effective treatment (see pp. 105–6).

When the identity of the drug or chemical is not known, as in cases of homicide and true suicide rather than unintentional suicide or accidental poisoning, treatment is more difficult. There are standard tests that Poisons Units/Medical Toxicology Units can use on urine or blood samples to identify common drugs and poisons quickly. Sophisticated analytical equipment can be used to try to identify the chemical responsible, but precious time is lost while this is being done. Other evidence may need to be collected from the scene of the poisoning. Another problem the forensic toxicologist or clinical toxicologist may face is that the victim may have taken a mixture of chemicals. Suicides often take a handful of pills (whatever they can find in the house) and perhaps wash them down with whisky or vodka. This makes the treatment and diagnosis of the poisoning much more difficult.

The crucial facts to be determined are the identity, and then the level(s), of the chemical(s) in the blood and when they were taken. The poisoning due to the most dangerous chemical that is at the highest level in the blood can then be treated. For example, if someone has taken an overdose of the drug combination Distalgesic, which contains paracetamol and dextropropoxyphene, the immediate treatment would be for the latter drug which can rapidly cause death from depression of the respiratory system.

If someone has taken a mixture of drugs, a particularly dangerous drug may constitute only a small proportion of it, in which case the treatment would be for poisoning with a drug present at dangerous levels. Drugs and chemicals interact with one another in various ways, as discussed earlier in the chapter, and this could add another layer of complexity to the treatment of the true suicidal victim who takes the contents of their medicine cabinet.

A knowledge of the science underlying toxicology is essential for the treatment of poisoning and necessary for the design of new antidotes.

3

Keep Taking the Medicine

There are No Safe Drugs, Only Safe Ways of Using Them

DRUGS are substances that many of us take for granted, at least in the industrialized, affluent West. We expect them to be safe and we get upset if they occasionally cause side effects, especially serious toxic effects. These are called adverse drug reactions. Drugs are the chemicals most commonly used for suicide and often feature in accidental poisoning cases. To some, the word 'drug' is synonymous with drugs of abuse. Others may include medicines they are prescribed by the doctor or buy from a pharmacy. What is a drug? It is a chemical of any origin, natural or synthetic, which is taken or administered in order to effect a particular beneficial change in the body. This may be to treat a disease or to alleviate its symptoms. The definition also includes some chemicals often used by many people such as alcohol, tobacco, and caffeine, which most people don't regard as 'drugs'. They are drugs, however, and can be at least as hazardous as some of the more notorious drugs of abuse. In some parts of the world different drugs are part of the culture, for example coca leaves and khat which are chewed by the peoples of South America and north-east Africa respectively. Drugs are an important part of our society, and yet we have an ambivalent attitude to them.

There are several ways in which drugs can be poisonous, but they are chemicals, to which the Paracelsus Principle applies, that at some dose they will be poisonous or toxic. Indeed it has been said that there are no safe drugs, only safe ways of using them. Although the toxic and even lethal effects of drugs taken in overdoses for suicidal purposes may be accepted by most people, the adverse effects that occur at normal doses are not as readily accepted. Unfortunately these adverse effects do occur. They are usually detected as a result of patients reporting them to their doctors who then report the effects to the relevant authorities (see pp. 86–8 and Figure 9).

A glance through a reference book like the *British National Formulary*, which is used by doctors and pharmacists, shows that there are hundreds of drugs for dozens of ailments, diseases, and conditions. It is not possible in this book to do more than mention a few examples in relation to overdoses and adverse drug reactions. I shall therefore use particular drugs to illustrate the kinds of problems that can arise with drugs and to discuss some of the more well-known examples.

The adverse, toxic effects of drugs can be divided into one of two types, A and B. Type A adverse reactions are due to the pharmacological (therapeutic) effect of the drug and are therefore usually predictable and related to the dose. They are usually exaggerated therapeutic effects. Type B are unexpected, idiosyncratic effects.

By their very nature drugs are molecules that have effects on the body, as they are intended to. If they are synthetic they have been designed this way; if naturally occurring they will perhaps have been extracted from a plant for a particular activity. They are therefore potent molecules which usually work through particular receptors in the body (see box, pp. 37–9). Consequently, after excessive doses there will inevitably be unwanted, adverse effects.

Many drugs will cause adverse effects if the dose is excessive but some cause adverse effects at the correct dose in certain individuals. This can be due to an increased sensitivity of the patient because of genetic or other factors (see pp. 29–31). There may also be side effects which are not apparent at low or therapeutic doses but which become more important at higher doses. Occasionally, unexpected and serious adverse effects occur in a few patients. Sometimes this is due to an interaction between two drugs or between a drug and a food constituent such as occurred with the drug terfenadine (used to treat hay fever) and grapefruit juice (see case note, pp. 33–4).

Following an overdose, not only will the expected effects often be exaggerated but other different toxic effects may appear. Some may be unrelated to the pharmacological effects of the drug.

Paracetamol

This is one of the most popular and readily available drugs around the world. It is an analgesic, which means it is used to reduce pain, and also an **antipyretic**, which means it reduces temperature, as in a fever for example.

Paracetamol owes its development to serendipity. In 1888 a pharmacist in Vienna made a fateful error. In response to a request from two physicians for a chemical called naphthalene to treat a patient with a parasitic infection of the gut, he mistakenly gave them a chemical called acetanilide. This was found to be very effective at reducing the temperature of the patient due to the fever. Although it proved to be too toxic for large-scale use as a drug, this discovery eventually led to the development of paracetamol, a related chemical.

Acetanilide, or antefebrin as it became known pharmaceutically, was cheap and very effective as an antipyretic and analgesic. However, acetanilide was recognized as being rather toxic because it altered (**oxidized**) the haemoglobin in the blood. Similar substances were tried, including paracetamol, but the chemically related compound phenacetin, also found to be effective as an analgesic and antipyretic, became the drug of choice.

The drug phenacetin became very popular in certain countries because of particular circumstances. For example, the Spanish flu epidemic of 1918 led to widespread use of the drug for the treatment of the high temperature and aches and pains. It also became a favourite with Swiss watchmakers who suffered badly from headaches as a result of the close work required in watchmaking.

The widespread use of phenacetin brought about a change in the way the drug was perceived and eventually used. It became more of a confection, so much so that people would take phenacetin powders regularly and **prophylactically**. They would give it, like a sweet, as a 'present' to friends. The nature of phenacetin as a potentially dangerous drug was temporarily forgotten. The widespread use of this drug resulted in the appearance of adverse effects, in particular damage to the kidneys. Although this was never proven to be solely due to phenacetin, rather than the combinations of drugs often taken which included phenacetin, the latter drug was the common denominator and so became associated with the adverse effect of 'analgesic nephropathy'. This paved the way for the reintroduction of paracetamol. It was found that in the body both antifebrin and phenacetin were changed mainly to paracetamol and that this was responsible for the therapeutic effects. Consequently paracetamol was investigated, found to be effective, and then developed as a safer alternative to antifebrin and phenacetin. Paracetamol is one of the most successful drugs ever marketed. At the same time, it is also the drug most commonly associated with overdoses.

The drug has two effects, reducing both temperature and pain. It works by inhibiting the production of substances that are involved with pain and temperature in the same way as aspirin (see box).

How paracetamol and aspirin work

Both these drugs, used to treat minor aches and pains and to reduce fever, have a similar mode of action. They inhibit an enzyme called cyclo-oxygenase, or COX. This produces substances called **prostaglandins**, which are mediators produced in response to damage to cells and which tell the body that there is damage, which we feel as pain. Blocking the production of the prostaglandins therefore reduces the sensation of pain. They also cause relaxation of blood vessels in the brain which causes headache; drugs such as aspirin reduce this by reducing the amount of prostaglandin present. The prostaglandins can also increase the temperature by acting on a particular area in the brain. If there are fewer of these mediators, the temperature will remain more or less normal and fever will be reduced. The prostaglandins, along with other mediators such as kinins and histamine, are also involved with the inflammation that occurs when tissue is damaged and infected by bacteria. Therefore aspirin will reduce inflammation by reducing the amount of prostaglandin available. Paracetamol does not do this because it is less effective at inhibiting COX; it mainly inhibits this in the brain, allowing the pain relief and temperature reduction. Therefore paracetamol does not reduce inflammation.

At first paracetamol (acetaminophen in the USA) was considered a safe and effective drug. It became increasingly popular and was used by millions worldwide. A problem then began to appear, as some people began to take overdoses of the readily available drug.[1]

The drug, unfortunately, causes serious toxic effects at high doses. Although two tablets, the normal recommended dose for an adult, are perfectly safe, twenty-two could be a lethal overdose. The number of deaths worldwide amounts to hundreds and many more are made seriously ill. However, paracetamol is a safe drug, provided it is used at the correct dose. (Remember there are no safe drugs, only safe ways of using them.) Here we have yet another illustration of the Paracelsus principle: 'All substances are poisons; there is none that is not a poison. The right dose differentiates a poison from a remedy.' Thus even a drug like paracetamol, which is safe enough to be sold over the counter in pharmacies, is capable of becoming a poison. The number of deaths caused by this relatively very safe drug, as well as the illness it has caused, have, unfortunately, led to calls for it to be withdrawn (see case notes).

Any drug taken in overdose is likely to cause death and illness, and there are many other drugs readily available over the counter that are

CASE NOTES

Paracetamol overdose

In 1966 a young man, an inmate of a mental institution, was admitted to hospital in Edinburgh. He had taken a quantity of paracetamol tablets, his estimate being 150, along with a quarter of a pint of vodka, some seven hours earlier. During the next twenty-four hours he was reported to be well but obstreperous. On the second day in hospital he complained of pain in the abdomen and nausea. On the third day he was breathing abnormally (hyperventilating) and appeared **jaundiced**. On the fourth day his condition worsened, and he died 80 hours after admission. At **post-mortem** the liver was found to have signs of extensive damage as did the tubules of the kidneys. The levels of paracetamol in the blood were found to be very high, consistent with an overdose of the drug. The victim died as a result of liver damage and probably kidney failure. Although the patient had been taking and was given other drugs, paracetamol was the only one taken at an abnormally high dose.

This case of fatal liver damage due to paracetamol overdose was reported in the *British Medical Journal*. It was one of the first such cases to be described. Another case was reported in the same article, also from Edinburgh, and another in the same volume of the journal. Overdose cases now number in the hundreds and possibly even thousands each year throughout the world.

If the patient had taken the overdose in 1996 he might well have survived due to the availability of an effective antidote.

CASE NOTES

Doctors demand curbs on 'killer' paracetamol

■ By Lois Rogers, medical correspondent

Doctors are demanding controls over paracetamol sales after research has revealed that poisoning incidents from the painkiller have reached record levels.

Results from the first analysis of patients receiving emergency treatment showed that as many as 40,000 people suffer serious overdoses.

Dr Chris Owens, a senior pharmacology lecturer at University College Hospital, London, studied 54,000 emergency patients last year. He discovered 167 paracetamol poisoning cases compared with 129 heroin overdose cases. Nationwide more than 500 deaths a year are associated with the drug. Owens believes the figures reveal a national scandal. 'It ought to command attention that we have this number of deaths from something you can purchase over the counter,' he said.

Sunday Times (14 November 1993)

potentially hazardous under these conditions. There are also many other substances that people intent on suicide can acquire and take, some of them much more unpleasant than paracetamol. Fortunately, with paracetamol there is an antidote, whereas with many chemicals that can be used for suicide there is no antidote.

Why does a drug that is safe when taken at the recommended dose become potentially lethal? The answer lies in the metabolism. The toxicity of paracetamol is not due to the drug itself, nor is it related to the therapeutic effect it has on the body at low doses (lowering of temperature and reduction of pain). Paracetamol becomes poisonous to the liver after large doses and may cause liver failure.

As discussed in Chapter 2, after a drug (or other chemical) is taken by mouth, it first reaches the stomach and then the small intestine. It is most likely to be absorbed from the small intestine, where it enters the bloodstream which will transport it directly to the liver (see Figure 3, p. 15). In the liver paracetamol is changed into two other substances, which are relatively harmless but another metabolite, which is produced in only small amounts, is potentially poisonous. This metabolite is normally safely removed by further metabolism, which employs the substance glutathione (see box, p. 24). After a large overdose, however, this process of detoxication is overwhelmed and the poisonous substance attacks the liver.

The result, if the damage is extensive, is liver failure and possibly death. The liver is so critical to the body that when its function becomes compromised, other parts of the body are affected. For example, the brain will be affected by the excess ammonia which builds up as it is not removed by the liver, and the patient may suffer convulsions and fall into a coma; and the blood won't clot properly as the clotting factors are made in the liver. Fortunately, research in the USA carried out on experimental animals, and then in human patients, led to an understanding of how paracetamol is toxic to the liver and to the development of a successful antidote, Parvolex.

The development of an antidote

Knowledge that paracetamol was detoxified by the thiol glutathione led the development of an antidote. Glutathione itself cannot be given to the patient after an overdose, so a number of similar substances were initially tried as possible antidotes. There was some success but some had unpleasant effects. It was then found that a substance called N-acetylcysteine would help to regenerate the glutathione in the liver.

This was without major side effects and seemed to be successful. It is given by injection directly into a vein. If given early enough after an overdose, up to ten or twelve hours, the antidote is usually successful.

When a patient has a suspected paracetamol overdose the first stage of treatment at an Accident and Emergency department of a hospital will be to wash out the stomach to remove any undigested tablets (commonly called pumping the stomach). Sometimes this can reveal the substance that has been taken, which can be useful with true suicides who won't disclose what drug or drugs they have taken. Blood and urine samples will be analysed for various common drugs and other chemicals including paracetamol. The analysis may identify the drug taken by the patient. Then a specific technique will be used to measure the amount of the chemical(s) in the blood. This will indicate the severity of the overdose, that is how much has been absorbed and whether an antidote is necessary (see also pp. 46–7). The symptoms of paracetamol poisoning after overdose can initially be quite mild and therefore misleading. The patient will probably be sick or complain of nausea and some pain in the abdomen, but this may recede and the patient can feel better for a day or so before they deteriorate and die. The poisoning can develop over several days and therefore it is not a rapid exit for would-be suicides.

The outcome, however, can depend on the individual patient. The recommended dose of paracetamol is 1 to 2 tablets (500 mg each) but the size of dose necessary to cause damage depends on the individual. The average individual who is not taking other drugs regularly, may have to take about twenty tablets to suffer liver damage, but with some individuals as few as ten tablets can have the same effect. There is individual variability in the sensitivity to paracetamol due to variation in the metabolism. A major factor is the use of alcohol and certain other drugs such as barbiturates (see case notes).

Barbiturates and alcohol both increase the activity of (induce) the enzymes that metabolize paracetamol to the toxic product. A chronic alcoholic or someone who has been prescribed barbiturate drugs will be at greater risk when taking more than the recommended dose of paracetamol because a greater proportion of the dose will be converted into the toxic product. It would be equivalent to taking a larger dose. In the case of an alcoholic the liver may already be compromised by repeated and long-term alcohol abuse and so be more vulnerable and less able to detoxify paracetamol; it may also have less protective substances like thiols.

In contrast, other drugs could inhibit the metabolism and so reduce the toxicity, as is believed to happen when a large amount of alcohol is taken

CASE NOTES

Beer and paracetamol

A 43-year-old man was admitted to hospital suffering hallucinations. He had fallen off his bike, fractured a bone in his shoulder, and was prescribed one to two tablets of Tylenol (paracetamol plus codeine) every four to six hours for two days. He continued to suffer occasional hallucinations and vomiting and from jaundice. Once in hospital liver function tests on his blood indicated that he had liver damage. He died in a hepatic coma thirty hours after being admitted to hospital. It was later revealed by relatives that the patient had also treated himself with nine Tylenol tablets plus ten tablets of another preparation after the bicycle accident. Another important factor was that he regularly drank half a case (twelve bottles) of beer each day.[2]

at the same time as paracetamol. Other factors such as food constituents and genetic variation can also be important. For example, it has been shown that paracetamol metabolism (by the toxic route) varies by as much as tenfold between individuals. Clearly the toxic effect(s) of paracetamol after an overdose varies between individuals, as can be seen in the next case.

If an overdose of paracetamol combined with other drugs is taken, the situation may be very different. For example, the combined preparation Distalgesic contains not only paracetamol but also a drug called

CASE NOTES

Remarkable tolerance

A 58-year-old woman was being treated in hospital for rheumatoid arthritis. It transpired that she had been taking large amounts of paracetamol for this condition for the past five years. It seemed that she was in the habit of swallowing 15 to 20 g of paracetamol, 3 g at a time, throughout the day, apparently without any problems. She also was drinking significantly (8 units of alcohol per day, which is more than the recommended daily amount). Usually, any more than 10 g of paracetamol is considered to be likely to cause toxic effects. Chronic alcohol intake also tends to increase the likelihood of toxic effects from paracetamol and to reduce the dose likely to cause these effects.[3]

So why was the woman immune to these effects? It seems that there were three reasons for her tolerance: first there was a slower rate of absorption of the drug from her gut; secondly detoxication was increased and there was less toxic metabolite formed; and finally metabolism overall was slower than seen in average individuals.

dextropropoxyphene. When overdoses of this are taken death can occur fairly quickly as a result of the effects of the dextropropoxyphene which causes depression of respiration. The antidote for paracetamol is useless in this situation.

Paracetamol is now sold in smaller packs to make it more difficult for suicidal people to acquire sufficient drug for an overdose. An understanding of why the drug is toxic has stimulated the development of other possibly less toxic analogues of this drug.

Thalidomide

There is much that is interesting about the drug thalidomide both historically and scientifically. Thalidomide became infamous and a symbol for all that was bad about drugs and the drug industry, but even this dark cloud has turned out to have a silver lining and there are a number of lessons that can be learnt.

It all started in 1962. Thalidomide was a sedative and was marketed as a drug to counter the effects of morning sickness in pregnancy. Originally made and marketed in West Germany as Contergan, it was later marketed in the UK as Distaval. After a period of use it became apparent that in a number of countries, including the UK and West Germany, babies were being born with unusual malformations. The malformations were, mainly, an absence of arms (amelia) or shortened arms (phocomelia). After some time an astute physician in Australia connected the appearance of this rare malformation with the use of thalidomide by the women to whom these babies were being born. It occurred only if the women had taken the drug between days 20 and 35 of their pregnancy, during which time the limbs and other structures are being formed in the embryo (see pp. 41–3). As well as the absence of arms, other abnormalities were also detected.

Thalidomide is a relatively safe drug for adults, including pregnant women themselves, but it may have adverse effects on the nerves in the extremities (**peripheral neuropathy**). This drug is therefore a classic teratogen, being *selectively* toxic to the embryo, which is exceptionally sensitive to thalidomide. The embryo is always potentially more sensitive than the adult simply because it is developing and the processes involved in development are easily disturbed by chemicals. Fortunately, the mother acts as a protective barrier to a certain extent so there is usually a threshold below which chemicals will not affect the embryo. With drugs, however, the threshold may be too low for them to be effective in the adult.

After the drug was withdrawn, litigation was started by victims' groups against the manufacturers. While there were, of course, many articles in the press about the drug and the plight of the babies born with the terrible malformations, few publications appeared in the scientific literature on thalidomide, as a result of the legal proceedings which went on for a number of years.

There were several results of the thalidomide tragedy. First, it was a major factor in the setting up of the Committee on Safety of Drugs (later to become the Committee on Safety of Medicines) in the UK. This committee reviews the data available on drugs that have been developed before the manufacturers are given a licence allowing the drug to be marketed. Secondly, the medical profession was shocked by what had happened, and it emphasized that drugs could be toxic and hazardous. This had a permanent impact. Finally, the tragedy also led to improvements in the safety evaluation of drugs, such as demanded by the Food and Drug Administration (FDA) in the United States for example, which have continued to this day (see Chapter 12).

How did it happen? Prior to the introduction of thalidomide, drug safety evaluation was fairly limited and nowhere near as rigorous as it is today. The animals used for these studies, rats and rabbits, have different processes of development of the embryo from humans, and so administration of thalidomide to pregnant rats or rabbits did not cause malformations. These species seem relatively insensitive to thalidomide, although often the rat can be more sensitive to chemicals that are teratogens because of differences in the placenta and process of development. Rhesus monkeys were later found to suffer the same malformations as humans. Thalidomide was therefore unusual. Furthermore, it seemed to be well tolerated as a drug, causing apparently few if any side effects or toxicity in adults. It was generally believed that the mother was a natural barrier and so protected the developing baby, and so the use of thalidomide in pregnancy was accepted. The disaster led to improvements in the testing of drugs for teratogenicity, including the development and use of *in vitro* methods. The methods now used to test drugs are much more likely to detect teratogenic effects. Much greater caution is also exercised in the prescription and use of drugs in pregnancy.

As we now know, teratogenesis does not occur only as a result of exposure to a chemical that is particularly toxic to cells or animals. This particular type of toxic effect simply depends on the chemical interfering with the processes of development (see Figure 7, pp. 41–3). Because the interference that underlies the teratogenic effect can be a very small or subtle change, it may have no significance or may not even occur in the mother, but in the embryo the effect may be huge. Thalidomide is a potent teratogen with specific effects on development in the embryo. It therefore has very potent effects on the embryo when taken at the crucial time but not at other times, and it has no effect at all in the mother.

Chemicals with such subtle effects on the embryo will not be detected in general safety evaluation studies designed to detect toxicity and performed using animals. There are now studies designed specifically to optimize the detection of teratogens. The study of the mechanism by which thalidomide is toxic did not progress very far while litigation proceeded, because of the difficulty of publishing the results. The court cases precluded publications that might influence the legal proceedings. However, once these were over, work was published which led only quite recently to an elegant theory explaining how the drug caused the effects it did.

How thalidomide is teratogenic

The thalidomide molecule exists in two forms called **isomers** (R and S), which are mirror images of each other. Only one (the S form) is teratogenic, and the other (R) is responsible for the sedative action. Unfortunately the liver converts the R form into the teratogenic S form and so the R form could not be used as a safe drug. It now seems that the three-dimensional shape of the molecule of thalidomide is crucial for it to be a teratogen. Only the S form of the thalidomide molecule fits into a particular section of DNA in the nucleus of the cell. It is thought that this interferes with the production of substances called growth factors. Some of the growth factors affected are those involved with the growth of new blood vessels. This is crucial in the development of an embryo: once structures such as limbs start to form, they must be supplied with blood. Other growth factors will be required for the development of limb buds. This particular effect explains why thalidomide is so potent but also so specific to the embryo.

The discovery of how thalidomide causes birth defects has suggested a potential use for it as an anti-cancer drug. The growth of blood vessels is an important part of the development of tumours, which if it were stopped would limit their invasion of tissues and therefore the spread through the body.

The case of thalidomide therefore illustrates another principle in relation to our use of chemicals: there are no safe drugs, only safe ways of using them. In addition to the Paracelsus principle regarding the dose of the chemical, we have to consider how chemicals are administered or the circumstances of exposure. There are certain conditions such as pregnancy or during breastfeeding when exposure to chemicals of any type is unwise or should be reduced as much as possible. There are patients such as the elderly or very young children whose dose may need to be reduced or to whom the drug may not be administered. There are also some genetic factors that make individuals sensitive to a particular drug or group of drugs, which should not be administered to such individuals (see pp. 71–2 on hydralazine). We now have the information to be able to adopt a more flexible approach and make the use of chemicals much safer, at least in some cases.

The final chapter in the thalidomide story is not yet closed. Further research with the drug has revealed that the mechanism that underlies its teratogenic action may be used to advantage in the treatment of cancer. It also has other effects which make it potentially a unique drug for the treatment of leprosy, HIV, and certain serious diseases affecting the mucous membranes (such as those lining the gastrointestinal tract and vagina). Its use in the treatment of leprosy, unfortunately, resulted again in malformations because of inadequate control. It is also being investigated for use in the treatment of tuberculosis.

Thus despite the fact that thalidomide caused such terrible effects in newborn babies, it may yet prove to be a very useful drug under the right circumstances for the treatment of serious diseases and not in women who are liable to become pregnant.

Aspirin

Like paracetamol, aspirin is an analgesic and antipyretic drug which is commonly available for sale over the counter at many shops and supermarkets, as well as pharmacies. However, it also has a third action, to reduce inflammation—it is an **anti-inflammatory** drug. It belongs to a group of drugs called non-steroidal anti-inflammatory drugs or NSAIDs.

Discovered to be a useful drug in 1899, Aspirin is chemically described as a salicylate, a group of chemicals that are also found naturally in plants. The ancient Greeks knew to treat arthritis by applying a preparation of crushed willow bark and leaves in oil to the inflamed joint. Hippocrates recommended it for gout and the removal of warts. Much later, in 1763, in England it was noted by the Revd Edmund Stone that people living in the country knew to treat aches, pains, and fevers with similar preparations which would lower the temperature in a fever. The bark of the willow tree (*Salix alba*) was later found to contain a chemical called salicin by an Italian chemist, Raffaele Piria, in 1830. It was discovered that salicylic acid was the ingredient in the bark preparations that reduced temperature, pain, and inflammation. Because salicylic acid had unpleasant side effects (upsetting the stomach and making the patient sick), the German chemical company Bayer set about making derivatives. Acetylsalicylic acid was one of those made but not tested. In 1899 a chemist who worked for Bayer, Felix Hoffman, administered acetylsalicylic acid to his father who had arthritis and could no longer tolerate salicylic acid. It not only worked better but did not irritate the stomach as much as salicylic acid. So aspirin was born and continues to be a major drug.

The main side or adverse effect of aspirin is to cause stomach bleeding, especially when taken in large doses or repeatedly. The reason for this is related to the way in which aspirin acts to reduce pain, stopping the production of substances in the body called prostaglandins.

How aspirin causes stomach bleeding

Aspirin is readily absorbed into the cells that form the lining of the stomach. Some of the cells in the stomach produce acid (hydrochloric acid) which aids digestion. Substances called prostaglandins reduce the amount of gastric acid produced and also help in the production of mucus, which protects the stomach against the acid produced.

Aspirin reduces the production of prostaglandins. It inhibits an enzyme involved in their synthesis called COX-1; this is how it acts to reduce pain and inflammation. Therefore aspirin causes changes that increase the likelihood that the acid will damage the lining of the stomach. Furthermore aspirin is an acid itself, and may also contribute to the natural acidity. If repeated doses of aspirin are taken, the result can be bleeding and ulceration which may be severe in some patients. Therefore aspirin should always be taken with food.

Some related drugs that have been developed do not have the same toxic effects as aspirin on the stomach as they cause greater inhibition of a related enzyme (COX-2) which exists only where inflammation occurs. These are as effective but without the adverse effects on the stomach.

In young healthy adults, however, there are few problems associated with the occasional use of aspirin in single doses as described on the packet, although the possibility of stomach bleeding may lead some people to prefer paracetamol. The newer drugs, NSAIDs, which have less effect on the prostaglandins in the stomach, and therefore cause less damage and bleeding and a lower likelihood of ulceration, may also be preferable.

Problems emerge when aspirin is taken repeatedly, and care must be taken especially if relatively large doses are taken repeatedly. When people repeatedly dose themselves, for example when they have flu or a prolonged cold, problems can potentially arise. The disposition of aspirin is unusual because the elimination can be saturated. This means that as the dose is increased, the time taken to eliminate the drug into the urine increases. Therefore if someone, perhaps with flu, takes the drug continually, that is several times a day for several days, the drug can accumulate in the body and start to cause toxic effects. Aspirin is also an anti-inflammatory drug, and for this reason it is sometimes used in elderly patients for the treatment of arthritis where other drugs are not tolerated or effective. As the function of the kidneys decreases with age, elderly patients eliminate salicylate less efficiently and therefore are more at risk of the drug accumulating. Repeated large doses of aspirin over long periods, which may lead to some accumulation, can also cause an adverse effect called salicylism whereby patients get severe headaches, ringing in the ears, and tremors in the hands.

If the drug accumulates there is the danger of overdosage, with similar effects to those seen after a single large dose is taken all at once, as might happen in suicide.

Reye's syndrome

Aspirin can also cause a serious condition in children, known as Reye's syndrome, when it is used in young children who have had certain viral infections. This syndrome includes accumulation of fat in the liver and it may be fatal. The mechanism that leads to it is unknown but it is not advisable for children under 12 to be given aspirin.

Overdoses of aspirin

As with paracetamol, overdoses of aspirin are unfortunately taken by people as a means of suicide, although the amount that has to be taken for a fatal dose is greater than for paracetamol. There are probably several hundred deaths per year from aspirin overdose worldwide, and in a recent year 10,000 calls to Poison Control Centers relating to aspirin overdose were reported in the USA.

Accidental overdosage can also be a problem especially in children, and the introduction of childproof containers has reduced the risk of this.

Problems sometimes arise if children regard drugs like aspirin tablets as sweets or candy, perhaps as a result of their parents suggesting this in order to convince them to take the medication. As indicated above, accidental overdosage can also occur if people treat themselves repeatedly and too frequently with aspirin.

Another problem is that often different preparations contain aspirin along with other drugs, and unless people carefully read the packet to discover the drugs contained in a cold cure, for example, they may take more aspirin than intended without realizing it. For example, someone may take aspirin for their aching and temperature, along with a decongestant preparation which also contains aspirin.

The toxicity of aspirin is an interesting example of multiple biochemical abnormalities. The drug is metabolized to salicylic acid which is responsible for both the toxicity and the desirable effects. After an overdose the salicylic acid causes effects on the breathing rate and on metabolic processes in cells, with the result that the temperature rises, the breathing rate increases, and the metabolic rate increases. A further result of some of these initial changes is that the acidity of the blood changes, first falling, then rising, and the level of glucose in the blood goes down (ie. the PH rises then falls). The increase in the acidity of the blood allows more of the drug to enter the brain where the main toxic effects occur. The symptoms are nausea and vomiting; a high temperature; rapid, deep breathing; ringing in the ears (in some people this may occur after normal doses); and headache. Serious overdoses lead to coma and death.

Fortunately there is a means of treatment but, in contrast to paracetamol, there is no antidote. As with many drugs (and other chemicals taken by mouth in overdose), the hospital Accident and Emergency department will wash the stomach out to remove any tablets that are still there. Sometimes undigested tablets can be present many hours after an overdose is taken (see pp. 46–7). The key to the treatment is increasing the elimination of the aspirin from the body and changing the acidity of the blood. This can be done by giving the patient bicarbonate of soda along with glucose. They would also be sponged down to reduce their temperature.

Apart from the symptoms, how would a doctor know that someone had taken an overdose of aspirin? As already described, there are simple tests that can be done on urine to find out if someone suspected of an overdose has taken any of the common drugs. After the drug taken is known, it is important to measure the level of the drug in the blood to find out the likely dose and therefore the necessary treatment (see pp. 46–47).

Beneficial effects of aspirin

Apart from the widespread use of aspirin to relieve pain, fever, and inflammation for the past 100 years, more recently other beneficial effects have been found. It has been known for some time that aspirin will reduce the clotting of blood, as a result of its interference with the production of prostaglandins, which are involved with the clotting process (see box above p. 51).

When and why would it be beneficial to slow the blood clotting process? Although we need to have a blood clotting system to prevent the loss of blood on injury, a blood clot (a thrombus) can be hazardous if it forms in a blood vessel and becomes dislodged into the circulating blood. This can move around the body and lodge in the heart causing a heart attack, or in the brain causing a stroke. For those who have had a heart attack, immediate treatment with aspirin reduces the chances of death by one-quarter (that is one person in four can be saved). Thus carrying an aspirin tablet to chew in case of a heart attack is suggested by some doctors. Subsequent treatment with low doses of aspirin also reduces the chance of having another heart attack.

Some doctors recommend patients take low doses (75 mg) every day to reduce the chance of a first heart attack. This is known as a prophylactic use of the drug. For the same reasons aspirin can also be used to reduce the chance of suffering a deep vein thrombosis, which can occur after prolonged inactivity, such as a long distance flight on an airplane or long periods in bed due to illness. Apart from damage to the surrounding tissue and pain, these can also lead to heart attacks and strokes. This practice may also reduce the likelihood of having a stroke, although this is complicated by the fact that some strokes are caused by a burst or leaking blood vessel in the brain. In this case the effect on clotting may increase the magnitude of the stroke.

Thus the effect on clotting might be seen as an unwanted side effect by someone who suffers stomach bleeding after repeat doses of aspirin, but in some circumstances it is desirable. This again illustrates a principle in the use of chemicals that is illustrated by the use of anti-cancer drugs (see below), the principle of selective toxicity, which we shall revisit in Chapter 4: what might be considered an adverse or toxic effect in some circumstances or in some organisms may be beneficial in others. The assessment and appreciation of toxicity sometimes needs to take this into account.

As well as slowing clotting, aspirin may also have another beneficial affect in relation to heart disease. It has been found that there is a relationship between inflammation, such as is caused by infections, and heart disease. As an anti-inflammatory drug aspirin may also have a beneficial effect here. Similarly the anti-inflammatory and antioxidant effects of aspirin have been suggested as being responsible for the apparent beneficial effect of aspirin in Alzheimer's disease. Two factors in this neurodegenerative disease are oxidative damage to the fats in the brain (see box, p. 23) and inflammation that can damage nerve cells. Aspirin could contribute to the prevention of both.

Finally, in this veritable list of beneficial effects, there is some suggestion that patients taking long-term doses of aspirin may be less prone to suffer from colon cancer. There is a possible basis for this, as some other drugs that interfere with the metabolism of prostaglandins (COX-2 inhibitors) stop the growth of blood vessels into tumours and kill tumour cells. Although aspirin is not as specific and potent an inhibitor of this process as the drugs investigated, it does have some inhibitory activity on COX-2.

As we have seen, this simple molecule, with its humble origins in willow bark, has a remarkable range of beneficial properties.

Penicillin

Antibacterial drugs

In earlier centuries millions of people died from infectious diseases, which were partly responsible for many early deaths and the shortened life expectancy. That this is no longer the case and this aspect of our lives has changed so much for the better is largely due to two things—sanitation and hygiene, better drinking water and living conditions; and antibacterial drugs. Many of these drugs were discovered by chance and their development is a shining example of the importance of chemicals in our society. They are also an important illustration of another principle in relation to the toxicity of chemicals: toxicity can be selective and can therefore be used beneficially.

Apart from arsenic compounds, used to treat infections such as syphilis, other antibacterial agents were discovered and developed in the latter part of the nineteenth century and the early part of the twentieth century. One of the earliest was sulphanilamide, the first of a long line of sulphonamide drugs. As so often in scientific research, this was discovered accidentally. Possibly as a result of his background as a chemist with an interest in dyes, Paul Erhlich had shown that some substances can stain or bind to bacteria and suggested that such a chemical might be active in killing that cell and so be useful in treatment. Chemical companies began to look at the properties of other substances such as dyes. One of these was Prontosil Red, which Gerhard Domagk, a bacteriologist working for the Bayer Company in Germany, discovered killed bacteria known as streptococci in infected mice but did not appear to harm the mice. Indeed Domagk chose to try it out on his own daughter who was suffering from a serious infection with the same bacteria. Although she turned bright red, she survived and was saved from likely death. It was later found that it was only part of the prontosil molecule that was active, as it was broken down by bacteria in the intestines into sulphanilamide which was then used in its place. Understanding how these agents killed bacteria allowed a range of new derivatives to be developed.

Another chance discovery was made by the scientist Alexander Fleming in London in 1928. He happened to notice, on his return from holiday, that an agar plate on which bacteria had been grown had become contaminated with a mould. The area around the mould was clear of bacteria and he wondered if this meant that the mould had produced an antibacterial substance. The mould was *Penicillium notatum* and the antibacterial substance was called penicillin by Fleming. In fact others had noticed this effect before, but it was neither them nor Fleming who developed the observation into what became one of the most widely used and successful drugs. Three scientists at Oxford University, Ernst Chain, Howard Florey, and Norman Heatley, isolated and purified penicillin ten years after Fleming's discovery. They were able to produce enough to use in human patients. Other scientists and clinicians also used early preparations and managed to produce enough to show dramatic effects on patients with life-threatening infections.

It was known even in the seventeenth century that mouldy bread could be used to treat infections. The mould that grows on bread is often a *Penicillium* type. Many other drugs have now been derived from moulds, or fungi as they are also known. However, as will be discussed in Chapters 6 and 10, fungi may also produce some highly toxic chemicals which have featured in human death and disease.

Antibacterial drugs are examples of chemicals that are selectively toxic to specific cells or organisms such as bacteria or fungi. These agents are useful to man because they are toxic to bacteria and have been 'designed' to be that way.

How penicillin kills bacteria

How is penicillin toxic to bacteria and why does this not affect human patients? This, in a nutshell, is due to its interference with the formation of the bacterial cell wall. Penicillin stops it being formed and so stops the bacteria from reproducing. Bacterial cells, unlike mammalian cells, have a rigid wall; penicillin and its derivatives act on this cell wall. The walls are made of molecules linked together into macromolecules. These macromolecules are known as polysaccharides as they are chains of sugar molecules (saccharides) linked together. There are also other chains cross-linked with the polysaccharides, comprised of chains of amino acids, or polypeptides. When the bacterial cell wall needs to be repaired or when the cell divides, which happens about every 20 minutes in bacteria, new polysaccharide and polypeptide chains form cross-links. This requires one of the amino acids at the end of the polypeptide chain to break off and the end to combine with that on another polypeptide chain. The process is catalysed by an enzyme. Penicillin is similar to the two end amino acids of the chain and the enzyme binds that instead, which stops the process so that the cell wall cannot be repaired or built when the bacteria divide. Damaged bacteria die, while those dividing are stopped in their tracks. Because human cells don't have a cell wall or the polysaccharide-polypeptide structure, they are not affected by the penicillin. Hence penicillin is selectively toxic.

Penicillin is a natural chemical, presumably produced by a mould to protect itself against bacteria. It is selectively toxic to that type of organism, and we use it in the same way to target invading bacteria. This is one of many examples of the positive uses of the specific toxicity of a chemical for the good of humankind (and other animals too). Penicillin has saved the lives of millions of people since its discovery. We have a humble mould to thank for producing it, and serendipity and Alexander Fleming and others for interpreting its effects. Serendipity plays an important part in scientific discovery and progress, but chance happenings must be observed and their importance recognized. As the famous French scientist Louis Pasteur once said, 'Chance favours the prepared mind'.

Is penicillin completely harmless to humans? Unfortunately it is not completely safe but, as you may appreciate by now, no drug is completely safe. Penicillin does have a very low **acute toxicity**, so relatively large doses are tolerated well, but it does cause adverse effects in sensitive individuals even at therapeutic doses.

Adverse effects of penicillin

Penicillin and related drugs do have some adverse effects and have caused deaths and serious illness. The number of lives these drugs have saved and continue to save, however, far outweigh the deaths and illness caused. This is an example of the important concept of risk–benefit in relation to chemicals in general and drugs in particular. We shall revisit this in Chapter 12.

Penicillin was developed at a time when safety evaluation of drugs was limited. Furthermore the Second World War occurred shortly afterwards during which penicillin was used extensively and saved thousands of lives. This probably gave the drug impetus. Had the drug been tested by today's rigorous standards it might not have been accepted, certainly if it had been evaluated in guinea pigs, a species which is particularly susceptible.

Penicillin is, in chemical terms, quite an unstable molecule and it interacts with proteins. These interactions can lead to allergic reactions (hypersensitivity) of various types, all of which are toxic effects involving the immune system (see Chapter 2 and box).

Penicillin and allergic reactions

Penicillin is an unstable, reactive molecule which in the body can easily react with constituents of the cells and tissues such as proteins. The reaction with proteins will alter their structure. They may become sufficiently different to be recognized by the immune system, and the body then mounts an immune attack on these altered proteins (known as **antigens**). This process can take one of several forms. For example, if the penicillin reacts with the proteins on the outside of the red blood cell, these may be recognized by the immune system. The immune system then reacts by producing **antibodies**, specific proteins which will recognize and bind to the altered red cells. The next stage is when a type of white blood cell called a natural killer cell sees and binds to the antibody on the red cell and destroys the red cell. The result of this is haemolytic anaemia, in which the number of red blood cells is depleted, possibly to dangerous levels.

The most serious adverse effect of penicillin, which is potentially life-threatening, is anaphylaxis. This only occurs in very few individuals and is the result of antigens circulating in the blood coming into contact with antibodies (immunoglobulins) of a particular type (IgE) which are attached to the surface of a particular type of cell (mast cells). When the antigen binds to the antibody the mast cell releases various 'chemical mediators' which cause various effects in the body such as a rapid fall in blood pressure and a constriction of the airways in the lungs, making it difficult to breathe. This **anaphylactic shock** is also responsible for the deaths that have occurred as a result of peanut allergy.

Penicillin will also cause the more common type of allergic reaction which results in skin rashes, reddening and itching of the skin. This is also due to antibodies but of a different type from IgE, and therefore the response is different. The mechanisms underlying allergic reactions are extremely complex and difficult to predict.

Penicillin is the most common cause of drug-induced allergic reactions and causes around 7 per cent of all adverse drug reactions. One per cent or more of patients receiving the drug may develop an adverse immune reaction. There are four general types of hypersensitivity reaction to chemicals, and penicillin can cause all four. The particular effect caused depends partly on the size of dose and partly on the individual concerned. Some of the hypersensitivity reactions are mild whereas others can be life-threatening. The most mild form of adverse reaction after relatively low therapeutic doses of penicillin is a skin rash, skin eruptions, and joint pain. High doses can in some cases cause damage to red blood cells. The most serious type of adverse effect is anaphylactic shock which occurs in about 0.004 to 0.015 per cent of patients. In an anaphylactic reaction the patient collapses when the blood pressure falls, and breathing becomes difficult. This can be fatal.

All of these types of reaction occur after repeated exposures to the drug. Once sensitized, the patient is likely to have a reaction when he or she is next given a dose of penicillin. Those who have had a reaction often wear a tag, which identifies them as being sensitive to penicillin, to alert medical staff should they intend to administer the drug when the patient is perhaps unconscious and unable to tell them of their sensitivity.

Other drugs from moulds

Many other drugs as well as penicillin have been derived from moulds or other similar organisms. The drug griseofulvin, an antifungal, was also isolated from a *Penicillium* mould. Cephalosporins, another group of antibacterial drugs, were isolated from other moulds. The antibacterial drugs streptomycin and tetracycline and the antifungal drug nystatin were all derived from yeasts. Using these naturally produced drug molecules, chemists in pharmaceutical companies have synthesized many derivatives with improved or different properties. A large number of other drugs have thus been created as a result of the original observation and development of penicillin for example.

Mutations and bacterial resistance

Mutations occur in all living organisms, including humans, and sometimes these are caused by chemicals. Sometimes the mutations are harmful, sometimes they are beneficial. Bacteria reproduce very rapidly, the cells dividing every 20 minutes under favourable conditions. As the process is not always perfect mistakes occur, and external influences such as ultraviolet light or the presence of chemicals may result in damage to the DNA. These mutations can cause changes in some aspect of the biochemistry of the bacteria. One of these changes might be beneficial, such as the ability to withstand penicillin by producing an enzyme that destroys the drug. Bacteria with this mutation will consequently survive in the presence of the drug and rapidly become established at the expense of other bacteria. This is very similar to the process that underlies evolution.

Thus, unfortunately, bacteria become resistant to drugs and new drugs have to be produced. In the case of penicillin it seems that some bacteria have developed the ability to break down the penicillin molecule by producing an enzyme to do this (penicillinase or β-lactamase). This necessitated the design by chemists of other drugs which were resistant to this enzyme such as flucloxacillin, or which inhibited it such as clavulanic acid. In time bacteria will become resistant to these too and new drugs again have to be developed.

Hydralazine and isoniazid: genetic factors in drug toxicity

Isoniazid: genetic factors in toxicity

Although not well known by the general public worldwide, this is one of the most widely used drugs. It is used to treat tuberculosis, previously known as consumption, a disease which has killed a thousand million people over time. Although improved living conditions and drug treatment probably led to it becoming less prevalent in the West in the past century, it is now becoming increasingly common. This is partly because of the occurrence of AIDS which increases vulnerability to infection. Tuberculosis is usually treated with a combination of drugs including isoniazid, streptomycin, rifampicin, and pyrazinamide. The treatments proved effective, as I know from my own personal experience, my mother having been treated successfully for this disease with a combination of three of these drugs in the 1950s.

Isoniazid was developed in 1951 as a result of the observation that vitamin B6 was weakly active against the bacterium that causes tuberculosis. Isoniazid has a related structure to vitamin B6 and acts by stopping the production of substances used by the bacteria for the growth of cell walls. The drug is effective although, as with other drugs, resistant bacteria have appeared.

Not long after isoniazid was introduced a number of curious observations were made. It was noticed that when the level of isoniazid in the blood of patients was measured, there seemed to be much variation between individuals in the actual maximum level achieved despite their having been given the same dose. It was soon realized that patients fell into two groups, and that these groups not only had more or less isoniazid in the blood, but they also had more or less of a metabolite. The two groups were called fast or slow metabolizers. It was later found that this was genetically determined, and that in the UK or USA, for example, about half the population were slow metabolizers and the other half were fast metabolizers. Apart from being the first known example of genetic variation in disposition/metabolism of a drug, this genetic factor proved to be a very important factor. The genetic predisposition was termed the **acetylator phenotype** and it is now known that the slow metabolizers have a faulty enzyme that slows the breakdown of certain drugs.

It was noticed that slow metabolizers would respond better to treatment, that is, the drug was more effective in these patients. This was because the level of the drug in blood and tissues was higher than in the fast metabolizers, and so the bacteria were more effectively destroyed. Unfortunately there was a negative aspect to this when adverse effects began to be detected.

Adverse effects of isoniazid

The drug was given to patients regularly for periods of at least one year and after a while some patients noticed tingling in their fingers and toes. These were some of the first signs of damage to the nerves leading to the extremities. This is called peripheral neuropathy. It was then discovered that this occurred more commonly in the slow metabolizers. Thus one group was genetically more susceptible to the adverse effect. This was found to be due to the higher level of the parent drug in the blood and tissues of this group, the slow metabolizers. Just as the effectiveness of the drug was greater in this group, so was the toxicity. The parent drug was responsible but what was the cause? Later it was noticed that the patients suffering from this adverse effect were also becoming deficient in vitamin B6, a condition known to cause a similar effect on the nerves. Treatment with vitamin B6 supplements proved to be a successful way of stopping this adverse effect. The mechanism is now partly understood, and so isoniazid therapy was able to continue.

However, a second adverse effect became apparent later, after many more patients had been treated: this was serious toxicity to the liver. It was noticed that a significant number of patients (maybe as many as 20 per cent) showed signs of mild liver dysfunction, which usually subsided without even stopping the drug. In a small number, certainly fewer than 1 per cent and possibly only 0.1 per cent, the liver dysfunction was progressive and could end in fatal liver failure.

The severe liver dysfunction also seemed to be more common in slow metabolizers, and it was proposed that this due to a deficiency in detoxication of a metabolite. Monitoring patients with liver function tests is one way to avoid this adverse effect. There may be other factors as well as the acetylator phenotype that are important, such as other drugs being taken at the same time. Excessive and sustained alcohol intake may also be a factor, as this is quite commonly associated with the occurrence of tuberculosis. Apart from weakening the liver it will also increase the amount of the enzyme that produces the toxic metabolite of isoniazid.

Hydralazine (apresoline)

This drug has some similarities in its chemical structure to isoniazid but acts in a completely different way. It is used to treat high blood pressure which it does by dilating blood vessels. This means that there is less resistance to the blood flowing through the small vessels, the capillaries, and so the pressure in the blood system (cardiovascular system) is lower.

Hydralazine has been in use since the 1950s and is usually used in combination with other drugs such as diuretics and β-blockers. In a significant number of patients, and typically after 18 months, adverse effects started to appear. These included joint and muscle pain (arthralgia and myalgia), a rash on the face and inflamed blood vessels (vasculitis). The rash on the face made afflicted patients look wolf-like, which gave rise to the name for the syndrome, Lupus erythematosus (*Lupus* is Latin for wolf). This disease can be caused by other drugs, such as isoniazid very occasionally and procainamide more frequently. It may also have other, unknown, causes and has some similarities with rheumatoid arthritis.

What makes this example particularly interesting is the factors that determine susceptibility. As with isoniazid, there are slow and fast metabolizers of hydralazine and it is only the slow metabolizer that suffers from the adverse effect, the lupus syndrome. However, this is not the only factor that affects the appearance of the syndrome as a result of taking hydralazine, as it also occurs more commonly and after a lower dose in women than in men, the ratio being perhaps as high as 4:1. The dose and how long the patient has been taking the drug also seem to be important. The final factor is that there is a greater incidence of the syndrome in those with a particular tissue type (DR4).

It would appear that a group of factors are necessary for this adverse effect to develop. So patients in which all these factors are present will be likely to experience this adverse effect if they take the drug. It is as if a series of windows all need to be open in order to see something. Women with the **slow acetylator** phenotype and the tissue type DR4 would almost certainly develop this syndrome if treated for long enough with hydralazine, even with the lowest dose. It also means that this syndrome can be avoided or at least reduced if the phenotype and the tissue type of the patients were determined before they were given the drug.

The mechanism by which hydralazine causes these effects is not known but it seems that the immune system is involved. Slow metabolizers may be more susceptible because they will maintain higher levels of the unchanged drug in the body which, like penicillin, is fairly unstable. There is other evidence that alternative routes of metabolism may be more important in the slow metabolizers, producing other reactive metabolites.

There are now other examples of drugs that affect one group particularly because of genetic factors, raising the possibility that in the future, if genotyping or phenotyping were generally available, doctors could avoid giving susceptible patients particular drugs. This will, however, require further understanding of how drugs cause toxic effects and of the factors that govern the appearance of adverse effects in susceptible individuals.

Many drugs are taken in overdose and those commonly used vary with what is available and in use at the time. For example, barbiturates were at one time more commonly prescribed and therefore often featured in overdose cases. Similarly tricyclic antidepressants have been popular and also featured in overdoses. Barbiturates also cause another problem shared by a number of drugs, that is, addiction. They also increase their own metabolism (they are enzyme inducers) and so ever increasing doses are required for the desired effect. They are used as sedatives and also to treat epilepsy. Another drug that has been widely used and may be addictive is codeine. This is changed by the body into morphine and it is this that has both the desired effects and addictive properties (see below). As well as such drugs, which are used for legitimate reasons but happen to be addictive, there are addictive drugs of abuse which both feature in overdose cases and cause adverse reactions.

Drugs of abuse

The most well-known drugs of abuse are heroin, cocaine, cannabis, LSD, amphetamines, and **ecstasy**. They all cause different kinds of effects and their toxic effects are also different. The first three of these are derived from or closely allied to substances found in plants. LSD is similar to substances found in a fungus which affects crops (see pp. 244–7). It is appropriate to consider these substances here as they are drugs in the accepted sense of the word, and some of them have legitimate uses as well as being drugs of abuse.

Cocaine

Cocaine has been used by humans for at least 3,000 years, initially in its natural form in the leaves of the coca plant and only more recently in pure form. The plant, a shrub (*Erythroxylon coca*), grows in South America and South-East Asia, and has been extensively cultivated in the foothills of the Andes in Peru and Bolivia. The leaves, which the natives of South America chewed to combat fatigue and hunger, contain cocaine. Once its effects were known, the drug began to be used in religious festivals and rites and its use became more controlled. After their conquest by the Spanish, the conquerors exploited the natives by making them work long hours in places such as silver mines, sustained by coca. When the Spanish conquistadors returned to Europe they brought coca with them. The *Gentleman's Magazine* in 1814 suggested that the famous scientist Sir Humphry Davy investigate coca to see if it could be used as a temporary substitute for food.

Its appearance in Europe really became prominent a bit later in the nineteenth century when coca leaves were used as part of a medicinal concoction (Vin Mariani) prepared by Dr Angelo Mariani in France in the 1860s. It was claimed to have anaesthetic, analgesic, and carminative properties and soon became very popular with patients. Other preparations containing extracts from coca leaves appeared, perhaps the most famous being Coca-Cola which was invented by a pharmacist in Atlanta, Georgia in the USA. Originally called Pemberton's French wine coca, the drink also contained caffeine as an important ingredient, as well as wine. The wine was later removed during prohibition in Atlanta and replaced by sugar, and the drink was renamed Coca-Cola. The effects of cocaine as a drug were becoming known, and in 1904 the extracts of coca leaves had to be removed from the drink.

The potentially useful effects of cocaine as a local anaesthetic were discovered by an assistant to Sigmund Freud. Freud, who was using the drug himself, asked his assistant Carl Koller to find out how cocaine reduced the feelings of hunger and tiredness. Koller found that the drug numbed his tongue, and went on to demonstrate its potential as an anaesthetic for eye operations.

Despite the fact that South American Indians have been using cocaine in a limited way by chewing coca leaves for hundreds of years, cocaine has now become a major drug of abuse. The chemical was isolated in about 1860, and individuals like Freud began experimenting with it. It can be injected, 'snorted' into the nose, or smoked. When smoked, as 'crack', it causes a particularly intense high followed by profound depression. After the effects have subsided there is a craving for more. Cocaine is highly addictive, especially when smoked, as it enters the brain very readily by this route. The drug causes pleasant feelings of euphoria with increased confidence, optimism, and energy but the feelings are short-lived and successive doses become less effective. Therefore larger and larger doses need to be taken to achieve the same result with the danger of increasing toxicity. In the UK there are probably more than 100,000 regular users of cocaine with three times that number of occasional users of the drug. In the USA there are 3 million to 5 million regular users with possibly 1 million suffering serious dependence. It is also known as coke, gold dust, snow, lady, and crack.

There are also large numbers of regular crack cocaine users (between 100,000 and 200,000 in the UK). Millions of South American Indians still chew coca leaves, mixing them with lime which helps to release the base drug. They probably absorb several hundred milligrams of cocaine, but also nutritious substances such as vitamins from the leaves. Whether there are any long-term effects to this usage of the drug is unclear, but taken in this way it may, like other such drugs, be a useful prop for these individuals.

How cocaine works

Cocaine acts as a local anaesthetic, that is at the site of application, probably by interfering with transmission of nerve impulses by changing the permeability of the nerve to sodium **ions**. While the drug is no longer used for this purpose, many of the newer local anaesthetics are based on the chemical structure of cocaine.

Its action on the brain to cause feelings of euphoria is due to it causing increased levels of the substances **dopamine** and 5-hydroxytryptamine (neurotransmitters) in the brain. Dopamine is particularly important in areas of the brain associated with pleasure. By blocking the uptake of dopamine into the nerves, cocaine potentiates its effects by making more available to interact with receptors. This is probably the basis of the addiction. The effects of reduced hunger and fatigue which the South American Indians experience is due to its ability to block the uptake of another neurotransmitter, noradrenaline.

The toxic effects of cocaine

The toxic effects are a combination of psychological effects and effects on major organs and tissues. Thus higher doses cause the addict to experience feelings of paranoia, hallucinations, and crawling sensations on the skin. The addict may also become obsessed with certain subjects and may suffer fits. One of the physiological effects of cocaine is that it causes blood vessels to contract, which consequently reduces the blood flowing to tissues. When this occurs the tissue, such as heart muscle, becomes starved of oxygen and nutrients and can suffer permanent damage. The heart and lungs may be affected, leading to chest pains and difficulty in breathing. When cocaine is snorted through the nose this effect on blood vessels leads to permanent damage to the nasal tissues, and addicts typically have a runny nose and sniff repeatedly.

Constant and increasing doses will lead to more serious, permanent damage to the heart called cardiomyopathy, arrhythmias, and possible heart failure. Strokes and acute heart attacks can occur. The reduction of blood to the kidneys can also affect their function.

Heroin

Heroin is a derivative of morphine, which is obtained from extracts of the opium poppy and has been used for several thousand years. Opium was known in 4000 BC when it was used by the Sumerians and is mentioned in the Ebers Papyrus. The early toxicologist Dioscorides was probably the first to describe the preparation, in AD 50: 'But it behoves them that make opium . . . to scarify about the asterisk with a knife . . . and from the sides of the head make incisions in the outside, and to wipe off the tear that comes out with the finger into a spoon, and again to return not long after, for there is found another thickened, and also on the day after.' Here Dioscorides describes the technique of scratching the pod of the opium poppy and allowing the juice containing the opium to exude onto the outside, then it can be harvested repeatedly. This is the process still used in countries where it is grown. The juice that dried on the outside of the seed capsule contained a mixture of **alkaloids**, of which morphine was the most important, but codeine, a closely related chemical, was also present.

Paracelsus prepared opium dissolved in alcohol which was known as laudanum and used it as an analgesic for pain relief. Opium was also used to induce sleep and for the treatment of diarrhoea (morphine is still used for this). As a result of trading and invasion, the use of opium spread throughout the Middle East, parts of Europe, and the Far East. Opium taken in preparations by mouth as tinctures, or even in cakes by the Chinese, proved to be very popular and to have many uses. It was widely used both for medicinal purposes and for the euphoric effects it induced as described by writers and poets, for example Thomas De Quincey in *Confessions of an Opium Eater*.

However, after the introduction of tobacco and the smoking of the leaves, the smoking of opium began. This occurred in China, for example, after the smoking of tobacco was banned in 1644. When exposure occurs by inhalation of the smoke, the constituent morphine enters the body much more rapidly through the lungs compared to absorption through the stomach and intestine. The use of opium in China increased so much that by the end of the seventeenth century about one-quarter of the population was using it. The British East India Company began supplying this growing need, in spite of the Chinese government's objection, and the tension eventually resulted in the first Opium War in 1849. The trade continued and increased but when opium use began to be a problem in other countries its export was curbed. Its use in the UK also increased such that by the middle of the nineteenth century the average consumption of the population was several grams a year. But it was not until 1920 that the Dangerous Drugs Act made opium and other drugs such as morphine and cocaine available only with a doctor's prescription and from a registered pharmacy.

By the early part of the nineteenth century the main constituent, morphine, had been isolated by Wilhelm Serturner in Germany and shown to be a powerful analgesic and narcotic, able to induce sleep. He called it morphine after Morpheus, the Greek god of dreams. The invention of the hypodermic syringe allowed morphine to be injected which made it more rapidly effective but also increased its addictive properties.

Then, in 1874, Frederick Pierce and his team at St Mary's Hospital in London made a derivative, diacetylmorphine, which proved to be more powerful. The German company Bayer eventually made and marketed this drug as 'Heroin'. It was soon found to be addictive, much more so than morphine, and its use was prohibited in many countries. The reason the new compound was more powerful and more addictive is because it enters the brain more easily. Therefore when heroin is injected by addicts the entry into the brain causes a more intense 'rush' or 'high' than occurs with morphine. Despite the limitation on heroin use in the early days after it was marketed, it is now a major drug of abuse. Heroin can be either smoked or injected. When smoked, a practice called 'chasing the dragon', the heroin powder is heated on tin foil and the fumes inhaled through a tube. When injected, usually dissolved in water, it causes a more powerful high. In the UK there are several hundred deaths per year from heroin abuse, and about 40,000 registered users, although some estimates put the number of regular users as high as 270,000. Each addict probably takes on average 750 mg per day. Many addicts suffer from infections such as Hepatitis C contracted from infected syringe needles. Heroin, mostly made illegally, is usually diluted or 'cut' for sale on the street. It may be cut with glucose, chalk, flour, or talcum powder. Some of these substances are not soluble in water and therefore a suspension rather than a solution is injected which can cause damage to blood vessels.

Heroin: morphine in disguise

When morphine is taken by mouth it passes through the liver, which metabolizes and inactivates a significant proportion. When morphine is injected into a vein, it bypasses the liver and so is more potent. In either case, only about 20 per cent of the dose of morphine enters the brain. Heroin is a derivative of morphine but is more soluble in fat than morphine. In order to gain entry into the brain, drugs generally need to be fat-soluble rather than water-soluble. So heroin enters the brain more easily, and once there some of it is converted into morphine and acts on receptors. Most of the effects are due to morphine rather than heroin. Heroin is like a Trojan horse, allowing more morphine to get into the brain, and is hence more potent. Morphine and similar drugs bind to specific receptors known as opiate receptors which are present in the brain and other tissues. The receptors in the brain are found in the areas concerned with pain and also with emotional behaviour. These receptors also bind substances normally found in the body called endorphins, which help to reduce the feeling of pain naturally by inhibiting the release of a chemical called substance P. Drugs such as morphine, known as agonists, bind to the same receptor and cause the same effect, reducing the feeling of pain. Binding to receptors in other parts of the brain such as the limbic system, which is associated with emotional behaviour, may be associated with addiction. There are several types of opiate receptors to which such drugs can bind, causing slightly different effects.

The toxic effects of heroin

Clearly addiction to or dependence on the drug is an adverse effect. Although the drug, heroin, itself may not be especially hazardous in the doses needed to cause a pleasurable effect or 'high' in a naive subject, repeated use of the drug will lead to both addiction or dependence and tolerance (see below) which increase the likelihood of toxic effects occurring. For example, the increasing doses needed due to tolerance may lead to suppression of breathing and a reduction in the sensitivity of the brain to carbon dioxide so that breathing may stop, for example during sleep. Other drugs that have a depressant effect, such as alcohol, will increase the likely toxicity if taken together with morphine or heroin. Other effects are constipation and the possibility of causing asthma in susceptible individuals.

As well as these direct effects, the change in behaviour and lifestyle and the requirement of money to buy the drug can seriously affect the health of the addict. As already mentioned, many addicts suffer from infections such as hepatitis and HIV. Poor diet and poor living conditions will also take their toll.[4]

Addiction and tolerance

Addiction to a drug, or dependence on it, is both a psychological and a physiological phenomenon. The psychological part is the need to repeat the pleasurable experience, the euphoria and feelings of well-being caused by heroin, for example. Stimulation of certain areas of the brain creates the feeling of craving for more. However, after the drug has gone from the body, and certainly within two days, the body undergoes withdrawal symptoms which at the least are very unpleasant. These include fever, aches and pains, cramps, nausea, vomiting, diarrhoea, sweating, depression, insomnia, and dehydration. The onset and the memory of these symptoms drive the addict to another 'fix' so as to allay them.

Tolerance is the requirement for more drug to cause the same effect. This seems to be due to the fact that the cells on which the heroin/morphine acts, produce more receptors in response to the blockade of the existing receptors. The more receptors there are, the more drug is needed to occupy them and the resultant effects will be less. If the drug is not taken or is withdrawn, the transmitter that occupies the receptors in the absence of the drug (for example, substance P) causes symptoms such as pain. The transmitter will have more receptors to react with and so will cause a greater degree of discomfort—the withdrawal symptoms. Because the tolerance to the different effects of heroin and morphine vary, toxic doses may be needed to achieve some of these effects. Hence the danger of long-term addiction and the possibility of damage and death.

There are other drugs available which can be used to wean an addict off heroin, such as methadone which interacts with the same receptors and causes a similar although less intense effect. It is more slowly removed from the body and so the addict adapts to the loss of the effect of the heroin more gradually. Methadone was developed in Germany during the Second World War as an alternative to morphine which became difficult to obtain as supplies of opium from traditional sources in the East were reduced by the conflict. Methadone was called Dolophine in honour of Adolf Hitler. Addicts and occasionally others do, however, take overdoses of these drugs. In the case of regular users this may be because of variations in the purity of the street drug, with naive users it may be the result of simply taking too much by mistake.

Treatment of morphine or heroin poisoning

Can toxicity due to an overdose of heroin be treated? The answer is yes, because understanding how the drug acts, by binding to specific receptors, allows treatment of poisoning to be carried out logically by using an antidote. The antidote used is normally naloxone.

Other drugs that bind to the same receptors as morphine and heroin but do not cause the effects have been devised. These are used as antidotes for the treatment of overdoses of morphine-like drugs. They are called antagonists, because they antagonize the effect of the agonist, binding to the receptor, but do not have the same effect. Naloxone is one such drug which binds to the receptors and displaces heroin and morphine, so stopping the toxic effects of excessive amounts of these drugs. This antidote works very rapidly.

The toxic effects of overdoses of heroin or morphine are related to the action of these drugs. The drugs affect receptors in the brain including the area where respiration is controlled, and so excessive amounts depress respiration and brain function. Hence victims can suffer symptoms ranging from lethargy to coma depending on the size of the dose. The heart rate may be slow and the blood pressure will drop. The lack of respiratory control and possibility of complete cessation of breathing is a particular hazard. If the antidote is given, breathing will be improved in one or two minutes. Most addicts inject heroin into a vein, and so when overdosage occurs rapid treatment will be essential. As well as heroin, other opiates may also feature in overdoses, and these cases can also be treated with the same antidote. For example, some common analgesic medicines available over the counter contain the opiate drug dextropropoxyphene along with

CASE NOTES

Was it heroin?

In 2000 Rachel Whitear, a 21-year-old heroin addict, was found dead in her bedsitter in Exmouth, Devon, UK. She had a syringe in her hand as if in the process of injecting the drug. It was assumed that she had died of a heroin overdose and her story, together with the photograph of her body, were widely publicized. However, an open verdict was recorded as forensic evidence indicated that the amount of heroin in her blood was below the expected fatal level. Furthermore, there had been no post-mortem. The case is being reinvestigated. Information about the dose or exposure to a drug or chemical is, of course, crucial in determining whether poisoning was a cause of death or injury. It should be shown that the injuries are consistent with the amount and type of drug or other chemical taken.[5]

paracetamol. When overdoses of this combination are taken, respiratory depression is the major toxic effect of immediate concern and needs to be treated with the antidote naloxone.

Ecstasy

In contrast to heroin and cocaine, ecstasy is a more widely used but non-addictive drug. It has been estimated that more than a quarter of teenagers have used it at least once, making it the most widely used illegal drug. Ecstasy, often referred to simply as 'E', is methylene dioxymetamphetamine, abbreviated to MDMA. It is an amphetamine derivative and is related to the naturally occurring drug mescaline. It was first made in 1912 in Germany and intended for use as an appetite suppressant, but its properties prevented its being used. It was reintroduced in the 1950s for use by psychotherapists, but its use was restricted in 1977 in the UK and 1985 in the USA.

It is reported to cause feelings of self-confidence, great energy, and drive. Users also experience a feeling of warmth toward others and increased sensitivity to sound and touch, as well as a reduction in anxiety and a general feeling of euphoria. The drug causes these effects by affecting the type of neurones that produce the neurotransmitter 5HT (5-hydroxytryptamine), which leads to an increase in the release of this neurotransmitter. This substance is associated with feelings of euphoria and plays a role in making us feel gregarious.

Ecstasy is not without negative effects, however, and some time after use there may be feelings of irritability, tiredness, and depression which can last for several days. Repeated users can experience depression, panic attacks, hallucinations, and psychiatric disorders. More seriously, there have been cases of fatal toxicity following, in some cases, a single tablet of the drug.

CASE NOTES

Leah Betts died of drinking water to counter drug's effect

▪ By Jeremy Laurance, health correspondent

Leah Betts, the teenager who collapsed after taking an ecstasy tablet, died as a result of drinking too much water, which made her brain swell.

Doctors who treated her at Broomfield Hospital, Chelmsford, Essex, where she was taken after lapsing into a coma at home during her 18th birthday party, will tell the coroner that 'water intoxication', and not an allergic reaction to the drug, was the cause of death.

The Times (22 November 1995)

This tragedy happened on 16 December 1995 in Essex in the UK. The girl took one ecstasy tablet at 19:45. By 20:30 she was thirsty and drank seven glasses of water. By 20:50 she was screaming, had a headache and blurred vision. At 21:00 she collapsed, and by 21:30 she was at the Accident and Emergency department of the hospital. She died five days later. Why? (See below.) It seems in this case, which is by no means the first, that the drug was only part of the cause.

The toxic effects of ecstasy and water toxicity

Like other amphetamines, the toxicity of MDMA appears to be increased in overcrowded and overheated conditions. This has been shown in experimental animals and appears to occur sometimes in humans too. Thus when taken by an individual in isolated conditions the drug slightly lowers the body temperature, when taken in overcrowded or hot conditions it raises the body temperature. As the drug is often taken by young people at all-night 'raves' in order to dance and socialize for many hours, these conditions and the attendant dehydration sometimes cause overheating and the body temperature may rise from normal (37°C) to 40–43°C. The use of the drug can result in disturbed heart, brain, and liver function and in death.

Case reports on adverse reactions to ecstasy show that it can cause altered mental states, increased heart rate, and increased rate of breathing. Failure of the kidneys and liver, coagulation of blood in the blood vessels, and raised temperature have all been seen in cases. Whether all of these effects are due to ecstasy or whether some are due to contaminants, to substances with which ecstasy is mixed, or to other drugs taken at the same time is not clear.

Because of the raised temperature, many takers will drink large amounts of water (see case notes). At the same time, ecstasy can increase the secretion of antidiuretic hormone (ADH), which is involved in controlling the production of urine. Therefore, after ecstasy is taken, urine production is decreased despite increased water intake and the water is retained in the body. If the accumulation of water is extreme the blood becomes so dilute that the blood cells are destroyed—they take in water and explode. Some of the cells in other tissues such as the brain also swell, putting pressure on particular parts of the organ such as the brain stem. In the case of Leah Betts this was probably the cause of her coma and eventual death as her blood level of sodium ions (a measure of the dilution of the blood) had fallen to 126 millimoles per litre of plasma compared to a normal range of 134 to 145. When the normal concentration of substances such as salts in the blood is decreased water flows into the cells by a process called osmosis, whereby water moves in an attempt to equalize the concentration, with the result that the cells swell and burst. (This is also one of the consequences of drowning when water goes into the blood through the lungs.)

It should be noted that, given the very large number of people using the drug, the incidence of acute toxic effects is low (see discussion on alcohol, pp. 198–212). However, some evidence suggests that ecstasy may have potentially damaging effects on the brain. Studies in experimental animals and humans have suggested that nerve cells producing serotonin (5HT), which are stimulated by the drug, were damaged by ecstasy.[6] This could lead to effects similar to those seen in Alzheimer's disease, such as mental confusion and changes in sensory perception and thought patterns.[7] The design of the human studies has, however, been questioned.[8] More recent studies, which claimed to show that single doses of ecstasy in monkeys damaged nerve cells producing dopamine, were eventually retracted by the authors, apparently after they discovered that the wrong drug had been given at doses high enough to be lethal to some of the animals![9] Thus there is cause for concern but it is not clear if there is a significant risk from ecstasy taken at 'normal' doses. Properly constructed studies in humans need to be done and, given the huge number of people taking the drug, such studies are vital as there could potentially be an epidemic of mental disability in future years.

Herbal medicines

Many traditional medicines and herbal remedies are used throughout the world and in many countries there is little, if any, regulation. In the USA alone this industry is worth a billion dollars per year. In the UK some herbal medicines are licensed, others do not require a licence when sold or supplied within particular guidelines, while yet others (such as herbal remedies) are sold as food supplements and are controlled by food regulations. In the case of the latter, no medicinal claims are allowed to be made.

Although plants are by definition natural, they contain many chemical substances, some of which can be used as drugs but others may be very toxic. Unfortunately many of the general public think that, as they are natural, herbal remedies and herbal medicines must be safe. But just as with man-made synthetic drugs, herbal remedies can and do cause harm to patients taking them.

A significant number of drugs in use today, or which have been used, have been derived from plants. Apart from cocaine and heroin, perhaps the most well known is **digitalis**, derived from the foxglove, and the pure drugs developed from this: digoxin and digitoxin. There are many other drugs derived, directly or indirectly, from plants, a number of which have been used recreationally such as mescaline, LSD, psilocybin, hyoscine, and cannabis.

There are so many herbal remedies and medicines that it is not possible to do more in this book than make some general points and look at one or two in greater detail, in particular some that have caused problems. There are several problems that arise particularly with herbal medicines and remedies. First, they have rarely been tested for safety, and therefore little toxicological information about them is available. Secondly, the concentrations of active ingredients within a particular plant can vary, and hence dosage is not precise. Thirdly, herbal products are mixtures, and therefore there are possible interactions between the components that may increase the toxicity. Inadequate knowledge on the part of the person preparing and prescribing the herbal medicine can result in incompatible mixtures being given. Finally, the plants from which the preparations are made may be contaminated with fungal toxins or pesticides. Sometimes the wrong plant is used because it has not been correctly identified, giving rise to toxic contaminants in the whole preparation. Whereas with synthetic drugs the exact chemical identity and purity must be known, this is not the case for herbal medicines. Often some or all of the constituents of a particular herb are unknown. Therefore the exact mechanism of the action of herbal medicines is rarely known, in contrast to synthetic drugs. This means interactions between components cannot be predicted.

Originally, herbal remedies were often infusions of herbs, but nowadays they may also be produced in more concentrated form, as capsules for example. The possibility of overdosage is therefore present. Because many herbal remedies are available for people to buy and administer themselves without the involvement of a licensed practitioner, some of them may increase the dose on the grounds that, if the recommended dose works, an increased dose will be even better. Herbal remedies may be prescribed by practitioners who may be medical herbalists, Chinese herbalists, or Ayurvedic or Unani practitioners, for example. These practitioners tend to use herbs from Europe, China, or the Indian subcontinent. In the case of the latter two the remedies may contain constituents such as minerals (arsenic, for example) and parts of animals.

Aristolochia fangchi

This Chinese herb has been associated with damage to the kidneys, called nephropathy, and in some individuals it has led to cancer of the kidneys.[10] In one case in Belgium a number of people were poisoned with this herb. The Belgian Ministry of Health estimated that about 10,000 people had taken Chinese herbs between 1990 and 1992, of whom at least 70 had developed end-stage kidney failure.

It seems that in one particular case the herbal preparation was used for weight reduction in dieters. One of the herbs that was intended to be used was replaced by *Aristolochia fangchi* (apparently the Chinese name for the correct herb was fangji and that for aristolochia is fangchi, clearly very similar names). The mistake originated with a wholesaler who imported the product from Hong Kong as a powder made from the root of the plant. The powder was then combined with other drugs in the final preparation. After some time patients began to suffer renal failure which occurred between three months and seven years after they had stopped taking the tablets.

Fortunately, in Belgium herbal medicines are available only on prescription, and so it was possible to trace the preparation back to the pharmacies that had prepared it and to analyse the constituents. Of 105 patients studied, 43 had complete kidney failure. In 39 of those examined, 18 were found to have kidney or bladder cancer. It was possible even to detect the constituent of the herb in samples of kidney from the patients affected.[11]

Aristolochia damages DNA

In the study of the patients with kidney cancer associated with the use of aristolochia, samples of kidney tissue were taken from patients. When analysed these were found to contain aristolochic acid, a known constituent of the herb. Molecules related to aristolochic acid were found bound to DNA from all the kidney samples analysed. When molecules of a chemical become bound to DNA, dysfunction of the molecule can result in various ways. The binding to DNA can lead to mutations or other disturbances in DNA function which underlie cancer, such as switching on cancer genes or oncogenes. Many carcinogens (or their active metabolites) have been found to bind to DNA and this is believed to be part of the process by which chemicals cause cancer.

This particular tragedy arose because of a mistake in a country with good regulatory processes, but at the time of the incident it was still possible to buy this poisonous herb in the USA, for example. Kidney failure associated with this herb has also been reported in the UK, where it had been used to treat a skin condition and also in France, Spain, Japan, and Taiwan.[12]

St John's wort

This herb (*hypericum*) is widely used for the treatment of mild depression; indeed, in Germany it outsells the synthetic drug Prozac. It has been shown to be effective and adverse effects do not seem to be a problem. However, it does interfere with the disposition of other drugs. In particular, it will generally reduce the activity of certain other drugs which may lead to a failure of treatment. This occurs because St John's wort is able to induce the enzymes that metabolize other drugs. The amount of the enzyme is thus increased, which leads to increased metabolism and excretion of the other drugs. Therefore a drug given after the herb has been used may not reach the required therapeutic concentration, or will not be active for so long, or conversely more of an active or toxic metabolite could be produced (see p. 34 for further explanation of this effect). Thus drugs used for the treatment of blood clots (warfarin), asthma, and heart disease, and the contraceptive pill, can all be rendered less effective if taken with this herbal medicine.

Herbal teas and pyrrolizidine alkaloids

In the West Indies especially, pyrrolizidine alkaloid toxicity is a continuing problem, as plants containing pyrrolizidine alkaloids such as *Heliotropium*, *Senecio*, and *Crotolaria* species are used in traditional medicine to make herbal teas. Chronic exposure to low doses of these alkaloids causes **liver cirrhosis**, and it is estimated that these alkaloids account for one-third of the cirrhosis cases detected at autopsy in Jamaica (see Chapter 6 for more details of these natural toxins).

Safety evaluation of medicines

The safety evaluation of chemicals in general will be discussed in more detail in Chapter 12, but some basic points relevant to drugs should be mentioned here. Before a new medicine can be licensed it needs to be shown to be effective and safe. Both of these involve studies of the effect of the drug in animals and humans (see Figure 9). After a potential new drug has been made it will be tested *in vitro*, for example, to see if it is active against bacteria or binds to a receptor. Even if it is active it may not work in an animal such as a human, so first laboratory animals are used to show that the proposed drug causes the desired effect, such as lowering blood pressure or reducing inflammation, and to find the necessary dose. Then laboratory animals are used to evaluate the safety before it is given to human volunteers. This involves carrying out various specified studies to look for any toxic effects at the dose expected to be used and at higher doses. If the drug proves to be safe and effective in animals, then a few human volunteers are given small single doses of the drug (this is a Phase I clinical trial). This allows the fate and safety of the drug in humans to be studied. If the drug proves to be absorbed in humans and without adverse effects, a small number of human patients are given the drug (Phase II clinical trial). If the drug is effective and without significant adverse effects, a much larger number of patients are given the drug (Phase III clinical trial). If this is successful, the drug will be granted a licence and can be sold. The drug is then available for use (either via prescription or sold directly to the public) but the evaluation continues. This last stage is called post-marketing surveillance and is Phase IV of the clinical trials. During this time doctors will report any adverse effects they have diagnosed or which have been described by patients. If serious adverse effects occur and can be related to the drug it will probably be withdrawn. According to the FDA[13], between 1971 and 2002 about 2.6% of new drugs

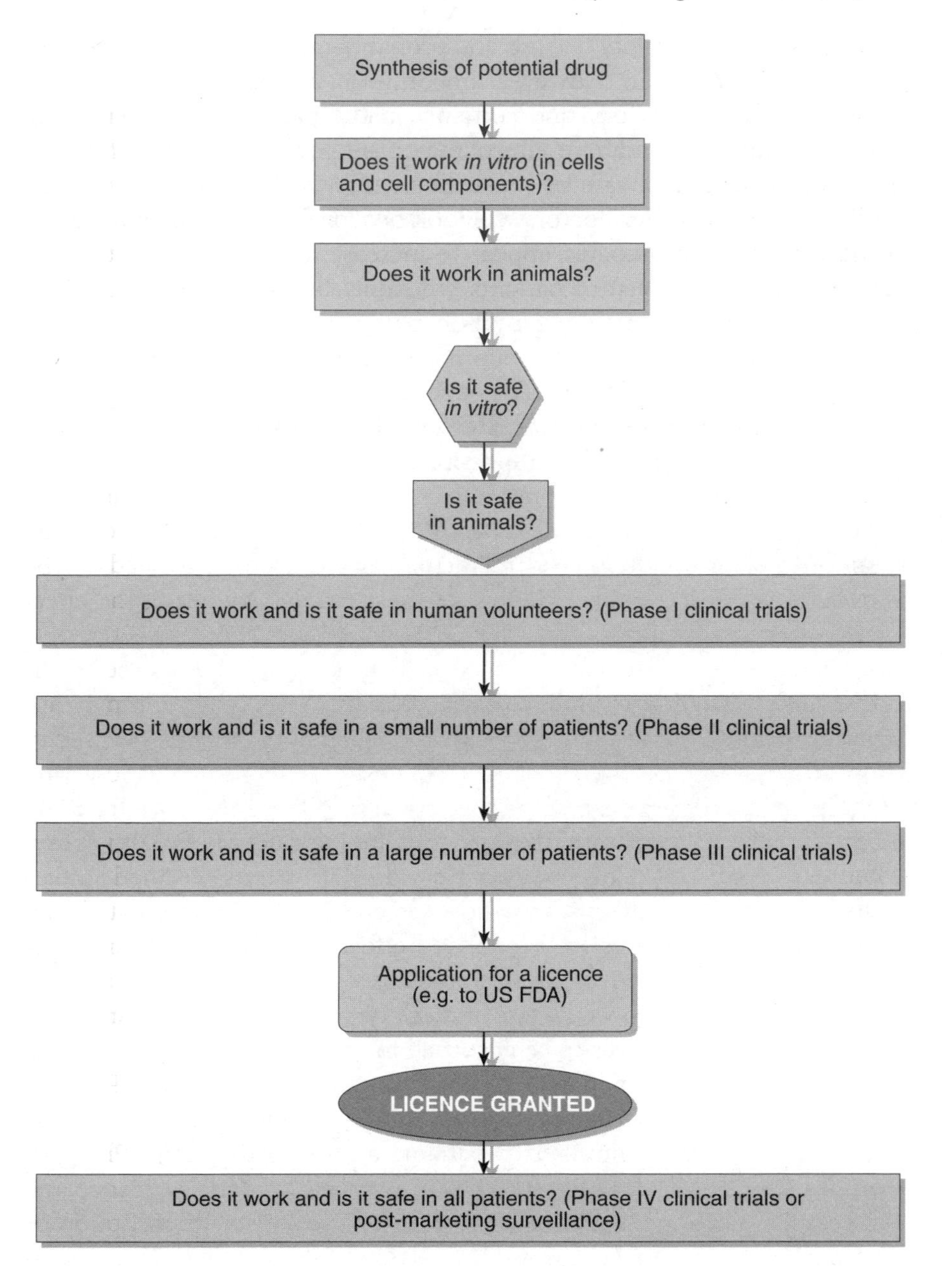

9. The process that chemicals must go through in order to become medicines.

were withdrawn for safety reasons and probably the most common reason is toxicity to the liver. However, uncommon adverse effects may go undetected for some time, and certainly in the past, before this system was well established, adverse effects may have gone unreported. Once a few doctors become aware of such effects, they will be reported in the medical literature, other doctors will look out for them, and the incidence of these effects will probably appear to increase. Because of the possibly huge numbers (millions) of patients who may be taking the new drug, rare adverse effects (say 1 per million patients) can become apparent. These will not be seen in the animal studies or earlier clinical trials because the numbers of animals or patients exposed are too small. This is why post-marketing surveillance is essential and why absolute safety of medicines before marketing is impossible.

4

Blood, Sweat, and Tears

Pesticides

> The clinical features are those of haemorrhage—blood in the urine, blood in the vomit, and bleeding from the nose.
>
> THE EFFECTS OF INTENTIONAL OR ACCIDENTAL INGESTION OF A RAT POISON (AN ANTICOAGULANT)

> Symptoms develop within twelve hours of exposure. The victim commonly presents with a headache, blurred vision, tremor, twitching, and convulsions which may be followed by coma. Incontinence and vomiting develop; other symptoms include sweating and excessive production of saliva and tears.
>
> THE EFFECTS OF POISONING BY ORGANOPHOSPHATE INSECTICIDES

DESPITE the fact that some of the substances used as pesticides have unpleasant and potentially lethal effects on humans, they are of great importance both in economic and human health terms. Humans have been changing their environment for thousands of years and in many different ways. First, attempts to control what were seen as pests probably simply involved killing them by force (shooting foxes, for example), fire (burning fields after harvesting to kill both animal and plant pests), or using manual methods to remove plant pests such as weeds. Only relatively recently have we become more sophisticated. In the nineteenth century farmers applied chemicals such as Bordeaux mixture (copper sulphate) to plants to kill fungus. There is evidence that the antifungal properties of sulphur compounds were known in very early times, and people with fungus infections (ringworm, for example) were encouraged to bathe in sulphurous springs.

Early herbicides were sodium chlorate and sodium arsenate; arsenic also features in one of the first insecticides, Paris Green, which contained copper and arsenic and was used in 1867 to kill the Colorado potato beetle. Kerosene was a readily available substance which was also used to kill insects, and cyanide gas was effective as a fumigant (and until recently was still used to dispose of wasps' nests).

Plants were known to contain insecticidal substances, and one of the first of these to be used, in the middle of the eighteenth century, was nicotine. A solution made from soaking tobacco leaves was found to be effective. Pyrethrum, contained in the flowers of the chrysanthemum plant, is a very effective insecticide and may have been known for centuries.

It was, however, only in the twentieth century that the production and use of pesticides became an industrial activity, with chemicals being specifically synthesized, tested, and targeted towards a particular pest. Although the word 'pesticide' is often thought to be synonymous with 'insecticide', in fact the word covers any substance used for destroying a living organism conceived as a pest, which could be a plant (using a herbicide), an insect (an insecticide), a mammal (a **rodenticide**), or a fungus (a **fungicide**). Unlike the other chemicals considered in this book (with the exception of antibiotics and anti-cancer drugs), pesticides are all substances specifically designed to be toxic, to kill particular living organisms. Pesticides are also designed, with varying degrees of success, to be selectively toxic only to the target species.

DDT and the organochlorine insecticides

The name of this insecticide is to many people synonymous with environmental pollution, and it is therefore an important example in this chapter but the story is not simple. DDT is an abbreviation for the chemical name of the insecticide dichloro-diphenyl-trichloroethane. It contains five **chlorine** atoms and is hence called an **organochlorine compound**. It was first made in 1874 but not found to be an insecticide until 1939 by Paul Müller. It was used extensively in the Second World War for the control of insects such as lice and mosquitoes, carriers of the diseases typhus and malaria. It was very effective in controlling these pests, and the diseases they carried, and undoubtedly thousands of soldiers' lives were saved. Since then, the lives of millions of people throughout the world have also been saved by this insecticide both as a result of the reduction of these and other diseases and as a result of the improvement in crop yields which has reduced starvation. Indeed, in 1953 it was estimated that the use of DDT for malaria eradication had saved 50 million lives and averted more

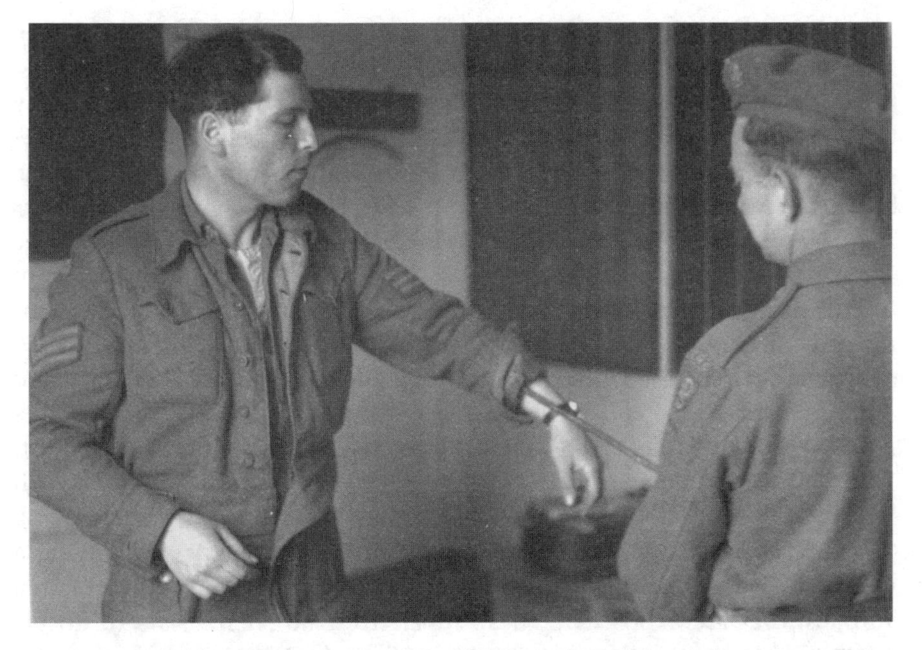

10. Dusting soldiers in the Second World War with DDT powder to control lice. This use of DDT dramatically reduced death from diseases such as typhus.

than 1 billion human illnesses. In 1971 the World Health Organization has estimated that more than 1 billion people had been freed from the risk of malaria, in the previous twenty-five years, mostly as a result of DDT use.[1] The soldiers who were dusted with DDT powder, wore underclothes treated with DDT, or sprayed or were contaminated with DDT did not suffer ill effects. Volunteers who were subjected to exposure to DDT reported no ill effects either. Indeed, some volunteers were induced to take half a milligram of DDT by mouth daily for over one year and yet showed no signs of toxic effects. In the UK there has been little evidence of harm to either humans or wildlife.

So why is DDT perceived as such a dangerous chemical? In the USA DDT was used extensively after the war for the control of various insect pests. It was used in some cases in much greater quantities than it had been previously, and possibly such quantities were unnecessary. For example, 1 lb of DDT per acre had been sufficient to eradicate the malarial mosquito from swamps, yet in the control of the bark beetles, which carry Dutch Elm disease, 25 lb per acre were used in the USA for spraying the trees in some cases. The disease was devastating the trees, and perhaps some overkill was understandable, but this use of DDT led to effects on wildlife, with birds being noticeably affected. For example, the population of the American robin plummeted and was even wiped out completely in some areas. The reason for this was that earthworms and caterpillars became contaminated and in turn contaminated the birds that fed on them. Fish were also found to be especially susceptible and died in some areas of DDT use. Before 1954 there had been little complaint about DDT, but such events started to attract comment. The reason for the concern and for the appearance of effects on wildlife was that DDT accumulates and is persistent in the environment (see below).

This meant that even the use of lower levels than the excess sometimes applied could, with time, lead to levels that were toxic or had particular effects on certain susceptible animal species such as predatory birds.

Accumulation of chemicals in the food chain: bioaccumulation and biomagnification

Chemicals such as DDT and other organochlorine pesticides are not very soluble in water but do dissolve in fat. Therefore in animals, including humans, that are exposed to these substances the chemicals tend to be distributed into tissues containing fat. Although DDT is metabolized to some extent and some is excreted (see Chapter 2), some of it remains in animals. If exposure is repeated or continuous, accumulation occurs. For example, fish or other organisms swimming in contaminated rivers may accumulate substances like DDT directly from the surrounding water or from sediments. This is called **bioaccumulation**. Even the level after accumulation may not be enough to be hazardous to organisms at the bottom of the food chain such as plankton or small fish. But the larger animals eating these smaller fry and further up, or at the top of, the food chain could acquire sufficient levels of DDT for it to be lethal. Thus the concentration in the fat in these animals can be much higher. For example, in Lake Michigan it was found that although the sediment had DDT levels of 0.02 **ppm**, herring gulls had levels of 98 ppm. Therefore a **biomagnification** of more than 1,000-fold had occurred (see Figure 11). It has been estimated that there is an accumulation factor of 5–15 between a predatory bird and its prey. Add to this the possible greater sensitivity of certain species and one can see how a substance such as DDT can cause the death of certain animals. In humans exposure occurs mainly if we eat food containing DDT or similar substances, but a constant diet contaminated with DDT could lead to accumulation if the level of intake exceeds the detoxication and elimination. Consequently food is monitored for such chemicals. Because breast milk has a high fat content, these substances will appear in the milk and be

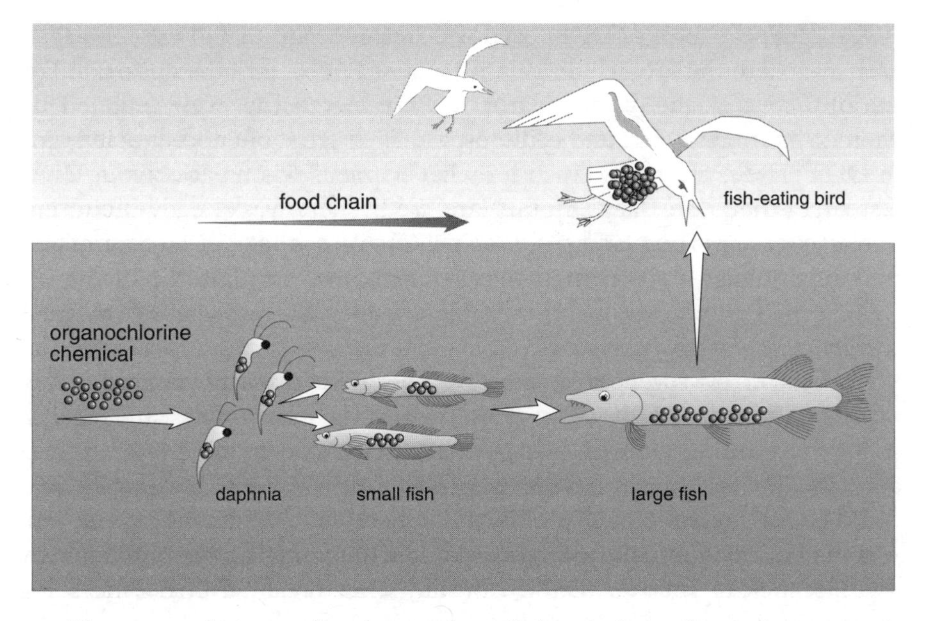

11. The accumulation and concentration of chemicals in the environment via the food chain. This is known as bioaccumulation and biomagnification, and applies especially to chemicals which are soluble in fat such as the organochlorine pesticides. The animals (e.g. fish-eating birds) at the top of the food chain acquire the highest concentration of chemical.

transferred to the suckling baby, which can then be exposed to a greater dose on the basis of its size. For example, lactating mothers exposed to DDT at 0.0005 mg per kg body weight per day were found to produce milk containing 0.08 ppm DDT, with the result that their infants were exposed to 0.0112 mg per kg body weight per day, twenty times greater than the mothers.

This principle is further illustrated by an example at Clear Lake in the USA. Here a very similar insecticide, DDD (also known as TDE), a breakdown product of DDT, was used, because it is less poisonous to fish than DDT. DDD was used to control midges which were a serious nuisance on the lake. Spraying over the water in which the larvae of these insects lived, in 1949, had a dramatic effect on the midge population for some years. There were no apparent effects on fish or, initially, on other wildlife. In later years more DDD was used but then effects on wildlife were detected, in particular on the bird the Western Grebe, whose breeding numbers were seriously reduced. Then in 1954 large numbers of dead grebes were found. Eventually it was discovered that the grebes had high levels of DDD in their fat.

Although the level of DDD used was low, it accumulated in organisms, each animal in the food chain gaining higher levels, until the animal at the top of the chain, the final predator, acquired levels that were lethal. This worried scientists when it was discovered. The same phenomenon applies to DDT. These examples illustrate what happens when excessive or even just larger than necessary amounts are used. There was already mounting public concern when, in 1963, Rachel Carson's book *Silent Spring* was published, which documented these occurrences and laid the blame on DDT. It had an enormous impact and the eventual result was that DDT was banned in many countries.

In the UK, however, it continued to be used for some time, perhaps more carefully. In 1963 it was used to control the pine looper caterpillar, a serious pest in pine forests, and large areas were sprayed with about 1 lb per acre. The caterpillar was successfully controlled and, although DDT could be detected in birds from the area, there were no deaths of birds nor was their population affected. This example illustrates how it is possible to use insecticides such as DDT carefully and avoid adverse effects on non-target species.

How DDT works and why it is toxic to insects but not mammals

The mechanism by which DDT acts is to disturb the function of nerves in the insect. Nerves in both insects and humans work by allowing an electric current to move down them. This action potential, as it is called, depends on the movement of two **metal ions**, sodium and potassium, across the membrane of the nerve, and involves channels for the sodium being opened very briefly. DDT interacts with the sodium channel in the insect nerve and retards its closure. This means that the flow of sodium and hence the electric current is prolonged and there may be several impulses instead of just one. The function of the nerves thus becomes uncontrolled. This effect of DDT seems to be reversible.

It seems that insects are more susceptible than mammals, for two reasons. The first is that insect nerves are more susceptible to the DDT. The second is that mammals have a more well-developed system of detoxication than insects and so remove it more effectively. Naturally, at high doses DDT will cause toxic effects in humans, due to effects on nerve function, but these effects are reversible.

As well as the direct and lethal effects of DDT on animals such as birds, there are also indirect and more subtle effects. For example, the decline in the numbers of predatory birds in the UK and USA, such as the peregrine falcon and the kestrel, was not necessarily due to lethal effects on the birds themselves but due to effects on their reproductive process. One such effect of DDT is to alter the production of eggs, in particular the eggshells. During the 1960s the eggs of birds such as the peregrine falcon and pelican were found to have thinner shells and were therefore liable to break, a phenomenon which had started in the 1940s. This was later found to correlate with the level of **DDE**, a breakdown product of DDT, in the egg. It is believed to be due to the effect of DDE on the disposition of calcium in the shell gland which is involved in the production of the eggshell. Other organochlorine compounds may also cause this effect but it seems that birds vary significantly in their sensitivity; the raptors and fish-eating birds being the most sensitive. However, recently the role of DDE in the decline in numbers of raptor birds, such as the California Condor, has been questioned.[2]

More recently, it has been found that substances such as DDT and other organochlorine compounds can have other effects on wildlife, again affecting the reproductive process and leading to reproductive failure, but also causing deformities in reproductive organs. This effect, now called endocrine disruption, was first noticed in fish in rivers in the USA and the UK where there may be many causes (see pp. 131–6). One of the most celebrated cases was in Lake Apopka in Florida. The population of alligators in this lake was found to be in decline and there seemed to be poor reproductive success among the animals. Both male and female animals had abnormalities in their reproductive organs. It was then found that there were very high levels of the DDT breakdown product/metabolite DDE in the lake. This was due to spillage of a related pesticide called dicofol, which was contaminated with DDE. The latter substance has been found to be capable of causing these effects under experimental conditions. DDE is now known to affect levels of the male hormone **testosterone** (it is known as anti-androgenic) rather than increasing the amount of female hormone (known as an **oestrogenic** effect). One of the effects of such endocrine disruptor chemicals is believed to be a reduction in sperm count as a result of early changes in the testicles. As there is some evidence that human sperm counts and fertility are declining (although some studies have shown an increase in sperm counts), it has been suggested that such substances as organochlorine compounds (for example, DDT) may be responsible. However, there is no direct evidence for this in humans. This will be discussed further in Chapter 5.

The real problem with DDT is its persistence in the environment and its accumulation in certain animals. It is degraded only slowly in soil and in some of the animals exposed to it. It may take between five and twenty-five years for the soil to lose 95 per cent of the DDT. Thus over the years of its use it accumulated in the environment, and certainly in wildlife there is a problem of continued exposure to the substance at levels higher than might be expected from the environment level.

Although DDT localizes in fat where it is probably relatively harmless, release from this fat when it is broken down in an animal's body to provide energy can release enough DDT to cause toxicity in susceptible species, for example in bats. As bats are insectivorous, they can accumulate DDT from their prey, which will become localized in their fat tissue. In the southern USA it was found that bats were dying during their migratory flights, when their fat was mobilized for energy, hence releasing DDT into the blood, which then caused toxic, in some cases lethal, effects.

What of its potential toxicity to humans? There have been no documented deaths and no established cases of illness in which DDT is the causal agent but it is still detectable in the environment and in some food. Because the levels of such substances in food are monitored and we have efficient detoxication systems, and because DDT tends to be stored in fat, it is likely that the potential toxicity is minimal.

What does the DDT story illustrate and what lessons can be learnt about the use of chemicals? When DDT was first used it was not only very effective but apparently also relatively harmless. This led to it being used in excessive quantities (the 'more is better' fallacy). The inevitable consequence was the death of wildlife and a public outcry which crystallized around the book *Silent Spring*. This was predicated partly on the fear, 'If it does this to birds, what is it doing to us?' DDT is designed to be especially toxic to insects, and other species such as birds are more sensitive than mammals such as humans. The development of a new technique for the detection of DDT which was extremely sensitive allowed traces to be found in many things such as breast milk and food as well as wildlife. It became easy to point the finger at DDT because many birds and other animals had detectable levels of the chemical. But just because a chemical is detectable in an animal does not mean that it causes either death or ill health or that the level is hazardous.

This story also serves to remind us that it is dangerous to assume that when problems don't appear immediately there is no problem. When the technique became available for the sensitive measurement of DDT it was found to be accumulating in the environment. And it has been only relatively recently that the effects on reproductive systems have been detected. This shows that it is necessary to have both the right tools and the inclination to detect potential problems.

It is possible to use chemicals such as DDT effectively and responsibly, as has been illustrated, and this particular substance has been of enormous benefit to humans. A little more care and respect for the chemical when it was introduced would have improved the risk–benefit balance. The example of DDT again illustrates that recognition of the principle of Paracelsus is vital in the use of chemicals, especially those intended as pesticides. We know also that the relationship between the dose and the effect is different in insects and in other species. Therefore using less DDT would still have been effective but have caused little, if any, harm to other species.

The effect of banning DDT

What has been the effect of banning DDT? First, other insecticides have replaced DDT, most, if not all, of which have been more toxic. No human deaths have been attributed to DDT, while hundreds of deaths have been caused by the organophosphate insecticides which succeeded it. Secondly, the control of the malarial mosquito was hampered when DDT was banned, and consequently many more people in countries such as India and Sri Lanka succumbed to malaria than might otherwise have done so. Millions of extra cases of the disease have occurred as a result. This was again illustrated recently in South Africa where the banning of DDT (and replacement with the more expensive and less effective pyrethroid insecticides) resulted in a huge increase in malaria cases in KwaZulu-Natal province (from just over 4,000 cases in 1995 to more than 27,000 cases in 1999). 'DDT represents a good example of how the misrepresentation and manipulation of toxicity data with little regard for risk–benefit can lead to the banning of a valuable chemical that has saved rather than threatened, millions of lives.'[3] Nevertheless DDT is still used in some parts of the developing world because it is both cheap and very effective.

The debate over DDT is still not over, and there are many toxicologists and other scientists who believe that the case in *Silent Spring* was probably exaggerated and possibly inaccurate in places. Other books, for example *Trashing the Planet* by D. L. Ray (1992), have appeared which have put forward alternative views. The detection of minute amounts of DDT in the environment or even in the fat tissue of birds (measured with exquisitely sensitive equipment) may not be relevant to the death of wildlife, that is, the association may be coincidental rather than causal. Because DDT became ubiquitous in the environment many people believed that they were accumulating it in their bodies and that eventually it would be hazardous. However, the amounts people have accumulated in their bodies in the USA, for example, are tiny in comparison to the doses volunteers have taken without harmful effects. There is no evidence of harm. Indeed a report by the UK Committee on Carcinogenicity in 1999 concluded that there was no evidence that DDT is associated with increased risk of breast cancer.[4]

Having said that, the effects of organochlorine compounds such as DDT on hormones in humans (endocrine disruption) have yet to be fully explored, but the likelihood is that DDT at current levels of environmental exposure is insufficiently potent to cause effects in humans. The effects observed in alligators discussed above were seen at relatively high exposure levels (see p. 134).

One of the consequences of the banning of DDT was that other insecticides were developed which became more widely used, for example dieldrin and the organophosphates. Unlike DDT, these were found to be much more toxic to mammals. Dieldrin, an organochlorine compound, was also more toxic to other species such as birds and resulted in large numbers of deaths between 1956 and 1961. One reason for this was the use of dieldrin-treated seeds which were eaten by birds such as pigeons. This led to the deaths of large numbers of pigeons in treated areas and then predators, such as foxes and cats, which ate them were also poisoned. Dieldrin is persistent and remains toxic in animals that have consumed it. Thus other, later, organochlorine insecticides were in some cases much more toxic than DDT.

Organophosphates

Organophosphates (OPs) are a group of chemicals with similar structures and similar modes of action, which includes the nerve gases. Only the pesticides will be considered in this chapter. Organophosphates were developed during the Second World War as insecticides and chemical warfare agents. The first was **parathion**, which is very toxic compared to DDT and to other OPs and is no longer used in most countries (see box and Table 1).

The OPs were found to be non-persistent because they are biodegradable. Generally it was found that when OPs were used carefully animals other than insects were less affected by OPs than by DDT. This was because, provided the particular insect pest could be targeted, other organisms would not be affected as residues were quickly degraded or were sufficiently volatile to disperse. However, as a group OPs have still been responsible for the deaths of many birds, mammals (including

Organophosphates

Organophosphates are all chemicals based around an atom of the element phosphorous. Chemically they are known as phosphoric acid esters. Depending on the chemical groups that are attached to the basic molecule the organophosphate will be more or less water-soluble and more or less soluble in fat. Hence some will be more persistent than others but not as persistent as DDT, for example. Unlike the organochlorine insecticides, organophosphates are mostly readily **biodegradable** because of chemical instability or perhaps breakdown by bacteria or other organisms. The half-life in water may be as short as nine days, for example. There are now many different OPs used for a variety of purposes, but the early compounds such as TEPP and parathion were very toxic to other animals.

Table 1 Lethal doses of insecticides

INSECTICIDE	LETHAL DOSE IN RAT	LETHAL DOSE IN HOUSEFLY
TEPP	2 mg/kg	
Parathion	21 mg/kg	
Chlorfenvinphos	31 mg/kg	
DDT	2,500 mg/kg	14 mg/kg
Malathion	>4,000 mg/kg	17 mg/kg

Malathion became widely used, as it had even greater selectivity than DDT.

humans), and fish and other aquatic organisms. There have generally been more effects on wildlife in the USA, where for example they have been sprayed over large areas of water to kill mosquito larvae, than in the UK.

Awareness of the dose necessary is, as always, crucial. In Canada an OP insecticide called phosphamidon was used to control insect pests on conifer trees. Initially the use of this pesticide killed many birds as well as the insect pest, but the Canadians found that by reducing the amount of the chemical used they could control the insects without affecting the bird population. Problems have occurred in particular when OPs have been used to treat seeds and, while this reduces dispersion, animals that eat the seeds can then be poisoned. This has caused the deaths of large numbers of birds in some areas.

As a group OPs have been responsible for hundreds of human poisoning cases and a significant number of deaths. For example, parathion has affected at least 1,500 people worldwide and caused the deaths of over 200 in such countries as India, Malaya (as it then was), Colombia, Egypt, and Mexico. Even malathion, which is one of the least toxic of the OPs, has been responsible for poisoning cases and deaths.

All OPs cause effects on the nervous system of insects and also of other animals. Some of them need to be changed by metabolism in order to be toxic. The reason insects are more susceptible than mammals is because they activate the pesticide to the toxic product but do not effectively detoxify the organophosphate, whereas mammals generally detoxify the insecticide more readily than they activate it. However, mammals can also be susceptible when the dose is excessive. Poisoning has occurred in rare cases after normal exposure levels (see box below, p. 102). The toxic effects both in insects and mammals is due to interference with the breakdown of a substance called acetylcholine, which is produced by the nerves. The effects are the result of accumulation of acetylcholine (see Figure 12). The symptoms of acute poisoning in humans, for example, are similar for all organophosphates and are readily predictable from a knowledge of the effects of acetylcholine on the nervous system. The symptoms range from minor effects such as constriction of the pupils and increased formation of tears and saliva, anxiety and slurred speech, to unpleasant ones such as diarrhoea, vomiting, and uncontrolled urination. At higher doses the symptoms become more serious and include convulsions, difficulty breathing, and paralysis of the muscles. These effects can be reversible once the level of acetylcholine has declined, except at high doses when the outcome may be fatal.

The symptoms or their timing may not be identical for different OP compounds, however, because the organophosphates have different potencies and are detoxified and distributed around the body at different rates. For example, some, like chlorfenvinphos, are more soluble in fat and will persist in the body longer; they may therefore have a slower onset but a longer duration of effects. The rate of detoxication will also be a major factor, and different OPs will have faster or slower rates as well as different routes of metabolism (see pp. 18–20). As well as the level of exposure, the frequency is also very important with OPs. This is because of the possibility of accumulation with insecticides such as chlorfenvinphos which are fat-soluble. In addition, because the organophosphate binds irreversibly to the target (an enzyme), there may be an accumulation of the effect over many days of exposure, for example, as might occur with an agricultural worker spraying crops every day. Although the 'dose' or exposure level may be small on each day, over time it can accumulate to a toxic level. For this reason the level of exposure is monitored in such workers.

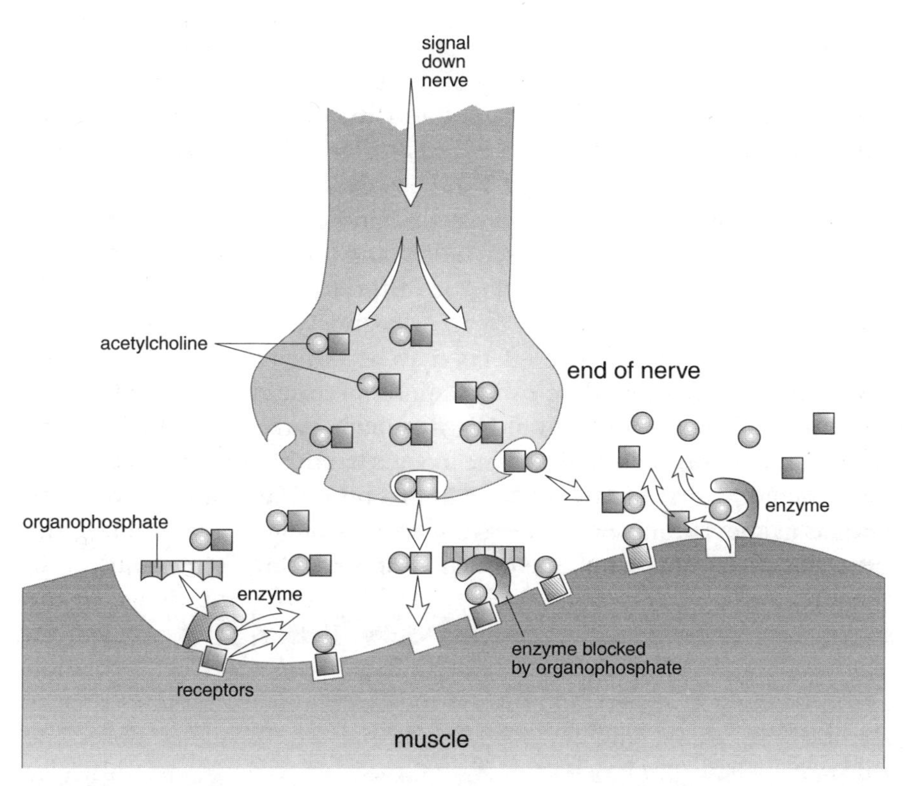

12. How organophosphates are toxic to nerves. Acetylcholine is a neurotransmitter chemical present in the ends of nerves. Acetylcholinesterase is an enzyme which breaks down acetylcholine so that it is no longer effective at causing muscle contraction. Organophosphates inhibit this enzyme allowing acetylcholine to accumulate.

Detection and treatment of organophosphate poisoning

Acute poisoning can be easily detected from the range of signs and symptoms that people exposed will show and describe. One of the most common signs observed is the constriction of the pupils. However, workers who are regularly exposed should be monitored by measuring a sensitive biomarker which is usually the level of inhibition of the enzyme **acetylcholinesterase** in red blood cells.

In the case of acute poisoning resulting from excessive exposure there are two antidotes that can be used. The first is atropine (which is found in deadly nightshade berries and is also known as belladonna), which antagonizes the effects of the acetylcholine that accumulates in the body as a result of organophosphate poisoning. The other, a substance called pralidoxime, was made specifically as an antidote, and this binds to and removes the organophosphate from the body. When combined, the two antidotes are many times more effective than when used separately, a phenomenon known as synergy. Neither of these antidotes will work if their use is delayed, nor can they be used against delayed neuropathy which some organophosphates cause (see below and for example, pp. 259–263).

As with drugs, an important factor in the toxicity of OPs can be the effect of simultaneous exposure to other chemicals, including other pesticides and contaminants. The effects may be difficult to predict as some chemicals can increase while others inhibit detoxication, and with some organophosphates there are multiple metabolic pathways. If two or more pesticides are sprayed by the same operator within a short time there could be dangerous and unexpected interactions. The same may apply to wildlife in the vicinity of the spraying. A poisoning incident in Pakistan illustrates this problem and the difficulty of predicting such events.

CASE NOTES

Pesticide poisoning in Pakistan

In 1976 there was a serious incidence in Pakistan of poisoning with the organophosphate malathion, which was being used to eradicate malarial mosquitoes. About 2,800 out of 7,500 workers spraying the malathion were poisoned, and there were five deaths. This happened because of poor storage, which led to a change in the malathion. The malathion was used as a powder dispersed in water, and it was stored this way. Perhaps the temperature in the building in which it was stored was too high, or the components of the powder were partly responsible. The malathion did not degrade as such, but a proportion of the chemical underwent a change called isomerization, in which two atoms in the molecule change places. The product remained almost the same but with one crucial difference: the isomalathion, as it was called, was a potent inhibitor of the enzyme that detoxified malathion in humans. The result was that the humans exposed to the mixture became as sensitive as insects to the malathion, with the result that the small amounts to which the sprayers were exposed became toxic and even lethal. Thus one of the safest organophosphates became very poisonous.[5]

Organophosphate-induced delayed neuropathy

There are other effects of OPs that are not predictable and are caused by only a small number of organophosphates. Of particular concern is the delayed neuropathy caused by some OPs. This has a different mechanism from the acute effects and may be irreversible. This is the effect described in Chapter 10 (see pp. 259–63), where the peripheral nerves, such as those in the legs and arms, are slowly destroyed, and consequently the limbs are paralysed, possibly permanently. Organophosphate insecticides associated with this are mipafox, leptophos, and methamidophos.

A few other organophosphates cause this effect, notably **tri-orthocresyl phosphate**, as described in Chapter 10. It seems to be due to the interaction between the organophosphate and a protein, which may be an enzyme, in the peripheral nerves. The protein seems to have a critical function and the binding to it is irreversible, causing the nerve to degenerate. The result is paralysis in the legs. The cause of this effect does not seem to be related to the interaction with acetylcholinesterase. The effect appears one or two weeks after exposure to the organophosphate.

There have been suggestions from experimental work in animals that there may be other effects, such as in the brain, that occur after low-level, long-term exposure to OPs and in the absence of other symptoms. This probably occurs by a different mechanism from that which causes the acute effects of high doses but which may be similar to that causing delayed neuropathy. If such effects were found to be occurring in humans, it would be of concern because of workers who are regularly exposed to OPs, for example farmers who use sheep dips. At present the situation is not clear and there is a debate about the presence or absence of such long-term subtle effects of OPs, especially in agricultural workers exposed to sheep dips and regular spraying.

It has been possible to devise antidotes for organophosphate poisoning because the underlying mechanism by which they cause acute toxicity is understood. These are very effective for the rapid treatment of acute poisoning episodes but are ineffective for peripheral neuropathy or if poisoning occurred some days previously.

Organophosphates illustrate several points. First, repeated exposure can be a problem not only because of accumulation of the substance (as can occur with other substances such as aspirin) but also because of accumulation of the effect, if it is irreversible. Therefore the dose–time relationship is important as well as the dose–response relationship, Secondly, understanding the mechanism allows effective detection and treatment; thirdly, other factors such as decomposition and exposure to other chemicals can have a large impact on toxicity; and finally, the dose is crucial, and it would seem, at least from the information available at present, that OPs can be used safely if they are used sparingly and carefully.

Paraquat: safe weedkiller or dangerous poison?

Paraquat is known to and used by many gardeners and householders as well as farmers. It is used in more than 100 countries around the world. In the UK it is sold as Weedol for home use and as Grammoxone for agricultural use. Paraquat is a contact herbicide, killing weeds it comes into contact with but not being absorbed into them sufficiently to destroy the roots. It is believed to interfere with and to disrupt photosynthesis, the essential mechanism by which plants use sunlight to manufacture their food supply. It is relatively safe when used as instructed and sprayed in solution because it binds strongly to soil, leaves, and other parts of the plant and is thereby effectively inactivated. This means that in areas that have been sprayed with paraquat new plants or crops can be planted or seeds sown very quickly. Therefore there is little, if any, risk of poisoning to livestock, wild animals, or humans. However, paraquat has proved to be very toxic to animals under certain conditions. If it is used carelessly it will irritate the skin and eyes, and if taken in by mouth it can be lethal. Hence the route by which it enters the body is a crucial factor, as well as the dose.

Weedol is sold as granules (2.5–8 per cent w/w) and when needed it is mixed with water, in which it readily dissolves, and then sprayed onto plants. Grammoxone is more concentrated (20 per cent w/v solution).

After Weedol was introduced it became notorious for two reasons. There were cases of accidental poisoning, and then cases of suicides and even homicides (see pp. 233–5) in which paraquat was used. Accidental poisonings seemed to occur most commonly in children, who mistook the solution for a soft drink. Gardeners, it seems, often dissolve the Weedol granules in water and then store it for use later. Sometimes the solution is stored in old lemonade or Coca-Cola bottles. Unsuspecting children, and even possibly adults, might mistake the Weedol solution for a fizzy drink. Sufficient paraquat could be taken in with a swig of the solution to be lethal to a child. A lethal dose is 3–4 g in an adult, which represents about 40–120 g of Weedol. With Grammoxone a lethal amount, about 10–20 ml in an adult, could be much more easily ingested.

When it became known through the press that Weedol could cause death when taken by mouth, those with suicidal intentions started to use the weedkiller for that purpose, and it was occasionally used for murder (see pp. 233–5). Deaths caused by paraquat rose as it became known how effective it was, and it was, of course, easily available. Between 1980 and 1987 there were more than 1,000 cases of poisoning and over 200 deaths in the UK, while in 1990–1 there were more than thirty deaths. Death from paraquat is not only unpleasant but also agonizingly slow, hence it is not like taking an overdose of sleeping pills. Victims can sometimes survive for two weeks, more than enough time for them to regret their action.

The toxic effects of paraquat are interesting and unusual, because the organ primarily affected is the lung, with the kidneys and sometimes the liver being damaged after particularly high doses. After a toxic dose the lungs are damaged and the cause of death is slow asphyxiation as the victim becomes progressively unable to breathe properly. The mechanism underlying the toxicity of paraquat is well understood (see box), but unfortunately it has not enabled an antidote to be devised, although a number of potential treatments have been tried. Only very early treatment can alter the course of the poisoning.

Why paraquat is poisonous

The paraquat molecule has some unusual features which account for the toxic damage it causes in the lungs. First, although only a small amount is absorbed from the gut, once in the blood paraquat does not distribute in the body evenly but is concentrated in the lungs. This is because it has similarities with a substance normally present in the body and for which there is a specific system to carry the substance into the cells of the lungs. Paraquat therefore reaches a high concentration in these cells.

Paraquat is also able to exist in a form called a free radical. These are normally very reactive forms of chemicals, but in the case of paraquat it is stable. Free radicals react with oxygen and convert it into a reactive form. There are high levels of oxygen in the lungs and hence paraquat is able to produce a lot of reactive oxygen. The result is that the lung cells are destroyed by this reactive oxygen and, while the lungs have a natural protection against the reactive oxygen that is always being produced in small quantities by normal processes, large amounts overwhelm the protective mechanism.

Treatment of paraquat poisoning

Although many antidotes have been tried, the only really successful treatment is removal of the paraquat from the gut before it is absorbed. This is done by the use of adsorbents such as activated charcoal or fuller's earth, and it must be done within an hour or so, as the paraquat is absorbed quickly. Once in the bloodstream it will start to accumulate in the lungs and damage them, and it may also damage the kidneys.

Haemodialysis or haemoperfusion, in which the blood is passed through a machine, may remove some of the paraquat but is not believed to be always particularly effective.

CASE NOTES

A lucky escape

In one case a 59-year-old man ingested 60 ml of Grammoxone, easily a lethal dose, and his blood level of paraquat was found to be above the threshold for lethality. Treatment consisted of sucking out the stomach contents, giving activated charcoal, washing out the stomach, haemodialysis, and giving an antioxidant (N-acetylcysteine, as used in paracetamol poisoning). Although the patient did survive, his lungs deteriorated after four days, and his liver and kidneys were clearly damaged. After sixteen days the lungs returned to normal, and he was eventually discharged.[6]

Because there is no antidote and the fatality rate is high, the manufacturers now produce packs for home use that contain less paraquat, so making it less likely that people will be poisoned. Another approach has been to formulate the paraquat with a substance known as an emetic, which will cause the victim to be sick. This will help rid the body of some of the poison. The addition of a strong-smelling substance and blue colouring to the paraquat formulation have also been used to reduce further the risk of accidental poisoning. Here again we have an example of a chemical substance that is safe when used properly for the right purpose, but which can be lethal when used for a different purpose. There will be other examples in Chapter 8 of chemicals found around the home that can, under particular circumstances, be lethal.

Other herbicides

There are other herbicides in use which cannot be discussed in detail here, but one particular pair of chemicals should be mentioned. These are 2,4-D and 2,4,5-T, which are abbreviations of 2,4-dichlorophenoxyacetic acid and 2,4,5-trichlorophenoxyacetic acid. These chemicals act like plant growth hormones but cause excessive growth. They are selectively toxic against weeds. Although these herbicides have low toxicity to animals they can be contaminated with the most toxic type of dioxin (2,3,7,8-TCDD), as a result of the manufacturing process. The mixture of 2,4-D and 2,4,5-T was

13. Spraying vegetation during the Vietnam War with the defoliant Agent Orange, a mixture of 2,4-D and 2,4,5-T. This herbicide mixture was contaminated with dioxin.

used in Agent Orange, a defoliant used during the Vietnam war by the American military. The contamination with dioxin led to American troops who were handling the herbicide being contaminated as well as the population of North Vietnam living in the areas that were sprayed. As a result of this, some Vietnam veterans have claimed that their ill health and that of their children were caused by their exposure to the dioxin in herbicides. There have also been reports of birth defects among Vietnamese children.

The fungicide pentachlorophenol can also be contaminated with dioxins. Dioxins will be discussed further in Chapter 5.

Other pesticides

Apart from the insecticides and herbicides already discussed, there are many other kinds of pesticides in use. These range from rodenticides (chemicals designed to kill mainly rats and mice) to fungicides, nematocides, and molluscicides (which target fungal infections of crops and animals, infestations of animals by nematode worms, and organisms such as slugs, respectively). Some of these have been associated with cases of human poisoning, both accidental and intentional, and also environmental contamination.

It is beyond the scope of this book to consider examples of the various kinds of pesticides, but one example of environmental contamination should be mentioned.

Anti-fouling paint and toxic tin

Barnacles growing on the undersides of boats are a problem because they reduce the speed of the boat and can damage the hull. Pesticides (molluscicides) have, therefore, been developed to combat this. One is tributyl tin (TBT). Paints containing TBT were applied to the underside of boats to discourage the barnacles from growing on the surface of the hull. However, small quantities of the substance would dissolve from the paint and disperse in the surrounding water. The result was that the concentration of TBT in the water was sufficient to start poisoning other molluscs and marine organisms, especially in enclosed areas such as harbours. One particular victim was the dog whelk, which suffered not from lethal poisoning but from a bizarre effect whereby the females developed male characteristics, so affecting breeding. This effect is called imposex. It was found to occur after exposures as low as 1 nanogram per litre of water (a nanogram is 1 thousand millionth of a gram). Although initially thought to be confined to areas around harbours and marinas, imposex has now been found in molluscs (such as the common whelk) in open seas like the North Sea. Some seventy-two species of gastropods have now been shown to be affected, and some populations have decreased to extinction. Other, higher, organisms such as fish, mammals, and birds are not at risk because they efficiently degrade and detoxify the TBT compounds, whereas the molluscs cannot. The TBT was found to accumulate in the whelks, and in some molluscs concentrations reached values of 500,000 times that in the surrounding water. Furthermore, TBT has a half-life of almost eight years. It seems that imposex was a result of the accumulated TBT interfering with the metabolism of the sex hormones in the female animals so that levels of the male hormone testosterone increased. Other, similar, examples are known to occur with organochlorine compounds (as mentioned above, p. 95 and see pp. 131–6) and various other environmental and industrial pollutants. This will be discussed in more detail in Chapter 5.

Pesticide residues in food?

As well as worrying about acute intentional and accidental poisoning by pesticides, the general public is concerned about some of these chemicals finding their way into food (see also Chapters 10 and 11). There have been cases of improper use of pesticides leading to serious illness and even death from food contamination. For example, the use of mercury-containing fungicides on seed grain which was subsequently used for food instead of planting led to the deaths of significant numbers of people in different parts of the world. Similar incidents have occurred with the pesticide hexachlorobenzene.

Is there a risk from pesticides contaminating food after normal use? Naturally, the use of pesticides to control insects, rodents, or fungi that threaten crops can potentially contaminate the crop or food made from it. Proper use of the pesticide will often result in minimal contamination and in many cases decontamination may simply be a matter of washing the fruit or other crop. However, because of the possibility of contamination, food is monitored by government agencies for the presence of pesticides and safety limits are set well below levels at which effects may be detected. If the level of a particular pesticide is below the limit set then the risk of any adverse effects will be extremely small (see Chapter 12 for a further discussion of this).

Although pesticides are designed to be toxic, they are designed to be selectively toxic to weeds or insects. As with any chemical, dosage is crucial, and pesticides will inevitably be poisonous to humans if they are exposed to sufficiently large quantities. If used correctly, the majority of modern pesticides are relatively safe. It is when they are misused that they cause toxic effects in animals or humans. The problems associated with DDT are perhaps the prime illustration of this principle. Thus chemicals are not equally toxic to all living things: some species are very sensitive, while others are very resistant.

5 First the Cats Died
Environmental Contaminants

The fishermen and their families living around Minamata City, in Japan noticed that their cats were ill and dying. They themselves also began to suffer strange symptoms: memory loss, ataxic gait, restrictions in their visual field and hearing difficulties. Post-mortem of victims who succumbed showed damage to neurones in the cerebral cortex of the brain.

THIS event occurred in the early 1950s and was the result of mercury poisoning caused by effluent from a factory nearby. It was one of the world's worst environmental pollution disasters.

Mercury: pollution and poisoning

The substance mercury has been described in many ways, but the following description is appropriate here: 'the hottest, the coldest, a true healer, a wicked murderer, a precious medicine and deadly poison, a friend that can flatter and lie'.[1] Named after the planet Mercury, it is a metal but, unlike all other metals, it is liquid at normal temperatures (that is, at room temperature or 21°C). Its Latin name is *hydrogyrum* (hence its chemical symbol Hg), meaning liquid silver, a name given to it by Aristotle.

Mercury can exist in several forms and all of them are toxic. Thus the element itself is poisonous because liquid mercury vaporizes and the vapour can be inhaled and absorbed. It can enter the brain in this state and cause damage there. The salts of mercury (inorganic mercury) can also be absorbed from the gut and can damage the kidneys. Finally, mercury can exist as organomercury, in which it is part of an organic compound. These are also very toxic and damage the brain and nervous system. Mercury is used in its various forms in scientific instruments, in battery manufacture, in chlorine production and other industrial processes, and in mining.

Metallic mercury is mostly a problem in confined spaces or where it is handled in industrial processes or in laboratories. It is not generally an environmental problem, more likely an industrial hazard. However, metallic mercury does occur naturally (it was mined in Spain and Slovenia, for example) and the use of mercury for metal reclamation is a potential environmental hazard in countries such as Brazil where miners use it to extract gold from river sediments. Inorganic and organic mercury may be produced from the metallic mercury during its use and subsequent release into the environment.

Metallic mercury was widely used in early experiments in alchemy, and more recently in scientific instruments such as barometers which use has continued until this day. Consequently scientists have been at risk from both the acute and the chronic effects of exposure to mercury vapour. The metal vaporizes very readily at room temperature and can be absorbed through the lungs by inhalation. The effects include loss of memory and appetite, insomnia, depression, and paranoia. All of these symptoms were suffered by the famous scientist Sir Isaac Newton, and indeed his hair has been found to contain significant amounts of mercury. Despite this he lived to the age of 85 and his mental abilities were hardly in doubt! Bleeding from the gums and gastrointestinal upsets are also effects.

Inorganic mercury, such as occurs in mercury salts, is mostly soluble or dispersible in water and therefore would disperse in the environment. At high levels it would certainly be poisonous to wildlife and humans alike but, dispersed in large volumes of water, it would be less of a risk. Under certain conditions, however, inorganic mercury may give rise to organomercury (see below). Inorganic mercury has been used in medicine as a laxative and to treat syphilis in the form of calomel (mercurous chloride) (see case notes). Corrosive sublimate (mercuric chloride) was used as a disinfectant and some traditional Chinese medicines still contain it.

Various forms of inorganic mercury are used in industry in battery production and as catalysts. Mercuric nitrate was previously used in tanneries and the fur trade (see pp. 166–7). Clearly these can cause both pollution and human poisoning. However, the form of mercury most likely to be associated with environmental pollution and to cause problems is organomercury. This is because in this form mercury is soluble in fat and can accumulate in animals in the food chain. The cases of serious environmental mercury pollution and poisoning illustrate this well.

CASE NOTES

The attempted murder of Benvenuto Cellini

That mercury is a double-edged sword is illustrated by the story of Benvenuto Cellini, the great Italian sculptor of the sixteenth century. He was the first person to produce life-sized works in bronze. He was also sexually very active and consequently contracted syphilis when he was 29. He refused treatment with mercury, commonly in use at the time, preferring to take guaiac, a preparation from the wood of a plant believed to be effective for the treatment of syphilis. This proved ineffective and his condition deteriorated. He seems to have had enemies who could not wait for him to slough off his mortal coil naturally from syphilis and wanted to dispatch him. They invited him to a dinner in which his food was poisoned. He became seriously ill and his physician realized that he had been poisoned with corrosive sublimate (mercuric chloride), which caused extensive damage to his gastrointestinal tract. However, Cellini recovered from this and, moreover, the syphilis had also receded. It appears that, luckily for art and culture, the poisoners had failed to administer a high enough dose to kill their victim but enough to kill the parasite that causes syphilis! This again is an illustration of the Paracelsus principle of the importance of dose.[2]

Minamata disease

Minamata is an industrial city on the Yatsushiro coast of Japan on the southernmost island (Kyushu). In the city there was a factory that manufactured the chemicals **vinyl chloride** (used to make the plastic PVC; see pp. 168–71) and acetaldehyde for many years. The processes used inorganic mercury (mercuric oxide) as a catalyst. The effluent from the factory contained inorganic mercury and perhaps also some organic mercury (methyl mercury), produced as a by-product in the chemical reaction in the plant. This effluent was discharged into the waters of Minamata Bay.

Inorganic mercury, although poisonous, might have been expected to disperse in the water, but it was mixed with other effluent and this sludge settled at the bottom of the bay. Here the mercury was converted by micro-organisms into organic mercury, which is quite different from inorganic mercury as it is not very soluble in water but very soluble in fat. Consequently it was readily absorbed by plankton and other organisms in the sea which were the food for shellfish and small fish. A constant diet of contaminated food led to increasing amounts of methyl mercury accumulating in the bodies of the shellfish and small fish. These would be eaten by larger, predatory fish and so the methyl mercury passed up the food

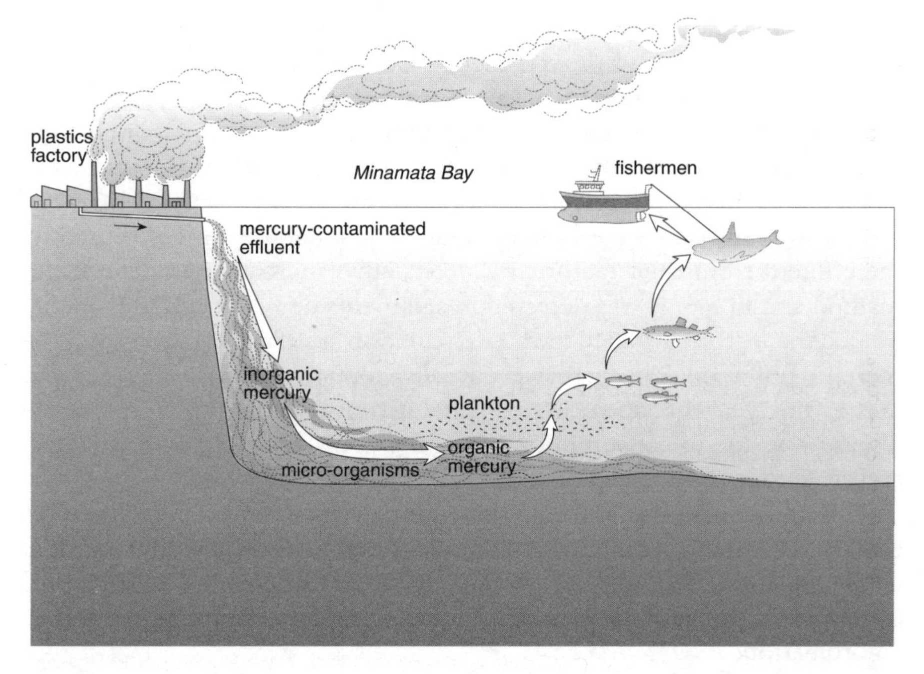

14. How mercury from a Japanese factory contaminated fish in Minamata Bay. The effluent, containing inorganic mercury and discharged into the bay from the factory, was not dispersed in the water, but formed sediment at the bottom where micro-organisms converted the mercury into the more toxic methylmercury. The methylmercury accumulated in fish and was consumed by people eating the fish.

chain. The result was bioaccumulation and bioconcentration of the methyl mercury. Fish can concentrate methyl mercury to the extent of 10,000 to 100,000 times that in the surrounding water.

The Japanese diet contains a lot of fish, and the local fishermen and their families would eat more fish than most. The fishermen were catching fish and seafood in the waters of Minamata Bay, waters which proved to be heavily contaminated with methyl mercury. In 1956 the first case of what became known as Minamata disease was reported and then other people started to present themselves to doctors and at hospitals with various symptoms such as muscular incoordination and difficulties in speech. Their pet cats, which were also eating the fish, suffered similar symptoms.

After a number of cases were seen, investigations started. At Kumamoto University, research suggested that the disease was a type of heavy metal poisoning transmitted by fish and shellfish. At the same time

Dr Hosokawa, director of the Minamata City hospital was conducting his own experiments based on the theory from the university. He fed cats waste effluent from the factory that was producing acetaldehyde and was able to produce similar symptoms in them, and he detected other changes by pathological examination at autopsy. The company that owned the factory, the Chisso Minamata Chemical Company, was aware of his work and by 1959 knew that it was likely that Minamata disease was caused by the effluent from their factory. In 1960 methyl mercury was detected in seafood and in 1961 it was detected in sediments derived from the factory. In 1966 the factory installed a water circulation system which removed the mercury pollution. The factory eventually stopped the process in 1968 and in the same year the Japanese government announced its opinion that the disease was due to consumption of methyl mercury in contaminated fish and seafood.

This environmental disaster led to over seventy deaths and some 2,265 cases of poisoning, yet it was many years before the company accepted liability and compensated the victims.[3] The victims suffered neurological effects such as memory loss, emotional instability, and loss of muscle co-ordination.

The teratogenic effects of methyl mercury

One of the particularly tragic aspects of the Minamata disaster was the effect methyl mercury exposure had on unborn children. Some of the mothers exposed to methyl mercury from the fish and seafood gave birth to babies who were severely affected with a disorder similar to infantile paralysis, suffering cerebral palsy and mental retardation. This occurred even in mothers who showed no symptoms themselves, a classic characteristic of a teratogen. Some babies were born completely paralysed.

Mercury can be transported into the brain and there interact with sulphur-containing proteins such as enzymes involved with nerve function. The fat-solubility of the methyl mercury would allow it easily to enter the embryo and its nervous system and inhibit processes that are essential in the development of the brain and spinal cord. Disruption of the activity of essential enzymes in the brain and nervous system would be crucial in a developing organism, in which a sequence of events is essential. The effect of methyl mercury on the babies born at the time was illustrated by a poignant photograph by the American photographer Eugene Smith.

A related incident occurred in another part of Japan, Niigata, north-west of Tokyo, where another factory, owned by the Showa Denko Company, was also producing acetaldehyde and discharging the effluent into the Agano river. More cases of the same disease, and symptoms as had been observed in Minamata, occurred in this area in 1965. In this case by the time victims were reported the factory had stopped using the process and was no longer polluting the river. As a result of this pollution a further 690 people were poisoned.

These tragic cases were the result of a disregard for the environment by the companies that were discharging effluent containing a known and very poisonous substance into the sea or a river. This pollution hazard was made worse by the fact that the mercury had been converted into a more toxic form, which was much more easily accumulated than inorganic mercury. Fishing in the polluted areas was restricted, leading to the loss of livelihood by fishermen, and these areas were subsequently dredged between 1974 and 1990 to remove the polluted sludge in the bottom of the bay. In 1961, fish in the Minamata Bay area were shown to have mercury levels of greater than 16 ppm, which declined to 0.4 ppm (the provisional regulatory level) in 1969. It has been calculated by the Department of Environmental Health, part of the Ministry of the Environment of Japan, that the annual cost of compensation to victims, the environmental clean-up, and the compensation to the fishing industry was approximately 12 thousand million yen, some 100 times the cost of putting in place measures to control pollution. These figures do not, of course, reflect the scale of the human misery caused by the failure to consider the consequences of the pollution but do show the futility of industrial companies trying to avoid their environmental responsibilities.

Other cases of organomercury poisoning have also occurred, the most serious and notable of which was in Iraq in the early 1970s. This was caused by the misuse of imported wheat and barley seed which had been treated with methyl mercury as a fungicide and was intended for spring planting. The warning that it was unfit for making flour was not heeded and the seeds were duly ground into flour, which was then used to make bread. In all there may have been more than 6,500 victims and some 500 deaths. The study of this large poisoned population allowed the symptoms to be defined—which included tingling sensations in the fingers and toes (paresthesias), lack of coordination of the muscles (ataxia), deafness and impaired ability to articulate speech (dysarthria)—and to be associated with a particular level of mercury in the blood.

As a result of this information a dose–response relationship has been established in humans which has been shown to be linear but different for each symptom, which means that the thresholds for the particular effects are different. Post-mortem examination of victims revealed degeneration of nerve cells in the brain, especially those involved with vision. As with the poisoning in Minamata, babies exposed via their mothers were born with cerebral palsy and mental retardation.

Similarly, feeding the treated seeds to livestock such as pigs and then eating the meat could also cause poisoning. This occurred in New Mexico in 1969 where several children in the family suffered neurological disorders and a baby exposed in the womb was born with brain damage. It was found that the level of mercury in the urine of the newborn baby was fifteen times that in the mother. It is known that the red blood cells in the developing baby concentrate methyl mercury thirty times greater than the same cells from an adult. Methyl mercury stays in the body for a relatively long time (the half-life is about seventy days), and is localized especially in the liver and brain.

The use of organomercury as a fungicide for the treatment of seeds has also caused poisoning of birds, with the mercury accumulating and concentrating in the food chain so that those at the top, that is raptors, become the victims. The deaths of birds in some countries (Sweden, for example) led to the banning of many organomercury compounds used in seed dressings. Data shows that over the hundred-year period from 1840 the mercury levels in birds such as the goshawk and crested grebe have increased about tenfold (see Figure 15).

Other sources of pollution include the use of organomercury-containing fungicides in wood pulp for paper-making. Effluent from such processing plants has contaminated rivers. Predatory fish (for the same reason as predatory birds) also tend to contain relatively higher levels of mercury than non-predatory fish, especially those from enclosed, relatively non-tidal areas like the Mediterranean Sea.

It is possible to calculate from the half-life and the known toxic concentration that the allowable daily intake of mercury, with the inclusion of a safety factor (see pp. 298–308), is about 0.1 mg per day. This would correspond to eating 200 g of fish with a mercury level of 0.5 ppm. At times the mercury level in fish from Lake Michigan in North America and from off the coast of Sweden has been found to contain ten times this level. Consequently, eating a lot of particular kinds of fish could increase the risk from mercury exposure, especially for susceptible groups such as pregnant women. Thus organizations such as the Food Standards Agency in the UK have advised pregnant women to restrict their intake of tuna and to avoid swordfish, for example. (Swordfish may contain about twenty times the methyl mercury level of fish such as cod and about six to seven times that of tuna.)

Finally, it should be pointed out that organomercury in our environment is not due solely to human activities, as micro-organisms can con-

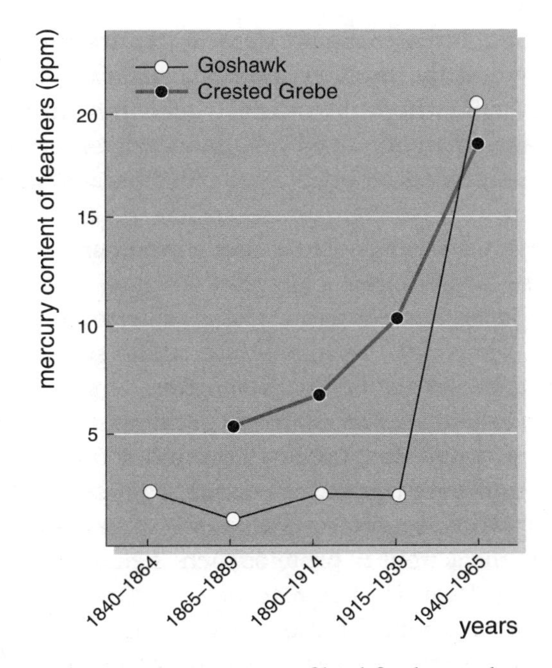

15. The increase in the mercury content of bird feathers (obtained from a Swedish museum) over a 100-year period. Data from Wallace *et al.* (1971), Mercury in the Environment: The Human Element. ORNL-NSF Environmental Program, Oak Ridge National Laboratory, Figure 4.

vert naturally occurring metallic or inorganic mercury, such as occurs in some rocks (for example, in Slovenia), into organomercury.

Why do the different forms of mercury cause different effects?

The answer lies mainly in their distribution and properties. Elemental (liquid) mercury is volatile and can be inhaled. If sufficient of it enters the body, it is able to enter tissues such as the brain and cause damage, giving rise to headaches and also damaging the gums and teeth, for example. Liquid mercury, if taken into the stomach (which could occur if a child chewed and broke a thermometer, for example), would not be especially hazardous as the metal would be eliminated from the intestines in due course in the faeces. Little would be absorbed into the bloodstream. Inorganic mercury, such as occurs in salts (for example, mercuric chloride, also known as corrosive sublimate), can be absorbed from the gut and sufficient enters the bloodstream to bind to sulphur-containing compounds. Although these initially detoxify the mercury, in the kidney they can degrade and the product mimics a natural substance which is transported into cells, where the attached mercury causes serious damage. The inorganic mercury does not, however, enter the brain as it cannot cross the so-called blood–brain barrier because it is not soluble in fat.

Organic forms of mercury, such as methyl mercury, which caused the Minamata disease, are different from the other two forms in that they are very soluble in fat and therefore readily able to enter the brain and nervous system (see p. 18). Again the interaction with natural sulphur-containing substances in the body is important. A product of this interaction mimics an amino acid for which a specific system of transport into the brain exists. So, rather like the Spartans in the Trojan horse, mercury enters the brain, and once there it can disrupt essential processes.

Mercury in all its forms interacts with atoms of sulphur which are often very important in the actions of proteins such as enzymes. One enzyme in particular (Na K ATPase) is known to be a particular target for methyl mercury. It is involved with the movement of ions such as sodium and potassium, which are crucial to the function of nerves. In severely poisoned victims nerve cells were destroyed in certain areas of the brain. Some victims had 50 ppm of mercury in their brains.

The differences in the effects of the three types of mercury are therefore due to differences in their distribution in the body.

Arsenic

This element, perhaps more than any other, has long been associated with poisoning, especially of the homicidal and suicidal type. Thus Agatha Christie called one of her plays *Arsenic and Old Lace* and the author Gustave Flaubert had his character Emma Bovary use an arsenic compound to commit suicide in the novel *Madame Bovary*. This aspect of arsenic poisoning will be discussed in more detail in Chapter 9 (pp. 221–2).

Unlike mercury, serious pollution due to arsenic is mostly caused by natural contamination of water. In further contrast, this contamination is with inorganic arsenic, which is generally more toxic than organic arsenic. Arsenic is found in various minerals in rocks, such as orpiment, realgar (arsenic sulphides) and lead arsenate, and iron pyrites may also contain arsenic. Atmospheric arsenic is produced naturally by volcanoes and micro-organisms which convert inorganic arsenic into volatile organo-arsenic compounds such as methylarsines. Organic arsenic does occur in the environment such as in fish and so may be ingested by humans. Indeed, some seafood such as mussels and prawns may contain more than 100 ppm of arsenic.

Pollution with arsenic can also be due to human, industrial activity, and most of the arsenic in the atmosphere is from the burning of fossil fuels. Arsenic has been used in fruit sprays, insecticides (lead arsenate), weed-killers (sodium or copper arsenite), rat poison, sheep dips, fly papers, wood preservative, and in glass-making. It is now also used in the semiconductor industry. Apparently, it was used as a pesticide by the Chinese centuries ago. Most of the arsenic used commercially in the world is probably contained in various pesticides. This inevitably leads to contamination of the environment.

Large-scale poisoning cases with arsenic have certainly occurred as a result of human activity, for example the poisoning of beer drinkers in Birmingham in 1900. About 6,000 people were poisoned and some seventy died. The cause was glucose (used in the preparation of the beer) that had been contaminated with arsenic. The glucose was manufactured in a process involving the use of sulphuric acid, which had become contaminated with arsenic from the iron pyrites used in its production. The beer contained about 15 ppm of arsenic, which meant that drinking six pints would give a dangerously high dose of 45 mg. At first the beer-drinking victims were believed to be suffering from peripheral neuritis due to alcohol, as arsenic causes very similar symptoms. An astute physician by the name of Reynolds recognized the symptoms as arising from arsenic poisoning.[4]

By far the most significant cause of arsenic poisoning in humans from an environmental source is contamination of groundwater in countries like Bangladesh, India, and China. The situation in Bangladesh has attracted particular attention and illustrates how one problem can be exchanged for another.

> Abdul Kasen lifted a greenleaf 'bandage' to reveal a large growth on the palm of his hand. The chances are that his cancer came from 20 years of drinking water from a village well laced with arsenic. Five of his six children also have skin blemishes and melanomas associated with the poison.[5]

Arsenic in groundwater results from the water flowing through rocks in which arsenic is a constituent of minerals such as iron pyrites. The arsenic present in the water is therefore primarily in the inorganic form, such as arsenite and arsenate. Apart from Bangladesh, the problem of arsenic contamination (that is, levels above the WHO guideline value of 0.01 mg/litre) occurs in many countries of the world, including the USA, Australia,

Finland, Argentina, Chile, Hungary, Mexico, Peru, and Thailand. In the USA some 13 million people are exposed to drinking water with levels at the guideline value.

The levels in groundwater in West Bengal in India are above 0.05 mg/litre in some areas, and more than a million people use this water. In Bangladesh the number of people exposed to arsenic levels above 0.05 mg/litre is between 28 and 35 million. Levels may be up to a hundred times the safe limit. This is a potentially huge poisoning problem which started when a solution was sought to the problem of contamination of drinking water with bacteria causing diseases such as cholera, typhoid, dysentery, and hepatitis. Over the past thirty years programmes aimed to provide safe drinking water by sinking wells to supply groundwater in preference to water from open dug wells and ponds which had often been contaminated with bacteria. The programmes were successful and there was a significant decrease in the infectious diseases.

Unfortunately, unbeknown to those who conceived and organized the programmes, in Bangladesh the wells drilled into the rock to provide safer drinking water tapped water contaminated with arsenic. Many of the people of Bangladesh had swapped their water-borne bacterial diseases for arsenic poisoning. Up to 90 per cent of the population of Bangladesh (130 million) drink well water. The number drinking arsenic-contaminated water has grown dramatically since the 1970s as a result of population growth and the continued drilling of wells. The problem is compounded by the lack of a sensitive and cheap means of testing for arsenic and uncertainty as to where safer water might be drawn from.

The effects of consuming arsenic over a long period of time, that is chronic exposure, range from initial skin lesions, including changes in pigmentation and thickening, to eventual skin cancer and other, internal cancers such as in the kidney and bladder which develop later. It has been estimated that 100,000 people have suffered skin lesions from drinking water contaminated with arsenic. The risk of cancer is increased if the water concentration is 0.05 mg/litre, which is the national standard in India and Bangladesh. In parts of China it seems that blackfoot disease also occurs, where there is damage to blood vessels, leading to gangrene and degeneration of the tissue.[6]

Dioxin (TCDD)

This substance is one of the most feared pollutants because of its great potency as a poison. Moreover, it is not a desirable chemical product: it has no known uses, it is simply a contaminant. There is no doubt that it is the most poisonous substance produced by human activity. It is not, however, the most toxic chemical known (this dubious honour is reserved for the plant product ricin, closely followed by botulinum toxin, which is produced by a bacterium; see pp. 151–2, 249–51).

Dioxins are produced naturally when wood burns, for example. They are a group of chemicals, about seventy-five in number, known chemically as dibenzodioxins. Only a few of these are regarded as particularly toxic, with TCDD (tetrachlorodibenzodioxin) being the most potent (see box, p. 122). A variety of human activities can produce dioxins, ranging from chemical synthesis to disposal by incineration of plastics such as PVC and the industrial bleaching of paper with chlorine. Burning waste of any description probably produces dioxins, as do car engines. The major sources of dioxins are:

chlorophenol production;
chlorinated herbicide production;
burning PVC;
burning wood;
car exhausts;
incinerators;
paper manufacture and bleaching;
manufacture of pesticides and wood preservatives.

Because of this dioxins are found everywhere in the environment and, with the advent of ever more sensitive analytical techniques, have been detected in paper, breast milk, food, air, soil, and water. As we saw earlier, just because a sensitive technique can detect a chemical in the environment does not mean that it is a toxic hazard at the level detected, but the ability to detect small traces everywhere has led to the profile of these chemicals being raised to become the number one hazard.

One of the first chemical processes to produce dioxins in significant amounts as contaminating by-products was the production of chlorophenols. These chemicals replaced the more toxic phenol as an antiseptic in the twentieth century and TCP (2,4,6-trichlorophenol) was widely used in the UK (my mother used it liberally in our home) for treating cuts and grazes, as a mouthwash, and in pastilles (the latter are still available). However, synthesis of this disinfectant produced dioxins as contaminants, including both TCCD and TCDF, the two most potent. It is likely that, at least until it was realized that this contamination took place, TCP contained some dioxin. Whether this was responsible for any ill health is unknown, but there was certainly no epidemic of birth defects or cancer after its introduction.

The synthesis of other chlorinated chemicals such as the herbicide 2,4,5-T also produce dioxins, including TCDD, as contaminants (see pp. 106–7 above).

The toxic effects of dioxin

Dioxin causes a wide spectrum of adverse effects in animals. The lethal dose varies enormously between different species. Thus, in guinea pigs a dose of only 0.5 μg per kg of body weight would kill them, whereas in hamsters 10,000 times that dose would be necessary to have the same effect. Animals exposed to dioxin undergo a number of changes, including loss of weight and changes in the liver; the thymus, which is particularly sensitive, undergoes atrophy. One reason the guinea pig is more sensitive than the hamster is because of differences in sensitivity of this organ. The half-life of dioxin in the guinea pig, 94 days, is also much longer than the 15 days measured in the hamster. Damage to the thymus leads to a decrease in the effectiveness of the immune system. Dioxin can also affect the reproductive system altering **oestrogen** levels and decreasing production of sperm at exposures as low as 0.001 μg per kg of body weight per day. It will also cause birth defects if pregnant animals are exposed. Finally, it has been shown to cause cancer in animals, although the fact that it increases the incidence of some cancers and decreases that of others suggests that it may be a co-carcinogen or **promoter** rather than an initiator of cancers. Some of the toxic effects of dioxin, such as those affecting the thymus, are mediated by an interaction with a receptor.

In humans the major and consistent effect is on the skin. Exposed people suffer from a severe form of acne, known as **chloracne**, after exposure to dioxin. Changes in the immune system have also been detected in the children exposed at Seveso in Italy and workers exposed at the Coalite plant in the UK. The latter had reduced levels of some immunoglobulins. There is no definitive evidence as to whether humans are more or less sensitive to dioxin than other mammals but the weight of evidence would suggest that they are less sensitive than many.

Human exposure to dioxin

There have been a number of cases of known human exposure to dioxin, mostly as a result of industrial accidents. The first reported human exposure to dioxin resulted from an accident at a plant making the herbicide 2,4,5-T in 1949 in Nitro, West Virginia. More than 120 workers were exposed and showed reversible changes to the skin known as chloracne, now known to be typical of exposure to this kind of chemical. This group of workers were monitored for thirty years by the Institute of Environmental Health at the University of Cincinnati. Far from showing adverse effects and ill health, the group appeared to live longer and have fewer diseases than non-exposed individuals.

In a later accident in Germany at the BASF factory in Ludwigshafen where trichlorophenol was made, 250 individuals were exposed and about half suffered chloracne. Follow-up of the individuals affected revealed that more died of cancer than expected. After another accident in the Netherlands in 1963, four men who had suffered chloracne after being involved with the clean-up may have died as a result of exposure, as their deaths occurred not long afterwards. Another accident occurred in the UK at the Coalite Chemicals Factory in 1968 following which about eighty people suffered chloracne.

These incidents had so far affected only the workers in the plants where the chlorophenols or 2,4,5-T was being made, but in 1976 an accident occurred in Italy where the surrounding population were exposed. This happened at the Icmesa plant at Seveso, near Milan, where the safety systems failed when the pressure in the reaction vessel making chlorophenols increased, with the result that the contents were released as a cloud into the air above Seveso. This cloud, mostly of 2,4,5-trichlorophenol, was contaminated with perhaps as much as 16 kg of dioxin, and at least 3 kg settled on some of the 17,000 inhabitants of the town. Levels of hundreds of milligrams per acre were detected in soil in the most heavily contaminated areas. A large number of animals, estimated at about 2,000 including pets, sheep, and rabbits, died following the exposure to dioxin. Reaction to the disaster was slow, partly because it had happened at a weekend in the summer and certain officials were away. Fortunately, no humans died although a number suffered chloracne, especially children. About 250 people were evacuated of whom about 180 had chloracne. There were a number of pregnant women exposed (150), some of whom opted to have abortions. None of the babies aborted or those born later had abnormalities that could be associated with dioxin; the incidence of abnormalities was no higher than usual.[7] A later study of more than 15,000 births in the area, including from the women living closest to the plant and most heavily exposed, again found no greater incidence of malformations than expected. A study by the Institute of Occupational Health at the University of Milan of the inhabitants over the ten years following the disaster found no dramatic changes in cancer rates, just two more than expected. There was an increase in some cancers, liver cancer in men and

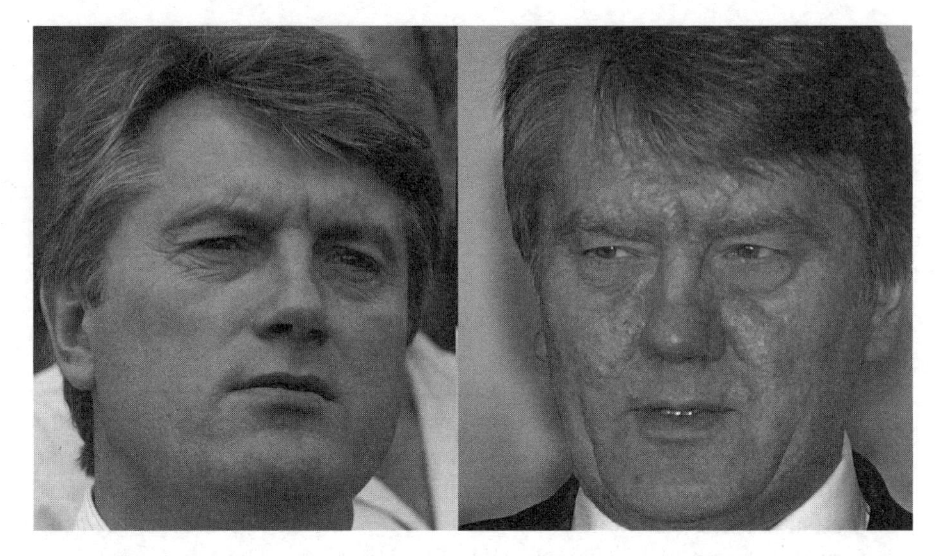

16. The pictures show the Ukrainian opposition candidate Viktor Yuschenko before and after he was apparently deliberately poisened. The changes in his skin, known as chloracne, are a well known effect of dioxin, which was detected in his blood.

cancer of the gall bladder in women, but a lower incidence of stomach and colon cancers in both sexes and of breast and uterus cancer in women. Most recently a larger study found that, compared with the general population, there were no more cases of cancer in the most exposed group and only a handful of extra cases of some rare cancers in those less exposed. However, dioxin became firmly embedded in the popular mind as an unimaginably dangerous pollutant and the *bête noir* of the environmental campaigners, taking the place of DDT.

At around the same time possible exposure during the Vietnam war was being raised as a cause of various diseases and ill health by veterans. The herbicide 2,4,5-T was widely used in Vietnam as a defoliant and was contaminated with dioxin (TCDD). The defoliant, called Agent Orange, contained a mixture of 2,4-D and 2,4,5-T and contamination with dioxin was on average 2 ppm (but ranged from 0.5 ppm to 47 ppm). The contamination of Agent Orange was reduced over time to levels as low as 0.02 ppm more recently. Agent Orange was sprayed from aircraft over tracts of jungle, with the result that soldiers and airmen who came into contact with the defoliant could have been exposed to the dioxin that contaminated it. The Vietnamese people, whose environment was sprayed with it, would of course have been much more heavily exposed as 2,4,5-T rained down on them and their homes. There is, apparently, evidence of an increase in birth defects in heavily sprayed areas of North Vietnam but the causes of this are not known.

From a knowledge of the amount of Agent Orange used, we can deduce that about 20 kg of dioxin would have contaminated around 2.5 million acres in four years, amounting to about 2 mg/acre/year. In contrast, the most heavily exposed areas around Seveso had levels of dioxin ranging from 400 mg/acre to 20,000 mg/acre. In Seveso human exposure at several orders of magnitude greater than was likely to have occurred in Vietnam did not appear to have led to long-term adverse health effects.

Of the Vietnam veterans who were potentially exposed, those who loaded the defoliant onto the planes and cleaned up the spills would have been the most likely to have been contaminated, and some of them did show the chloracne typical of dioxin exposure. However, the Center for Disease Control in the USA was unable to find any correlation between particular health problems and service in Vietnam, and Agent Orange exposure in particular. Analysis of the blood of veterans who had contact with Agent Orange did not show more dioxin than the general public. A study by the US Air Force did not find any abnormalities in the immune systems (a known effect of dioxin) of those involved with using Agent Orange in Vietnam.

The workers who have been involved in the manufacture of 2,4,5-T have, of course, also been exposed to dioxin. Some diseases have been associated with the herbicide, including some cancers as well as chloracne and another type of skin disease.

After Seveso a couple of other events took place in the USA in which dioxin featured. The first was Love Canal, a small community built on an old chemical waste tip in New York state, near Niagara Falls. Some of the chemicals present in the soil, including dioxins, were leaching into the canal. It is not known which dioxins these were, but the levels in the canal would have been likely to have been very low as dioxins are only very sparingly soluble in water. It has been estimated this might be about 0.2 **ppb**. In 1978, not long after Seveso and Agent Orange were in the news, people began to suspect that a number of ailments, which included cancer, brain damage, asthma, and deformities in children, could be blamed on dioxins in the canal. A report published by the New York State Commissioner for Health ('Love Canal: Public Health Time-Bomb') led to demands by householders living nearest the canal to be rehoused. The Environmental Protection Agency then increased the fear and panic by declaring that there was a real danger. President Carter declared Love Canal an emergency area, and the remainder of the 1,000 inhabitants were evacuated. The canal was dredged and contaminated waste removed. The Center for Disease Control in the USA investigated the residents but found no increased level of illnesses compared to a control population.

In Times Beach, Missouri sludge derived from a factory making chlorinated chemicals was sprayed onto roads, and dirt contaminated with dioxin was washed by the flooding Meramec River into homes and shops. EPA scientists arrived in the town with their protective clothing to test the contamination. This naturally caused some panic and the US government eventually bought out the town for $33 million. The locals were tested for dioxins but were found to have no higher levels than the normal population. Again, investigation showed no higher levels of cancer or birth defects than were recorded elsewhere.

After Seveso, the current information indicates that there were no deaths that could be related to dioxin and no real increase in cancers, birth defects, or other illnesses. At Seveso exposure was known to be high, and animals died in their hundreds. This suggests that humans are much less sensitive than most other mammals to the effects of dioxin. It is known that some types of animals are much less sensitive than others, for example hamsters are at least 5,000 times less sensitive to dioxin than guinea pigs. Perhaps we have a sensitivity more akin to that of a hamster. Rhesus monkeys are much more sensitive and therefore not good models for humans in this case. Although dioxin is a highly toxic chemical and we should make all efforts to reduce exposure, it may be that the excessive concern and panic recent accidents have caused, and the efforts expended in reducing amounts to unmeasurable levels, may not be necessary. Furthermore, at least some of the dioxin detectable is produced naturally. Dioxin can also be detected in breast milk, in some cases at levels higher than the limits set by regulatory agencies, but the benefits of breast-feeding outweigh the risk to the newborn child (see also p. 311). The last chapter on Seveso is perhaps not written, as some cancers take many years to develop, and future generations may yet be affected, for example by a decrease in reproductive function, which seems to be especially sensitive to dioxin. However, such effects look increasingly unlikely, and clearly dioxin is not as potent in humans as was once feared.

Dioxin poisoning has recently been in the news again, as it was believed to have been the agent used in an attempt to poison the Ukrainian opposition leader Viktor Yushchenko in 2004. It was reported that dioxin was detected in his blood at 6,000 times the level expected in a human. His face is now disfigured by chloracne, a known effect of this poisoning. It remains to be seen what other toxic effects become apparent.

PCBs and PBBs

Like dioxin and DDT, these chemicals are associated in the public mind with pollution and nasty chemicals. What are they? PCB is an abbreviation for polychlorinated biphenyls: this means the chemical biphenyl has various numbers of chlorine atoms attached to it. There are 200 possible substances, known as congeners, with the number of chlorine atoms ranging from one to ten in the molecule. The PCBs that are usually manufactured are a mixture of products with various numbers of chlorine atoms in the molecules. PBBs are similar chemicals called polybrominated biphenyls, in which various numbers of bromine atoms are attached to biphenyl.

This chemical description does not tell us what the chemicals are like and why they may be a problem in the environment. Like DDT and dioxin, PCBs and PBBs have very little solubility in water but are soluble in fat. Because they contain many chlorine (or bromine) atoms they tend to be resistant to breakdown by animals and other living organisms. Therefore they will accumulate in animals in the environment, as does DDT (both biomagnification and bioaccumulation will occur; see pp. 92–4). With PCBs the biomagnification factor can reach values of 10,000 or 100,000. For example, in the Great Lakes of North America the concentration of PCBs in plankton was found to be 0.0025 ppm, whereas in trout it was 4.8 ppm and in herring gull eggs 124 ppm (50,000 times greater).

Because of their properties as chemically stable, viscous liquids with low electrical resistance, PCBs have been used widely in electrical equipment such as transformers and capacitors, as plasticizers, and as fluids in pumps. They have trade names such as Aroclor 1242 (which means it is a PCB which contains 42 per cent chlorine). PBBs have been used as fire retardants (see p. 258). The wide use of PCBs has led to human exposure as a result of disposal of old transformers, and so on. Their persistence means that they are detectable in the environment and in the fat and tissues of animals and humans. Their manufacture and use is now restricted, but because of their persistence they will remain in the environment for a long time.

The PCBs have low acute toxicity, as shown by studies in experimental animals, but chronic exposure does seem to cause a number of toxic effects. The PCBs used commercially are often mixtures of compounds with various levels of chlorination. The greater the number of chlorines, the greater the toxicity. They are also often contaminated with other substances such as dibenzofurans, which are similar to dioxins and are very toxic. Chronic human exposure occurs mostly as a result of eating contaminated food, particularly fish, especially from lakes. This has been studied in the Great Lakes of North America where significant levels of contamination have been detected in fish.

The potential toxicity of PCBs to humans first came into the public arena in 1968, when over 1,500 people around Fukuoka in south-west Japan were poisoned by eating food cooked in rice oil contaminated with PCBs. This was not a single (acute) poisoning incident as the victims used the contaminated rice oil for cooking for some three months. The oil had become contaminated with PCBs as a result of a leak in the machinery in the plant producing the rice oil. People started to suffer from various diseases, and the illness became known as Yusho disease (meaning rice oil disease). Most prominent was chloracne, a severe effect on the skin which is also caused by dioxins. Babies born to women who were exposed also showed symptoms of the disease. Eventually some 2,000 cases of Yusho disease were recorded.

Eleven years later, a similar case occurred in the Chungwa and Taichung districts of Taiwan in which rice bran oil was contaminated with PCBs. Again around 2,000 people were poisoned and were found to be suffering from a skin disease that was called Yu-Cheng. In these incidents a knowledge of the level of contamination of the oil allowed estimates of human exposure to be made. Ingestion of more than half a gram of PCBs over the three months led to moderate to severe symptoms. The maximum total dose ingested was between 3 and 4 grams.

Studies of these populations have suggested that there are effects on the immune system, which have also been observed in experimental animals. These changes (decreased levels of antibodies and of certain white blood cells) result in victims having lowered resistance to bacterial and viral infections. Other changes observed were: delayed growth and development in children; middle ear disease was more common in exposed children; more behavioural problems were reported; and in some boys the length of the penis was decreased. Some of these effects were only slight.

Observations in experimental animals have shown that some PCBs (specifically 3,3′,4,4′-tetrachlorobiphenyl, TCB) can interfere with thyroid hormone levels. It seems that a metabolite of the TCB is similar to the thyroid hormone thyroxine and therefore competes with it for binding to a specific receptor. One of the consequences is loss of thyroxine and vitamin A from the blood. There is evidence for similar effects in humans in that higher levels of PCBs in breast milk correlated with lower levels of certain thyroid hormones in mothers and infants.

However, a confounding factor in both Yusho and Yu-Cheng diseases was the role of contamination with other chemicals, in particular penta-chloro-dibenzofurans (PCDFS). These chemicals are similar to dioxins and are known to be highly toxic. They were found to be present in the rice oil and at a higher level than normally found in PCBs because the PCBs were heated in the factory and then again later when the contaminated rice oil was used for cooking. The level of PCDFs in the rice oil was higher in the Japanese contamination incident than in that in Taiwan.

These PCDFs were still detectable in victims of the poisoning more than eleven years after the exposure, and they seemed to be retained more than PCBs. They have been detected in the livers of victims of Yusho disease some years after the exposure occurred. Furthermore, the isomers and congeners (see box) retained in the livers were the most toxic, it would seem because these are more resistant to metabolism and so are more persistent.

PCCDDs and PCDFs

Polychloro-dibenzodioxins and polychloro-dibenzofurans are basically contaminants which can be formed when chemicals containing the element chlorine are synthesized or subjected to heat. There are many of these substances (75 PCDDs and 135 PCDFs) which can have various numbers of chlorine atoms, when they are known as congeners. In some cases there can be compounds that have the same number of atoms but arranged differently, when they are called isomers.

TCDD-2,3,7,8-tetrachlorodibenzodioxin, usually called just dioxin, is the most toxic of these chemicals, and the toxicity of the others is expressed in relation to it. This is known as the toxic equivalency factor, or TEF, which can be applied to mixtures as well as single chemicals.

PCDFs are very similar to dioxins and, like them, are very toxic, especially 2,4,7,8-tetrachlorodibenzofuran, although this is not quite as toxic as TCDD. They have been found to be contaminants of PCBs and in the Yusho and Yu-Cheng poisoning incidents were more abundant in the PCBs than was usual. As with dioxins and PCBs, there are many different isomers and congeners.

Because some of the symptoms of Yusho and Yu-Cheng diseases (see Table 2) were similar to those caused by PCDFs rather than PCBs, it has been concluded that these contaminants are the causative agents. Indeed the level of PCDFs in the blood of victims correlates with the symptoms whereas the blood level of PCBs does not.

Although studies in experimental animals suggest that PCBs may cause

Table 2 Percentage of victims showing symptoms of Yusho disease

SYMPTOMS	VICTIMS SHOWING SYMPTOMS (where known) (%)
Acne, pigmentation of skin and nails	82–87
Discharge from eyes	83–88
Jaundice	10
Numbness of limbs	32–39
Hearing difficulties	
Vomiting and diarrhoea	17–39
Chronic bronchitis, reduced serum antibodies	40
Brown skin pigments, retarded growth, abnormal teeth (in newborn babies)	

cancer, it seems likely that they are promoters rather than carcinogens, which means that they promote the carcinogenicity of other chemicals which cause cancer. There is evidence of increased incidence of certain cancers in the people exposed to the contaminated PCBs in Taiwan and Japan. This could be due to the PDCFs contained in the rice oil or to a combination of PCB and PCDF action.

Exposure to PCDFs in Yusho and Yu-Cheng diseases

The contaminated rice oil responsible for the Yusho disease contained 5 µg of PCDFs per gram of oil. In Yu-Cheng disease, up to 1.7 µg per g of oil were detected. The daily intake by victims was therefore about 0.1 to 0.2 µg per kilogram body weight per day (µg/kg bw/day). Disease occurred in those exposed to between 0.05 and 0.1 µg/kg bw/day.

The total overall consumption of PCDFs by those with signs of disease was between 3.3 and 3.8 mg.

In this example knowledge of the level of exposure of the victims, and measurement of the levels in their bodies, were particularly important. This is often lacking in such situations, which makes it virtually impossible to relate effects to a particular chemical. It should also be noted that the exposure levels in these two incidents were relatively high. Levels of PCBs (and dioxins) that are present in the environment and might contaminate food, for example, are monitored and are generally very low and unlikely to represent a significant risk. This may even be the case after accidental contamination (see the 'Dioxin and the Belgian poultry scandal', pp. 257–8). Levels are likely to be highest in meat and particularly in fish, especially farmed fish and fish oil. The limits set for these chemicals incorporate a large safety margin (see chap. 12 and p. 311).

Drugs in sewage

One form of pollution that has increased and become a potential problem is the appearance of drugs in rivers. This can be observed particularly in periods of drought when the concentration increases. The source of these drugs is primarily the urine eliminated by people taking them. The drugs find their way into the sewage system, and rivers, reservoirs, and the sea become contaminated. As the water in rivers and reservoirs is partly derived from sewage that has been processed, unless the drugs are removed they could potentially appear in the drinking water. Drugs will not necessarily be removed by the processes involved in sewage treatment which is meant to remove bacteria primarily. The most commonly used drugs, such as aspirin and paracetamol, will naturally be more abundant than prescription drugs of more limited use. Some will not cause a problem in the low concentrations in which they occur. More potent are the steroids found in the contraceptive pill, which has been widely used by a large number of women since the 1960s. Residues of the contraceptive pill have been found in rivers, presumably derived from sewage. These have the potential to affect the endocrine system of sensitive animals exposed in rivers, although significant amounts have not been found in drinking water.

Gender benders or endocrine disruptors

Endocrine disruptors are chemicals that cause adverse health effects in an animal or its offspring due to changes in hormone function. In recent years some scientists have concluded from a number of coincidental observations that certain chemicals in the environment may be interfering with the hormonal systems of animals and possibly of humans. These endocrine disruptors *could* be responsible for a range of dysfunctions that have been observed in the reproductive systems of humans and other animals. The effects in wildlife have been well documented and can be reproduced experimentally. The human effects are more difficult to associate definitively with environmental chemicals, and some are controversial.

It is now widely accepted that chemicals that are capable of causing hormonal effects have been released into the environment. Such chemicals may mimic natural hormones, or act to antagonize either male or female hormones and so change the natural balance. Although chemicals with this activity can be detected in the environment, the question remains whether the concentrations are high enough to cause the effects observed. Early work in the UK showed that male fish kept in cages in rivers and close to the outfall of sewage treatment works underwent certain changes; they became more feminine and hermaphrodite (they had both male and female attributes). Although such effects occurred close to the sewage outfall, it is likely that dilution of such chemicals would occur as the sewage mixes with the river water and disperses.

One group of chemicals that have the potential to be sufficiently potent are the drugs found in the contraceptive pill. For example, it has been found that ethinyl estradiol, one of the synthetic hormones used in the contraceptive pill, and other related female hormones are detectable in the waste water from sewage treatment plants in a number of countries at levels around 6 ng/litre. In the UK, the outflows from some of these plants deliver biologically significant levels of such compounds into rivers. It has been suggested that this is the main reason for the appearance of feminized (intersex) male fish in these rivers. In some rivers all of the male fish are showing signs of feminization. Some of the effects include the growth of eggs in the testes of the fish, the appearance of female reproductive tracts in male fish, and a decrease in numbers of sperm.

These effects are irreversible and can be reproduced in fish in the laboratory. The male fish have now been found to produce a protein (vitellogenin, found in the yolk of eggs) that is normally produced only by female fish in response to the female hormone (oestrogen). Vitellogenin is produced by male fish both in polluted rivers and in the laboratory when they are exposed to known endocrine disruptors. The protein is used as a biomarker for this effect (see box).

Up to 50 per cent of some rivers consists of treated sewage (it could be more in the summer in south-east England) and a significant amount of drinking water can be derived from rivers. So far synthetic hormones have not been found at sufficient levels in drinking water to cause a biological effect in humans (see above). The contraceptive pill and its breakdown products are not the only chemicals that can cause these effects (although they are probably the most potent). Another important oestrogenic chemical is natural oestradiol, which is derived from the urine of pregnant

Endocrine-disrupting chemicals in English rivers

In 1994, researchers put cages of male rainbow trout into five rivers. In each case five sites were chosen, one upstream of a waste treatment plant, one at the point where the effluent was discharged, and three others at various distances downstream. It was found that the female protein vitellogenin was produced in male fish in four out of the five rivers at the site of the effluent discharge. The other waste treatment plant did not receive industrial waste. In one river, where the effluent was the most active, at all the other sites, up to 5 km downstream, the fish produced vitellogenin. It has also been found that male fish such as flounder in estuaries in the UK also have high levels of vitellogenin. In the Tyne and Mersey estuaries the increased levels of the female protein were 10,000 to 1 million times more than in control animals.

One of the groups of suspect chemicals in the most 'active' effluents in the survey was the alkyl phenols. However, there are probably several endocrine disrupting chemicals in the effluents and rivers, as different waste treatment plants may feed into the same river. As well as the alkyl phenols, a large number and variety of chemicals have been found to cause this effect, including organochlorine insecticides such as DDT, organotin compounds, phthalate esters, plant products, dioxins, polycyclic aromatic hydrocarbons, PCBs, and as expected natural and synthetic oestrogens.

These chemicals or their metabolites seem to act by binding to the oestrogen receptor by mimicking the natural oestrogen, thereby causing an effect. However, generally they are much weaker than the natural hormone. This knowledge has allowed simple *in vitro* assays to be devised which can detect if a chemical has the activity. Although not foolproof, they can indicate if a chemical should be further investigated.

women. The problem of intersex or feminization of male animals has been recognized as being due to many different kinds of chemicals as well as those occurring in sewage, both chemicals released by industry and other environmental contaminants such as organochlorine pesticides. Thus DDT and PCBs are both able to cause these kinds of effects, as are industrial chemicals called alkyl phenols, plasticizers, and hormones used in farm animals such as diethylstilboestrol. The alkyl phenols nonylphenol and octylphenyl have both been shown to cause male fish to produce the female protein vitellogenin.

Some natural chemicals such as those produced by fungi (zearalenone) and plants (genistein) may also be oestrogenic and have hormone-like actions. Such chemicals can be detected in the effluent from paper mills and are suspected of causing masculinization of female fish. However, all these chemicals are very much less potent than the natural hormones and the synthetic **ethynyloestradiol**.

One chemical which has been widely studied is tributyl tin, a molluscicide and constituent of anti-fouling paint on boat hulls. This has been shown to cause changes in female molluscs (such as the dog whelk) which become masculinized, growing male organs, a phenomenon called imposex. Tributyl tin causes this effect at extremely low concentrations (1–2 ng/litre). This was a particular problem for two reasons: (1) in harbours and marinas, because of the confined spaces, toxic concentrations were able to accumulate in molluscs from the many boats moored there (tributyl tin being quite soluble in fat); (2) the mollusc is exceptionally sensitive to the tributyl tin (mammals such as man degrade the chemical fairly rapidly). It was later found that even in areas of open sea, such as busy shipping lanes, there were effects on molluscs.

Probably the most well-known effect of endocrine disrupting chemicals in wild animals is the case of the alligators in Florida.

CASE NOTES

The case of the shrinking alligator penis

It was reported that the alligator population of Lake Apopka in northern Florida was declining, presumably because of lack of success in reproduction. The male animals were found to have abnormal testes, low testosterone levels, and small phalli (penises). The female alligators had abnormalities too and high levels of oestrogen. It is believed that high levels of organochlorine compounds, resulting from a spillage of the pesticide difocol, were responsible. The difocol was contaminated with DDE, a breakdown product of DDT, which may have affected the level of the sex hormones. Apparently such effects can be produced in alligators by treating them with DDE.

What of the effects in humans? While there is evidence, some of it controversial, that changes have occurred in the reproductive systems of humans in various countries throughout the world over the last fifty years or so, ascribing these changes to particular chemicals, or even to chemicals at all, has been difficult. There has been an increase in testicular cancer as well as breast cancer since 1945, particularly in certain countries. Data from studies of populations in some countries, for example Finland, have indicated that sperm counts and the quality of the sperm in males have declined over the same time period, but by no means all studies have found this and some have found the reverse. In some places there has been an apparent increase in disorders of the male reproductive system such as hypospadia (a defect of the penis) and cryptorchidism (undescended testicles).

There is, however, one example of a chemical known to have caused adverse effects in both males and females. The chemical is the synthetic oestrogen diethylstilboestrol. This was used in the 1950s to prevent miscarriages until its use was stopped in the early 1970s. Women were given the drug during pregnancy. However, the effects did not occur in the women who took the drug but in their offspring. In a significant proportion of girls born to women who had used the drug there was dysfunction of the reproductive organs, disruption of the menstrual cycle, and abnormal pregnancies. A few of the girls developed vaginal cancer but not until they had reached puberty. In boys born to the women who had taken the drug there was an increased incidence of cryptorchidism and a small penis. There was also some evidence of a decrease in sperm count and ability of the sperm to move. These effects and testicular cancer have been reproduced in experimental animals treated with diethylstilboestrol. Diethylstilboestrol has been used as a growth promoter in cattle and so residues in food may exist; the chemical and its breakdown products could also potentially appear in river water.

The oestrogen ethynyloestradiol is, as we saw earlier, known to be present in sewage treatment plant effluents and could potentially appear in drinking water. It has apparently been detected in drinking water in the UK but the data have been questioned. Even if it were present, it is not known whether the concentrations would be high enough to affect human males.

As we have seen, there are many other chemicals capable of causing these effects, although they are generally very much less potent that the natural oestrogens and the synthetic oestrogens used in the contraceptive pill. Mixtures of these chemicals, which includes natural substances from plants, pesticides, and industrial chemicals, may be more potent than the individual components. This phenomenon is called synergy (see p. 35) and is currently an area of intense scientific interest, as is the subject of endocrine disruptors.

Effects on the male reproductive system can also be produced by substances that interfere with male hormones, so-called anti-androgens. DDE is one such chemical, but attempts to correlate DDE levels in human breast fat tissue with the incidence of breast cancer have produced equivocal results. Dioxin shows significant anti-oestrogenic activity, but after the accident at Seveso a decreased incidence of breast cancer was observed in those who had been exposed.

Although all the synthetic chemicals and natural chemicals considered here have weak oestrogenic activity compared to natural oestrogens, there are several ways in which chemicals can affect hormones, and combinations of chemicals could be more potent than expected. Although there is little indication of this at present, future research may resolve the issue.

In conclusion, there is no doubt that the effects of endocrine disruptors can be reproduced in experimental animals. However, the effects in animals after environmental exposure, and in which levels were measured, have been detected mostly when contamination was relatively high and in most other cases evidence for a causal link is weak.

For humans, except for cases where high levels of exposure have occurred (for example, in the therapeutic use of diethylstilboestrol), the data is inconsistent and inconclusive. Information on levels of exposure is lacking and there is no firm evidence for a link between low level exposure and adverse health outcomes in the human population. For all the adverse health outcomes the evidence of association with environmental exposure to endocrine disruptors is weak (except in the case of minor effects on thyroid hormones, where evidence is moderate). In some cases, such as changes in sperm count, a possible connection is scientifically plausible. Further evidence is needed, and especially evidence that links actual exposure levels in humans to effects.[8]

Lead

The metal lead has been used for more than 2,000 years and has been associated with poisoning for much of that time. The ancient peoples who used it were aware of at least some of the symptoms of lead poisoning, for example in 300 BC Hippocrates described a case of lead poisoning. It has been responsible for poisoning in many different ways, including contamination of the human environment, which is probably the most important current source of exposure.

In earlier times the unfortunate workers involved with the mining and especially the smelting of lead were particularly vulnerable. Once the lead was being used, for example as glazes for pottery and paint, in the construction of lead pipes for water, and in lead cooking pots, large-scale poisoning was possible.

Because lead is eventually deposited in our bones and remains there (as a phosphate salt), it is possible to determine a person's exposure when the skeleton is all that remains. From this information we know that we are currently exposed to more lead than prehistoric man as we have twice as much lead in our bones. However, our level of exposure is not in the same

CASE NOTES

Too much lead in the beer

The victim had complained of abdominal pain for two to three years which had become worse about ten weeks before his death. His red blood cells were found to be abnormal and he had inflammation of the peripheral nerves. The victim was a publican who liked to drink the first beer drawn from the barrel every day with his customers. Unhappily for him, this had been lying overnight in lead pipes some 20 feet long which connected the barrel with the tap. The beer and drinking water in the pub were found to have a high lead content.[9]

league as that of the Romans or of some who came after them, including the people who worked in and/or lived near factories during the Industrial Revolution in the nineteenth century. It has even been suggested that lead contributed to the decline of the Roman Empire!

Where did the lead come from and what effects did it have? Lead is a cumulative poison. Like arsenic, it has many targets in the body and some of the symptoms of poisoning are easily confused with the symptoms of other diseases. Chronic, long-term lead exposure will cause effects on the nervous system which may become serious. The victim feels tired and listless, has constipation, is anaemic, and can become infertile. If, as seems likely, it was the upper levels of Roman society that suffered most from lead poisoning, some of these effects could have influenced the running of an empire.

As the Romans ingeniously introduced lead water pipes, it is likely that their water, carried in these pipes, was contaminated with lead as it still is today, especially in areas of soft or acidic water. The use of lead glazes in pottery and, even more importantly, lead cooking pots were other contributory factors. Thus the lead was dissolved from the surface of the lead pots or from the lead-containing glaze on pottery. Analysis of the bones of Romans from the time of their empire has shown high levels of lead, sufficient to cause lead poisoning.

The syndrome caused by lead poisoning was known as Saturnine gout (gout can be one of the symptoms of lead poisoning). It is similar in cause and effects to so-called Devonshire colic in eighteenth-century England and Colic Pictonium in late medieval France. These were due to lead dissolved by acids in cider and wine respectively. An English doctor showed that it was the lead used to line the apple presses that was the cause of the colic. More recently, those making and drinking 'moonshine' whisky during Prohibition may have fallen foul of lead poisoning by using lead pipes and lead-containing solder for the stills. Some of these individuals suffered from chronic kidney damage called nephritis.

Lead was also used in medicines, and in pastes for the treatment of skin diseases 2,000 years ago. In the nineteenth century some doctors recommended lead acetate and opium mixtures for the treatment of diarrhoea. This was described in early versions of the **British Pharmacopoeia** (*Pil. Plumbi cum Opio*). Certain lead salts were believed to be powerful astringents which would help in the treatment of wounds and promote healing. Thus Goulard's extract (*liquor plumbi subacetatis fortis*, or lead acetate) was still recommended in textbooks in the early twentieth century[10] in dilute form for the treatment of ulcers, acute inflammatory skin conditions, and eczema and as a gargle in tonsillitis. Lead compounds have also been used to treat inoperable cancer.

Both acute and chronic exposure to lead became more common with the advent of the Industrial Revolution and there were as many as a thousand cases per year of lead poisoning in the UK at the end of the nineteenth century. During the nineteenth century there were also many cases of lead poisoning due to contamination of domestic drinking water in areas such as the north of England where slightly acidic water was delivered to houses in lead pipes. Lead was also used in fungicides in the form of lead arsenate.

CASE NOTES

Too much canned food

An expedition in 1845 led by Sir John Franklin to explore the North-West Passage around Canada took two ships and 129 men. The ships became trapped in pack ice and disappeared later in the year without trace. The graves of three of the men who died in 1846 were found on Beechey Island. In 1988 their bodies were exhumed and analysed. High levels of lead were discovered, suggesting poisoning as a cause of death. Where would this lead have come from? The ships had been provisioned with supplies of food to last several years—in cans. At the time cans of food were sealed using solder, which contains lead. This is the likely source of the lead that may have killed many more of the crew as they subsisted on canned food.

Current lead exposure

Despite awareness of the dangers, lead poisoning is still an issue in the UK and many other countries. Occupational exposure can still occur and in

Table 3 Past and present sources of lead

PAST SOURCES OF LEAD	CURRENT SOURCES OF LEAD
Lead in pipes/equipment for drink manufacture	Lead in water from pipes
Lead paint	Old lead paint
Lead salts as food colourants	Electrical batteries
Toy soldiers, coloured pencils	Leaded petrol
Fungicides	Lead smelters

countries where leaded petrol is used extensively there will be more environmental exposure by inhalation. It has been estimated that half the atmospheric lead is from this source. People such as traffic policemen or mechanics in garages are particularly vulnerable.

There are three main sources of lead: water, food, and air. Absorption from food or water from the gut is relatively inefficient, with about 10 per cent being absorbed. Lead in water is mostly from old lead pipes, from which the lead will be dissolved particularly by soft and acidic water, or it may be present naturally, derived from rocks and soil. Absorption of airborne lead through the lungs is much more efficient and, although there is less lead in air than water, this is a more important route of exposure. Thus lead smelters, which produce lead particles, and car exhausts, which generate lead from the additive used in leaded petrol, are the main sources of exposure. The lead additive in leaded petrol is tetraethyl lead which is an organic form of lead and, as with organomercury, it has different toxic effects, being especially toxic to the central nervous system. Tetraethyl lead is also very readily absorbed into the body and the central nervous system. Although some is degraded in the engine to inorganic lead, some is still present in the emissions from car exhausts.

Reduction of lead in petrol in the USA to less than 1 per cent in 1971 and then to 0.06 per cent in 1977 has had a dramatic effect on levels of lead in people: in 1980, the average daily lead intake in male babies was 45 μg and 84 μg in adult men; by 1990 these levels were down to about one-tenth.

A significant source of exposure for young children has been leaded paint, which may be present in old houses and resulted in large-scale poisoning in children in the slum areas of the USA at the beginning of the twentieth century. This paint may contain as much as 40 per cent lead, and the habit of eating flakes of the paint, known as 'pica', has lead to some serious cases of lead poisoning, even quite recently in the UK. A study conducted by the European Union in Glasgow in 1979 revealed that 10 per cent of babies had a blood lead level of more than 300 μg/litre. Removal

FUTURES The world of science & technology

THE GUARDIAN Thursday October 1 1981 23

from the Guardian, June 27, 1967

Concern about the dangers of lead pollution is not new. Fourteen years ago the Guardian's science correspondent Anthony Tucker warned of the danger of official complacency. Today we repeat the message

Still pumping poison into our children

Drawing by Peter Clarke

There is no longer any argument about the danger of lead pollution. But, argues Robin Russell Jones, commercial interest has once again triumphed over public safety, for the arguments used to keep lead in petrol are the same as those used 70 years ago to keep lead in paint

How much more evidence is needed before the Government takes effective action to abate this appalling menace?

17. This headline reminds us that lead, such as that derived from the use of leaded petrol, is a potentially dangerous pollutant. Luckily this source has now diminished in many countries.

of paint containing lead can create dust which, if ingested or inhaled, is also a potential source of lead poisoning. Children are particularly vulnerable because they absorb as much as five times more than adults (from the gastrointestinal tract). Their developing central nervous system is also more susceptible to lead.

In 1991 a comprehensive review of available data resulted in the maximum tolerable blood level for children being reduced to 100 µg/litre in the USA. The Centers for Disease Control in the USA state that above this level there is a risk of adverse developmental effects in children such as reduced IQ, growth retardation, and lower hearing acuity. Lead levels above this in the blood in adults was associated with an increased risk of hypertension. The equivalent 'intervention level' for blood lead in the UK has been set at 250 µg/litre, but, apparently, 'cognitive deficits' have been detected at levels below this. Such effects are often difficult to measure, and three bodies (the Medical Research Council, a Royal Commission, and the Department of Health) which separately evaluated the data in the UK found that studies on this effect of lead have been inadequate and that the case was not proven. Studies are still being published which purport to show effects (see Table 4).[11] It should be remembered that environmental lead levels (and blood levels), at least in developed countries, have fallen and continue to fall.[12]

The average concentration of lead in the blood of people in the UK not exposed at work is probably at least 50 μg/litre. The level depends very much on where a person lives because of sources such as natural lead in water, lead pipes, local metal working industries, and so on. Biochemical changes in haemoglobin synthesis can be detected at a blood level of 100 μg/litre. Severe toxic effects such as renal injury and encephalopathy (brain damage) can start at blood levels of 1,000 μg/litre and above in adults, although this will depend on the length of exposure. At one time 800 μg/litre was the permissible level in the UK, but this has been steadily reduced, especially because of the sensitivity of children (see Table 4).

It is known with certainty that, at levels to which many of us are exposed, lead has biochemical effects such as causing a partial inhibition of enzymes involved in haemoglobin production. Some of these effects may be of no consequence to the health of the human exposed and cannot be considered as toxic effects. This is because the body has protective mechanisms and what is known as reserve functional capacity. This is

Table 4 Lead toxicity in children

BLOOD LEAD LEVEL (μg/litre)	EFFECT
100	Reduced growth, IQ, hearing[1]
200	Changes in nerve function
400	Reduced haemoglobin production
600	Colic
700	Anaemia
800	Kidney damage
850	Brain damage
1,300	Death

Note: These levels are only approximate, as the length of exposure will be an important factor.

[1] The effects on IQ have been questioned and some studies have not replicated these effects. The most recent publication (R. L. Canfield *et al.*, 'Intellectual impairment in children with blood lead concentrations below 10 μg per deciliter', *New England Journal of Medicine*, 348 (2003), 1517–26) suggested effects occurring at levels below 100 μg/litre and reported a correlation between blood lead levels and IQ. They found a decrease in IQ as blood lead increased from 10 to 100 μg/litre. However, measurement of IQ is not precise and has many potential confounding factors. In rats the no adverse effect level (NOAEL) is 50 μg/litre.

similar in principle to the modern jet aircraft, which may have two or more engines but can fly normally and safely with only one if the others should fail. By being aware of the very real dangers of lead and its sources it is possible to reduce exposure to levels at which toxicity is unlikely.

Another current but occasional source of lead poisoning that is still a potential hazard are some cosmetics and herbal remedies imported from countries such as India.

Effects of lead poisoning

Lead in the inorganic form, as in lead salts, causes a range of effects depending on the amount. General effects on the gastrointestinal tract lead to pain (colic), constipation, and diarrhoea; vomiting can also occur. There are sometimes pains in the joints (gout), and weakness in the arms or legs or hands (hence 'wrist drop') resulting from effects on the nerves. Headache and blindness are sometimes symptoms, as well as mental disturbances which in severe cases can reach insanity. Chronic exposure will cause damage and dysfunction of the kidney, leading to nephritis and possible kidney failure.

Lead distributes into the red blood cells in the body and consequently causes damage there, interfering very specifically with the means by which haemoglobin is synthesized in red blood cells. This leads to a loss of haemoglobin and therefore of functioning red cells which need haemoglobin to carry oxygen. The lack of red cells is called anaemia and contributes to tiredness, listlessness, and a pale appearance. Fortunately, this effect is reversible once the exposure is reduced, but the effect on the kidney, peripheral nephritis, is more serious and may not be reversible.

Lead poisoning can be readily detected and once detected it can be treated. It can be detected in a number of ways, from simply measuring the amount of lead present in the body to measuring specific biochemical markers (biomarkers) in the blood, for example. The metabolic disturbances involved in the interference with the production of haemoglobin can be used for diagnostic tests. Treatment of lead poisoning is by removal of the source of lead and possibly the use of chelating agents which bind to and remove the lead from the body (except the bone) by allowing it to be eliminated into the urine.

Aluminium

In July 1988, 20 tonnes of aluminium sulphate were accidentally tipped into a reservoir containing water already treated for drinking. The accident happened when a relief driver was making the delivery to the Lowermoor treatment plant, which was unmanned. The aluminium sulphate normally used in the water treatment process was put into the wrong reservoir. The result was that the people of Camelford, a small town in Cornwall, received drinking water that was heavily contaminated with aluminium and highly acidic (**pH** 3.9–5). There was a delay in informing the public and in detecting and remedying the source of the accident. The aluminium level in the water was reported as being up to 620 mg/litre. The acidity of the drinking water compounded the problem by dissolving other metals, such as copper, from pipes. The water issuing from taps in the homes of residents thus contained not only aluminium at high levels, but other metals such as copper as well.

The residents complained of foul-tasting water and later of various symptoms such as gastrointestinal disturbances, skin rashes, arthritic pains, sore throats, and loss of memory. Animals, like fish, that were exposed to the water died. As a result of this accidental contamination, two committees evaluated the available information, and their reports concluded that there was no convincing evidence of harmful accumulation of aluminium, or of a greater prevalence of ill health in those exposed.

Other studies have found evidence of harmful effects, however, although some of these have been criticized. For example, a study published some years after the accident reported that forty-two of fifty-five residents studied had poor psychomotor performance. There were acknowledged flaws in the design of this study, which were beyond the control of the authors, but the conclusion was that 'aluminium poisoning probably led to long term cerebral impairment in some people in Camelford'.[13]

Aluminium is the third most abundant element on earth but it is known to have toxic effects that cause brain disease, bone disease, and anaemia. There is concern especially in relation to its possible role in Alzheimer's disease. The changes observed in animals exposed to aluminium are similar to those observed in patients with Alzheimer's disease. Aluminium has been found in some areas of the brain of victims of Alzheimer's at levels not too dissimilar from that in the brains of the animals exposed. Some of the findings and their interpretation are controversial. Unfortunately, the amounts of aluminium absorbed by the residents of Camelford are unknown. A further review of the information available is currently being considered by the Committee on Toxicity in the UK.

We are all exposed to aluminium from the metal utensils we use and also from the occasional use of medicinal preparations such as antacids, but it is poorly absorbed and the risk is probably very small. Dialysis patients with renal disease were found to be at risk of brain damage due to the aluminium derived from the equipment. Realization of this led to a lowering of the exposure of such patients, which decreased the occurrence of toxic effects of aluminium. Patients on dialysis with end stage renal disease, in whom some accumulation of aluminium occurred, showed evidence of metabolic abnormalities and in psychomotor function.

6

Natural Born Killers

Poisonous Chemicals Designed by Nature

Many people think of chemicals as not only intrinsically dangerous but also exclusively man-made and hence unnatural. Chemicals are found throughout the natural world. The world is naturally a chemical environment, both the earth itself and the animals and plants on it being made from chemicals. As we have seen, minerals such as compounds of arsenic, cadmium, lead, and mercury all occur naturally and have at times been responsible for poisoning.

In addition to these substances and the normal constituents of living things, such as DNA, fats, and sugars, there are many toxic chemicals manufactured by plants, animals, and micro-organisms. Some of these will be explored later (see pp. 239–57). These natural but synthetic chemicals are often used by the plant or animal for protection, as they engage in chemical warfare against predators. Plants may be trying to stop animals such as birds or insects eating their berries, seeds, or leaves. For example, many fruit kernels and pips contain cyanides, while the deadly nightshade plant produces atropine in its berries. Both can be lethal to humans. Thus many plants produce natural insecticides.

There are a very large number of chemicals that have been identified in plants and many thousands that have not. Some are toxic, others benign. It is not possible within the limits of this book to consider more than one or two. It is important to note that some extremely toxic chemicals are found naturally and do contribute to human disease and poisoning (see also pp. 239–57). Some of these chemicals may have only a single target species; others will be toxic to most. The animals that are predators for these plants can become immune, presumably, through a process of natural selection.

Knowledge of the toxic effects of some of the chemicals in plants have been known to humans for hundreds, and probably in some cases thousands, of years. They have been used for warfare, murder, suicide, infanticide, political assassination, and execution. Indeed, as noted earlier, *toxikon* is the ancient Greek word for arrow poison. The use of these poisons was recorded in 1200 BC in the ancient poetic work from India, the Rig Veda. In Homer's *Odyssey*, Odysseus was 'seeking the deadly poison wherewith to anoint his bronze-tipped arrows'. The poison may well have been the natural toxin aconite, which is derived from the plant known variously as wolfsbane or monkshood. Ovid in his book *Metamorphosis* referred to it as *akonitos*. This and other natural poisons used for killing will be discussed further in Chapter 9 (see pp. 213–14, 227–30). Some have been unwittingly responsible for disease and death, and even now plants containing toxic chemicals contribute to disease when used in herbal remedies.

Poisonous chemicals produced by plants, animals, or micro-organisms are called toxins. Some have been used therapeutically and become the basis for widely used therapeutic medicines, and some have been used as recreational drugs (see Table 5).

Some of these plant toxins have been used as drugs, or drugs have been derived from them.

Table 5 Some well-known plant toxins and their sources

PLANT TOXIN	PLANT(S) FROM WHICH IT IS OBTAINED
Atropine	Deadly nightshade, henbane, thornapple, mandrake
Digitalis	Foxglove
Physostigmine	Calabar bean
Hyoscine (scopolamine)	Henbane, mandrake, thornapple, deadly nightshade
Antiarin	Upas tree
Oubain (strophanthin)	Strophanthus gratus
Tubocurarine (curare)	Chondodendron tomentosum
Toxiferine	Strychnos toxifera
Nicotine	Tobacco
Aconitine	Aconite

Hyoscine and the screaming mandrake root

Goe, and catche a falling starre,
Get with child a mandrake roote,
Tell me, where all the past yeares are,
Or who cleft the Divels foot,
Teach me to heare Mermaides singing,
Or to keep off envies stinging,
And finde what winde,
Serves to advance an honest minde.

From a song by John Donne (1572–1631)

Hyoscine, or scopolamine, is an alkaloid found in a number of plants but it is the mandrake, which grows in many Mediterranean countries, around which much folklore has grown. The chemical is toxic but in small doses it has a sedative, soothing effect. It has been used as a premedication before surgery. The root of the mandrake plant maybe Y-shaped and, because it can look like the lower half of the human body or the male organ, it was at one time believed be a lower form of living thing. It has for this reason always been associated with fertility, as can be seen in the poem above. In the book of Genesis in the Old Testament Rachel, who is childless, asked her fertile sister Leah, 'Give me, I pray thee, of thy son's mandrakes', after Reuben, Leah's son, found mandrakes in a wheat field.

Extracting the mandrake root from the ground was, by all accounts, at the very least an unpleasant experience. The root was reputed to give off a foul stench and apparently, when pulled from the ground, emitted a scream or groan. If one heard this one would surely die or go mad, as Juliet observed in *Romeo and Juliet*: 'And shrieks like mandrakes' torn out of the earth, / That living mortals, hearing them, run mad'. Dogs were employed to drag the root from the ground with the aid of a rope tied around their neck and attached to the plant. It is not clear what enticed the dog to do this or whether it survived the process, but the intrepid collector also needed to collect a sample of woman's urine and menstrual blood to sprinkle over the plant before it was removed from the ground!

The root was highly prized for its reputed aphrodisiac properties as well as its toxic and therapeutic effects, which presumably made the effort of collection worth while. In the Middle Ages it was often used for poisoning, the root being allowed to ferment to produce a poisonous brew. The infamous Dr Crippen used scopolamine to kill his wife. Some of the effects of scopolamine are employed to good effect by criminals, as it can cause the victim to enter a zombie-like state during which they are powerless to resist. After the attack or robbery the victim suffers from amnesia, allowing the criminal to escape detection. The drug has been used for the abduction of women as slaves in earlier times, and more recently by criminals in Colombia (where the drug is called burundanga). In the past women in Colombia dispatched unwanted children by smearing extracts of certain plants (various species of *Datura*) on their nipples before suckling the infant.

Scopolamine and plants containing it, such as thornapple, mandrake, and jimsonweed, can have a powerful sedative effect if used at the correct dosage. It has been used also for the treatment of epilepsy and asthma. Poisonous doses will lead to respiratory depression and symptoms such as dry mouth, dilated pupils, and restlessness. At high doses the victim suffers convulsions, delirium, coma, respiratory failure, and death. Jimsonweed, or Jamestown weed, was the cause of many fatal poisonings in 1666 in the American colony in Jamestown, Virginia, when soldiers suffering from hunger during a famine ate the weed.

Scopolamine is often found with another alkaloid, atropine, to which it is very similar. Deadly nightshade (*Atropa belladonna*) contains not only atropine but also scopolamine. When dropped into the eye, the juice of the berries was found to cause the pupils to dilate. It was used for this purpose by women during the Renaissance, leading to the name 'belladonna', because it made the women appear more beautiful. The berries were also successfully used by Roman poisoners like Livia and Agrippina, the wives of the emperors Augustus and Claudius. One deadly nightshade berry may be enough to deliver a fatal dose of the alkaloids. (The other part of the Latin name for deadly nightshade, *Atropa*, is from Atropos, meaning 'the fate that cuts the thread of life'.)

Pyrrolizidine alkaloids

While these are not the most poisonous chemicals produced by plants, they are widespread and have in the past caused, and still cause, serious illness. A significant number of poisoning cases have been due to the use in herbal remedies of plants containing these alkaloids.

Pyrrolizidine alkaloids are a large family of similar chemicals found in over 6,000 different species of plants in the *Leguminosae, Compositae*, and *Boraginacae* families. For example, plants of the *Senecio, Heliotropium*, and *Crotolaria* species, many of which occur as weeds throughout the world, produce these alkaloids. About half of the pyrrolizidine alkaloids have

CASE NOTES

Toxic tea

In Austria an 18-month-old boy became ill and was taken to hospital, where he was found to have liver disease of the type characteristic of pyrrolizidine alkaloid poisoning. He had been given herbal tea since he was 3 months old. The herbal tea should have been been made with the plant coltsfoot but was mistakenly made with alpendost. The boy had congestion of the blood vessels of the liver (veno-occlusive disease) which were also damaged and bleeding.[1]

A study in two hospitals in South Africa identified twenty children suffering from this type of liver disease which was thought to be due to the use of traditional remedies. Most of the children had fluid in the abdominal cavity and an enlarged liver, indicating liver damage and dysfunction. There was a high level of illness and mortality, and in those who survived the disease progressed to liver cirrhosis. In four cases pyrrolizidine alkaloids were detected in the urine of the children.[2]

been identified as toxic, and these plants are probably the most common cause of poisoning in the world for humans and both livestock and domestic animals. Poisoning may occur as a result of contamination of cereal crops, and so flour made from the crops will contain the poisonous substances. In some countries plants used in the preparation of herbal teas or in herbal remedies will contain the alkaloids.

Poisoning with these alkaloids has occurred in various parts of the world, especially where agricultural conditions are poor and the indigenous population may be forced to use contaminated crops. For example, in South Africa in the 1930s, poor whites suffered the toxic effects of these alkaloids as their staple diet was wheat which had become contaminated with plants producing the alkaloids, whereas their Bantu neighbours who ate maize which was not contaminated were not affected. More recently, poisonings have occurred in Tashkent, central India, and northern Afghanistan. In one incident, where 1,600 poisoning cases were reported, the threshed wheat was found to be contaminated with *Heliotropium popovii* seeds and pyrrolizidine alkaloids could be detected in the grain.

As we saw in Chapter 3, pyrrolizidine alkaloid toxicity is a continuing problem, as plants containing pyrrolizidine alkaloids are also used in traditional medicine to make herbal teas, especially in the West Indies.

Apart from humans, livestock may also be exposed and suffer the toxic effects. Where there is abundant vegetation for grazing, animals will ignore plants that contain the alkaloids, like ragwort (*Senecio jacoboea*). If, however, food plants are sparse, poisonous weeds will be eaten. In some

Dangerous chemicals in common plants

The pyrrolizidine alkaloids cause an unusual form of liver disease known as veno-occlusive disease. Studies in experimental animals have shown that these chemicals are very toxic, especially to the liver and lungs. The alkaloids, such as monocrotaline, undergo metabolism in the body to become products that are reactive and damage both the cells lining the blood vessels in the liver and the liver cells. This leads to destruction of liver cells and bleeding, and finally to the liver disease. The blockage of the blood vessels in the liver eventually stops the blood flowing to the liver, which means that it cannot function properly, and so the dysfunction increases. The blood supply is then diverted into new blood vessels which can be seen growing just under the skin of the belly of the victim.

countries, like Australia, widespread losses of horses, cattle, and sheep have occurred from heliotropium poisoning. Exposure of livestock may lead to indirect human exposure through the consumption of milk, as the alkaloids can be detected in the milk of cows grazing on such plants.

Ricin

CASE NOTES

Six arrested in poison terror alert

■ By John Steele and Sandra Laville

Six men were being questioned by anti-terrorist police last night after traces of the poison ricin were found in a flat in north London.

⟨http://www.telegraph.co.uk/news/⟩, accessed January 2003

The most toxic plant product, ricin, is also the most toxic chemical known to man. Unlike the pyrrolizidine alkaloids, ricin is a protein which most cleverly targets the interior workings of cells. The way in which this most potent poison works is quite elegant (see box, p. 151). It is made and found in the seeds of the castor bean (*Ricinus communis*).

The initial symptoms of ricin poisoning are gastroenteritis, in which there is bleeding, followed by **arrythmias** of the heart, depression of the central nervous system, coma, and then death. The toxin is especially potent when injected, whereas a single castor bean can be eaten without

CASE NOTES

The poisoned umbrella

Ricin is best known as the substance that is believed to have been used to kill Georgi Markov, a Bulgarian dissident journalist and broadcaster, in London in 1978. While walking across Waterloo Bridge, Markov was stabbed with a special umbrella which injected a small spherical pellet into his leg. This is thought to have contained a potent poison. The symptoms he suffered before he died were consistent with ricin poisoning. Ricin was also a poison known to be favoured by the Bulgarian secret police who, it was believed, killed him to put an end to his anti-communist broadcasts and writing.

effect provided it is not chewed. Even if it is chewed, the outcome may not be fatal (although this would depend on the number of beans eaten), however, because ricin is not well absorbed from the gut. If the seeds are finely powdered and the particles inhaled, it is more poisonous. As well as containing ricin, the seeds also contain a protein that causes red blood cells to stick together, but which is also poorly absorbed from the intestine when the bean is eaten.

This toxin has been used by natives in some countries for infanticide, the removal of unwanted babies by their mothers. In 1962 the US Army

Ricin: a molecular Trojan horse from the castor bean

The toxic ricin is a small protein molecule consisting of two parts, chains A and B. The B chain is similar to proteins called lectins which recognize and bind to the membranes surrounding the cells in our bodies. The B chain attaches the ricin to the cell membrane which then folds inwards so that the ricin molecule is taken inside the cell inside a bag called a vacuole. There is only one bond between the A and B chains and this now breaks. The B chain then makes a hole in the vacuole through which the A chain passes into the cell. Here it heads straight for structures called ribosomes, where proteins, many of which are vital for the functioning of our bodies, are made. The A chain then selectively removes a specific molecule (the base adenine) from the **RNA** in the ribosomes. RNA contains the information required to make proteins, and removal of part of the information blocks the synthesis of proteins. The cell therefore dies. One molecule of ricin may be sufficient to kill one cell. This makes it the most potent toxin known.

took out a patent on the powdered seeds for possible use in chemical warfare.

As with some of the other chemicals discussed in this book, the possibility of using ricin as a drug, in particular for the treatment of cancer, has been explored. The possibility of attaching the part of the toxin that is lethal to cells, to antibodies, which would then target cancer cells, is being studied.

Another selective and very toxic chemical that is produced by plants for the purpose of chemical warfare against predatory herbivores is sodium fluoroacetate. What is particularly interesting about this chemical is that it is one of the few fluorine-containing organic chemicals found naturally in a living organism. Fluorine is abundant in the earth but is mostly found in the inorganic form as fluorides, which are stable minerals. Plants found in South Africa and Australia produce this simple organic chemical, which is very toxic because it specifically blocks a major metabolic pathway that is essential in most animals for the production of energy in the body. The heart and brain fail through lack of energy and the exposed animal dies. Strangely, however, some animals have become immune to the effects and can eat these plants. The chemical has also found a use as a pesticide in New Zealand, where it is used for killing possums. The possum, which is not native to the country, threatens to overrun many areas and has become a major problem.

Hemlock: the executioner of Socrates

The poison hemlock contains two alkaloids that are poisonous, coniine and coniceine. They act to block the transmission of nerve impulses, which results in death by failure of respiration. The plant is famous for its part in the execution of Socrates in ancient Greece, who was found guilty of corrupting the young and neglecting the gods. The execution was described by Plato in *Phaedo*, written in 360 BC:

> And the man who gave the poison began to examine his feet and legs . . . then he pressed his foot hard, and asked if there was any feeling in it; and Socrates said no; and then his legs, and so higher and higher, and showed us that he was cold and stiff; and he afterwards approached him and said that when the effect of the poison reached the heart Socrates would depart.[3]

These symptoms are due to the blockade of nerves, those involved with feeling and with movement.

18. Socrates drinking hemlock, the Athenian state poison.

A similar plant is water hemlock, which contains a different toxin, called cicutoxin. This is very potent and exposure to it is often fatal. One study of poisonings with this plant found that 30 per cent of victims died. It affects primarily the brain and the spinal cord, causing seizures and epileptic fits, possibly by overstimulating certain nerves (cholinergic pathways).

Another toxin found in plants, in particular tobacco (which is similar to coniine and is another alkaloid), is nicotine. This substance, with which we are all familiar, is a very toxic chemical, and its presence in cigarette smoke is the essential ingredient that smokers crave. The tobacco plant and the habit of smoking the leaves, known as tobago, was probably first seen by Columbus and his crew in South America. Sir Walter Raleigh also saw the plant in his travels to the new continent of America. Leaves from the plant were sent back to Europe in the mid sixteenth century, and an explorer by the name of Jean Nicot de Villemain sent some seeds back to Europe. He helped to popularize the habit as a panacea, which became widespread in the sixteenth century. From the explorer's name and the name given to the practice of smoking, the plant was called *Nicotiana tabacum*. The active substance it contained, isolated in 1828, was called nicotine.

When the tobacco leaves are burnt the nicotine is released, and when the smoke is inhaled about 350 μg of nicotine is present in a single puff. As well as being a plant toxin, nicotine is a drug in the true sense of the word, in that it has effects on the body that are desirable, at least to the addict. Nicotine is very toxic and has even been used as an insecticide; there have been deaths from accidental exposure to the chemical. Its effects after moderate poisoning are similar to those caused by organophosphates. A high dose may lead to death by causing respiratory arrest as a result of blockade of the nerves that control breathing. That it is a potent and rapid-acting toxin is clear from the following: 'if a couple of drops of pure nicotine are placed on a dog's tongue the dog drops down dead in a few seconds'.[4]

Nicotine binds to a type of receptor, now called a nicotinic receptor, and causes the same effects as a release of the neurotransmitter acetylcholine at nerve endings. In this it is similar to organophosphates, which lead to excess acetylcholine at nerve endings (see pp. 100–101). Nicotine first excites and then inhibits the central nervous system; it first stimulates and then paralyses nerves. It reacts with receptors in muscle and nerves, and is able to enter the brain from the bloodstream and interact with nicotinic receptors. At first there is a stimulation, with the smoker experiencing alertness and decreased irritability, aggression, and anxiety. With higher doses there is depression of the brain as a result of saturation of the receptors.

Repeated exposure, as occurs in habitual smokers, leads to an increased rate of metabolism of nicotine and decreased sensitivity of the receptors to the nicotine. More nicotine is therefore required to satisfy the needed stimulation, and tolerance develops. This is a separate effect from addiction which seems to be due to increased levels of a substance called dopamine in the brain. This is due to the nicotine activating nerve cells which results in increased release of dopamine and causes a reduction in the amount of an enzyme that destroys dopamine. It seems that other addictive drugs may also work by increasing dopamine levels.

Strychnine: poisoned by the last dose in the bottle

One of the most well-known, infamous poisons, beloved of murderers and crime writers, is strychnine, which is derived from a tree found in India, *Strychnos nux vomica*. Strychnine may be found in the crushed seeds of the tree, along with other alkaloids, and hence it is often known by the name nux vomica. It is also found in the upas tree, along with antiarin (a cardiac glycoside), and the sap of the tree containing these toxins was used for making poison darts for the execution of criminals in Malaya. As Charles Darwin described it: 'A few moments after being wounded by the executioner, the poor victim trembled violently, uttered piercing cries and had frightful convulsions; death occurred in 10 to 15 minutes.'[5]

The easy availability of strychnine has led to some fatalities; about 50 mg is a lethal dose. As early as the sixteenth century it was used as a rat poison in Germany, and it is still used for this purpose in some countries.

Strychnine has also been used in medicines, originally in the sixteenth century, becoming more common some centuries later; preparations containing it were still described in some pharmaceutical books as recently as 1973.[6] Easton's syrup and tablets, which contained strychnine along with quinine, have featured in accidental poisoning cases. The drug heightens awareness, which led to it being used as a constituent of nerve tonics, and its action on the gut led to it becoming a component of purgative pills. The pills were sometimes sugar-coated, which occasionally led to children being poisoned who had mistaken them for sweets (there are cases from 1958 and 1961). Its medicinal use seems to have been fraught with dangers, and rarely have there been great benefits. Careless use by doctors has occasionally killed patients, mistakes in prescribing having led to accidental poisonings. For example, strychnine was once added to a preparation instead of quinine, and as a result seven patients died. Other cases were reported of strychnine having been either wrongly added or added at the incorrect dosage: 'The patient asked for a tonic and the doctor thought of strychnine . . . unfortunately he misread the dosage . . . and gave him 60 times the intended dose. The patient remarked that it was strong medicine . . . complained of stiffness in the muscles . . . had one convulsion and was dead 45 minutes after drinking the medicine.'[7]

The action of strychnine leads to violent convulsions in the victim, and to contortion of the muscles as they contract to an abnormal degree (see box). Sometimes the jaw muscles also contract, imparting to the victim's face the appearance of an apparently sardonic grin. Death from this poisoning is painful because of the extreme muscle contraction, and terrifying for the victim who is fully aware at all times of what is happening. He or she becomes exhausted by the muscle contractions and convulsions, during which breathing stops, and after perhaps five or six of them, respiration fails to restart and the victim asphyxiates. Death is often relatively rapid, but if the patient survives more than three hours, he or she is likely to make a recovery.

How strychnine kills

This natural toxin acts on receptors in the nervous system, in the spinal cord, which respond to a substance called glycine, an inhibitory neurotransmitter. When released from special nerve cells the glycine inhibits the action of nerves that control muscles. It is a control mechanism to stop the nerves sending messages too rapidly, thereby causing the muscles to react too violently. Strychnine thus diminishes resistance to nerve impulses and so heightens sensitivity to stimuli. In the presence of strychnine, this control is lost: the muscles contract excessively, and the patient suffers violent convulsions and contortions. All the muscles are contracted but the most powerful prevail, leading to extreme arching of the back. The victim dies from a combination of exhaustion and asphyxiation because the nerves controlling breathing are out of control. Breathing often stops during the violent convulsions and may or may not start again during the relaxation period.

Treatment involves sedating the patient and keeping them in a darkened room isolated from external stimuli. This is because the slightest noise or touch can be followed by a violent movement or convulsion. Sometimes an anaesthetic, such as ether or **chloroform**, was used to stop this reaction, but barbiturates seem more effective and are preferred. Helping the victim to survive the convulsions allows the body to metabolize and excrete the toxin.

Another cause of accidental poisonings was as a result of the strychnine sometimes collecting as a sediment in the bottom of the medicine bottle, either because of incorrect preparation or because the patient did not shake the bottle. Thus there have been several accidental poisonings when patients took the last dose in the bottle, which turned out to be their last as they overdosed.

As a result of such incidents, it was only a matter of time before strychnine was banned in countries like the UK. As we shall see in Chapter 9, there have been cases of suicide involving strychnine and many cases of murder. Homicidal fatalities were still occurring in some countries (for example, Romania) in the 1950s, and accidental poisonings still occur.

Bracken

The shoots of the bracken fern are eaten as a delicacy in Japan, but they may contain a poisonous substance called ptaquiloside. Ptaquiloside degrades into a chemical which can cause cancer. This may explain the high incidence of throat cancer among the Japanese. Animals that eat the fern as fodder also suffer from cancer, typically cancer of the bladder and intestine.

How ptaquiloside may cause cancer

A breakdown product of ptaquiloside has been shown to react with DNA, with the result that the DNA chain breaks. The ptaquiloside breakdown product reacts specifically with adenine, one of the four bases that form the code in the DNA molecule. The loss of this base from the DNA and the change in the DNA structure can cause mutations by altering the reading of the base pair code. Such genetic changes and mutations can lead to cancer.

Fungal toxins: toxic toadstools and magic mushrooms

There are many toxins produced by fungi of many different kinds, such as aflatoxin, a potent carcinogen which can contaminate food. (Some of these will be discussed in Chapter 10.) Fungi in the form of mushrooms can be eaten, but many fungi, including mushrooms and toadstools, produce potent toxins. Some of these are or have been used as drugs.

Probably the most poisonous mushroom in Britain is the death cap mushroom (*Amanita phalloides*), which is also found in other parts of the world (see Figure 19). It may occasionally be eaten by mistake, although poisoning with this mushroom is rare. The mushroom contains a number of toxins: several phallotoxins and several amatoxins. The phallotoxins produce violent gastroenteritis four to eight hours after the mushroom is eaten. The amatoxins have a delayed toxic effect, targeting the liver and kidneys, and causing destruction of the cells of both. After eating the mushroom one may experience few if any symptoms, apart from non-specific effects like nausea, followed by vomiting and diarrhoea. There can be a phase of perhaps two days in which the victim seems to recover. However, he or she can then suffer liver and kidney failure, as indicated by jaundice and alterations in the chemistry of the blood, such as low levels of sugar in the blood and high levels of nitrogen containing waste products.

The amatoxins and phallotoxins are cyclic peptides (peptides are made

19. The death cap mushroom, one of the most poisonous, which is sometimes responsible for human poisoning.

of groups of amino acids and are the building blocks of proteins). The phallotoxins act on the membranes of cells while the amatoxins, which are more potent, prevent the production of proteins from the information encoded in the DNA (this is the main function of DNA). About 10 mg of one of the amatoxins is sufficient to kill a human being.

Animal toxins

Snakes: rattlers and cobras

Ten per cent of the snakes in the world, about 300 different species, are poisonous, and snakes are an important source of natural toxins. Poisonous snakes belong primarily to certain groups, such as that including the mamba and cobra, the vipers, and the sea snakes and another group that includes the boomslang and the mangrove.

In a typical year there are some 45,000 snake bites in North America alone, but only 20 per cent of these are from poisonous snakes and only a proportion of these involve toxin or venom. This still leaves over 1,000 poisonous snake bites in which venom is injected, but deaths number only a handful, perhaps twelve to fifteen. The reason for this is that medical care is normally not too far away and the antidote or **antivenin** is available. Without treatment, for example, approximately 75 per cent of rattlesnake bites would be fatal. Similarly, in Australia, which has some of the world's most poisonous snakes, death from snake bite is rare because antidotes are prepared to the different toxins and are made available. Fortunately, many snake venom toxins are also relatively slow to reach target sites and therefore slow to act. The situation in other parts of the world, however, is very different, and snake bite can be a common cause of death. In India and Pakistan there are about 50,000 to 70,000 deaths per year from snake bite, and in Burma and Brazil the numbers also run into the thousands.[8]

Unlike the toxins in plants or fungi, of which there may be only one or two in each, snake venoms are not single toxins but a whole family. Venom from the North American crotalids (this family includes rattlesnakes) contains at least nineteen different toxins, some of which work in concert to increase the harm and speed of damage. For example, there are enzymes in the venom that break down the skin and other tissue at the site of the bite, allowing the toxins to enter the deeper tissues more easily; other toxins affect the permeability of the blood vessels and capillaries, allowing blood and fluid to seep into the tissues, which leads to swelling, and helping to transfer the toxins into the blood circulation.

In some snake venoms there are several substances that affect blood clotting (see box), with the result that clotting doesn't occur properly and the victim can bleed to death through the wound. The effect will have been increased by the action of the enzymes degrading the tissues around the site and internally. One purpose of the venom containing enzymes that degrade the tissues of the victim is, of course, to aid digestion later when the snake eats its prey. This degradation causes inflammation and can result in severe pain. There may also be breakdown of muscles as a result of the action of another specific toxin. There are also toxins that affect the nervous system and heart, but these are not normally as significant in the crotalids.

In the elapids, the group that includes the cobra, a toxin acting on nerves and muscles, a **neurotoxin**, is responsible for the death of the victim. It causes paralysis of breathing (see box). In addition, the victim will show speech incoordination, the eyelids will close, and there will be muscle weakness and lack of energy due to the lack of oxygen. About 6 mg of the cobra neurotoxin is sufficient to cause death, which occurs very rapidly, for example a woman in Sri Lanka died fifteen minutes after being bitten by a cobra.

Although treatment is possible using antivenins (antivenom), the combinations of toxins in a particular snake venom can vary, which means that they need to be as specific as possible. Antidotes to the effects can sometimes also be used, for example anticholinesterase substances which are used in treatment of the death adder bite. The antidote stops the breakdown of acetylcholine, a neurotransmitter, which is in short supply as a result of the actions of the toxin.

Rattlesnake and cobra venoms: thin the blood and dissolve flesh

Many snake venoms, including those from the rattlesnake, interfere with blood clotting in some way. Destruction or alteration of components of the clotting process, for example the protein *fibrinogen*, is a common method, with the result that clotting does not occur. At the same time enzymes are degrading tissue and muscle. The muscle is degraded by the action of a specific toxin, myotoxin, which increases the calcium level in muscle, precipitating the self-destruction of the muscle tissue. Destruction of blood cells also occurs, again through the action of enzymes, and there may also be coagulation of small blood cells. The victim may die within an hour, but the average is between 18 and 32 hours. This is usually due to a loss of blood pressure or shock from the loss of fluid in the circulatory system and blood cells.

With cobra venom, the major effect is due to a toxin that acts on the nervous system. This neurotoxin is a small molecule, which can distribute throughout the body rapidly. It acts like curare, paralysing the centre in the brain that controls breathing. By acting at the point where nerves control muscles it blocks the transmission of nerve impulses and causes muscle weakness and again affects breathing. The eyelids droop and speech becomes incoordinated.

In addition, cobra venom contains a toxin that causes changes in the heart rhythm and a loss of blood pressure. Finally, a further toxin and a combination of enzymes cause red blood cells to rupture. Thus a bite from the cobra will be rapidly fatal.

Toxic toads and frothing frogs

> Double, double toil and trouble;
> Fire burn, and cauldron bubble.
> Fillet of a fenny snake,
> In the cauldron boil and bake;
> Eye of newt, and toe of frog . . .
>
> Shakespeare, *Macbeth* IV. i

Whether Shakespeare knew of the toxins in some frogs and toads is doubtful. We don't normally associate frogs and toads with poison because they don't bite. However, some frogs and toads, for example the frog dendrobates, produce very potent toxins to ward off or even kill predators, and possibly harmful bacteria, and humans may fall victim to their effects. Frogs that produce such poisonous skin secretions are found in Colombia, Costa Rica, Panama, and also in Australia. Indians in Colombia used such toxins from frogs to coat the tips of their arrows, as was described by Captain Charles Cochrane when he was exploring Colombia in 1823–4:

> Those who use poison catch the frogs in the woods, and confine them in a hollow cane, where they regularly feed them until they want the poison, when they take out one of the unfortunate reptiles and pass a pointed piece of wood down its throat, and out of one of his legs. This torture makes the poor frog perspire very much, especially on the back, which becomes covered with white froth; this is the most powerful poison that he yields, and in this they dip or roll the points of their arrows, which will preserve their destructive power for a year. Afterwards, below this white substance, appears a yellow oil, which is carefully scraped off, and retains its deadly influence for four to six months, according to the goodness (as they say) of the frog. By this means, from one frog sufficient poison is obtained for about fifty arrows.[9]

More recent studies have identified the most potent toxin produced by the frog as batrachotoxin (although it is one of about a hundred toxins). This is one of the most toxic substances known—five times more potent than tetrodotoxin, the toxin found in puffer fish (see pp. 252–3)—and a dose of 200 μg would be lethal for a human. Interestingly, tetrodotoxin is also found in the Californian newt. Batrachotoxin is secreted through the skin when the animal is stressed, and so when animals lick or bite the frog they are exposed to the toxin, which causes death by stopping the heart beating.

A toxin produced by Australian frogs called uperolein causes inflammation of the skin, dilation of blood vessels, and a large fall in blood pressure. This toxin is a peptide and also affects muscle, which leads to diarrhoea and breathing difficulties.

Toads can also produce toxins, which are different from those produced by frogs, one of which, bufotenin, has similar effects to LSD, causing hallucinations and also lowering blood pressure. Another toxin, bufotalin, affects heart function like digitalis, which may lead to arrythmias and death. Cases of poisoning have occurred as a result of people handling these frogs and toads, as some of them can produce considerable amounts of the toxins.

Toxins from sea creatures

Those of us who have been stung by a jellyfish know the pain this can cause. In the case of the Portuguese man-of-war jellyfish, the pain is intense and is accompanied by many other symptoms, including pain in the muscles in the injured limb or even the whole body, pain when breathing, and even breakdown of red blood cells and kidney failure. The toxin, physalitoxin, is a large protein which is lethal to mice at a dose of 0.2 mg per kg body weight, that is a lethal dose in humans would be about 12 mg.

A related sea creature, the sea wasp, is particularly dangerous and has been responsible for several deaths in Australia. Other sea creatures can be unexpectedly poisonous, for example cone snails which produce conotoxins. Like tetrodotoxin and saxitoxin (see pp. 252–3), conotoxins stop the movement of ions such as sodium, potassium, and calcium, hence affecting the function of both nerves and muscles. Even humble seaweed can contain toxins, and that found in the Death Sea weed from the Caribbean and Hawaii, which is called palytoxin, is especially potent, being one of the most poisonous substances known. Based on the lethal dose in a mouse, the lethal dose in a human would be about 9 μg. Although this is more than ten times the lethal dose of botulinum toxin, it is still extremely poisonous. The toxin works by creating pores in the membranes of nerves, allowing ions to flow in or out, which changes the electrical activity of nerves. It also causes red blood cells to rupture by allowing a rapid loss of potassium. This poison has also been used to tip arrows.

Spiders

Most people own up to a dislike of spiders, even those that are harmless, but those that bite are in a different league of terror. Possibly the best known is the black widow, found in the USA, a small, unobtrusive spider but very dangerous. Three of the most venomous spiders, and those most commonly involved in poisonings, all have venoms that act in a similar way. Thus the black widow, red back, and funnel web all have effects on nerves and muscles, that is, they affect breathing and control of muscles. The black widow toxin, latrotoxin, causes a massive release of neurotransmitters, for example acetylcholine, at nerve terminals by binding to a receptor in nerves. The toxin causes pain and cramp in muscles, moving from the area of the bite to the muscles of the back and abdomen, which become rigid. Nausea and vomiting, sweating, tremors, speech defects, and insomnia can occur within minutes or hours of the bite. Paralysis of muscles, including those involved with respiration, high blood pressure, cardiovascular collapse, and coma are all serious effects.

The red back spider, found in Australia, which was responsible for deaths of adults (5 per cent of those bitten died) and especially of children before antivenin became available. One of the favourite haunts of the red back is underneath toilet seats, so bites on the backside and in the crotch are not uncommon! There are several thousand victims annually in Australia. The symptoms are similar to those caused by the black widow, and pain can be very severe.

In contrast, the brown recluse spider from the USA has a more complex venom, which causes damage to arteries and veins, and the formation of clots. This leads to pain and inflammation, and damage to tissues which may leave a scar. Victims can die of circulatory or kidney failure.

Another Australian venomous spider is the funnel web, whose toxin, atraxotoxin, causes effects on the nervous system leading to muscle paralysis and high blood pressure. A bite from the spider can be lethal to a human.

Scorpions

The first and only time I saw a live scorpion was in Morocco. These potentially dangerous and aggressive-looking creatures are found in many parts of the world, in deserts and tropics. They are nocturnal, remaining hidden during the day, for example under stones or in shoes! There are many types of scorpion (for example, thirty in North and South America), some of which can inject a fatal sting. The most deadly is the *Centruroides sculpturatus*, found in North America, which produces a toxin that acts on ion channels in nerves, causing symptoms similar to those in strychnine poisoning, with increased excitability. The nerves fire more impulses, with the result that victims cannot control their muscles, which contract, causing the limbs, eyes, mouth, and throat all to move uncontrollably. The body releases massive amounts of adrenaline, the heart rate and blood pressure rise, and the victim sweats. Finally, as a result of the uncontrolled muscular contractions leading to paralysis, breathing stops. Fortunately, there is an antidote (an antivenin) available for treatment.

Some scorpion stings do not contain neurotoxins and their venom is relatively harmless, producing mainly irritation and inflammation at the site of the sting as bee and wasp stings do.

There are many other poisonous animals and plants, but there is not the space in this book to deal with them.

Bacteria

Later in the book I shall deal with one of the most toxic substances known, botulinum toxin, derived from the bacterium *Clostridium botulinum*, which is occasionally found in food. A related bacterium, *Clostridium perfringens*, also produces a nasty toxin, called alpha toxin. Like *Clostridium botulinum*, the bacteria cannot live in the presence of air. The spores are found in soil and even in our intestines, but if they get into a deep wound, where oxygen is absent, they can infect it and cause gas gangrene. The toxin is an enzyme that destroys cell membranes. It is a particularly ingenious bacterium because it targets the immune system and the blood supply so as to reduce the supply of blood, and therefore oxygen, to the infected tissue. It improves its own chances of survival by making the environment conducive to it and by reducing the ability of the immune system to remove it.

Many bacteria produce toxins, some of which are exceptionally poisonous and cause various types of effect. Bacteria that cause diarrhoea may do this by producing toxins that act on the gut and alter the uptake of fluid from the gut or the secretion of water into it.

Some bacterial toxins are used experimentally, such as tetanus and diphtheria toxins. Diphtheria toxin kills cells by interfering with the synthesis of proteins in a similar way to ricin.

7

The Mad Hatter and a Bad Case of Acne

Industrial Chemicals

INDUSTRIAL diseases have existed ever since humans began manufacturing on a large scale, and during the Industrial Revolution they were common. Before that, certain occupations, like mining, were hazardous. In ancient times those working in arsenic or lead mines would undoubtedly have suffered from the adverse effects of the minerals they mined. Arsenic mines in Pantus in ancient Greece were described by Strabo, the Greek historian, as follows: 'the air in the mines is both deadly and hard to endure on account of the grievous odour of the ore, so that the workmen are doomed to a quick death.'[1] Some occupational diseases were common, and had well-known names, such as hatter's shakes (see below), farmer's lung (allergic lung damage/asthma from inhaling fungal spores in hay), and phossy jaw (fragile jawbone from phosphorus exposure).

As men and women can spend up to a third of their lives at work, what they are exposed to in the workplace is of great importance. One of the first diseases that was clearly recognized to be due to an occupational hazard was cancer of the scrotum, which occurred in chimney sweeps. This was first recognized and described by Sir Percival Pott in 1775. It was, apparently, due to the exposure to soot and coal tar. It was not until much later (in 1916) that it was shown experimentally that coal tar could cause cancer in rabbits exposed to it. In many cases of industrial diseases resulting from exposure to chemicals, especially in earlier times, the association has been made by studying populations (**epidemiology**), while the basic experimental work has provided the evidence only later.

Mercury

The phrase 'mad as a hatter' is in common English usage but its origin is not so well known. It is derived from the workers who used mercury in the preparation of felt for hats. Beaver and rabbit fur, which were used to make the felt hats, were treated with mercuric nitrate, a mercury salt, in order to matt the fur together, a process called carroting. After this process, the fur would be heated and the workers inevitably became exposed to mercury vapour. The symptoms developed by the workers became well characterized from observations of groups of workers in the industry. These consisted of tremors ('hatter's shakes'), mental disturbances ('mercury madness'), a jerky walk, stammering speech, and apparent pathological shyness.

Elemental mercury, which is a liquid, is found in the free state in rocks, for example the mercury mines in Slovenia and Spain. It was known as quicksilver and its Latin name, *hydragyrum* (from which its chemical symbol, Hg, is derived), means liquid silver. It is also found combined with sulphur in brightly coloured cinnabar, and with oxygen.

Liquid mercury was certainly known in 1600 BC in Egypt and probably also in China and India. Compounds of mercury such as cinnabar have been mined for more than 2,500 years in Spain, and it was known from around AD 300 that heating this produced metallic mercury. Mercury has been used in various ways for centuries and its use has often been associated with toxic effects. The Greeks and the Romans knew it was poisonous. Apart from the felt-making industry, exposure to mercury would have occurred in the mines where the metallic mercury was found. Mercury was later used extensively in scientific instruments such as barometers and thermometers. The alchemists used mercury in a vain effort to turn base metals into gold, and it is believed that King Charles II died of mercury poisoning as a result of his dabbling in alchemy.

One of the unusual properties of mercury is that it reacts with certain other metals to form amalgams. Amalgams of silver and tin became used for filling cavities in teeth in the nineteenth century and similar amalgams were used until very recently. This ability to form amalgams has also been utilized in the extraction of metals like gold and in plating. Thus amalgams of gold and mercury could be used to deposit or to plate gold onto other surfaces such as buttons. The process led to mercury vapour being released and the workers becoming poisoned, suffering from 'gilder's palsy'. The waste mercury also poisoned those living near the factories. It was reported that sixty Russian workers died as a result of exposure to

20. The Mad Hatter from *Alice in Wonderland*. The character in the cartoon reflects the fact that hatters at one time suffered damage to the nervous system as a result of working with mercury.

mercury when gilding the dome of St Isaac's cathedral in St Petersburg. Similarly, the making of mirrors involved using a silver amalgam, which also exposed the workers to mercury. A more bizarre case of mercury poisoning occurred when the liquid metal was being transported by ship in 1810. One of the flasks containing the mercury broke open during a storm, with the result that three of the crew died of mercury poisoning.

Various forms of mercury have more recently been used in industrial processes: as an electrode, in electrical equipment, in batteries, as a catalyst, and in fungicides. Some of these, especially its use in fungicides and as a catalyst, have caused serious human poisoning, as have have seen.

Paracelsus, the author of the first monograph on occupational diseases, published in 1567 after his death, described in detail the effects of chronic mercury poisoning, distinguishing between the acute and the chronic effects. As a physician, he also devised a use for mercury to treat syphilis, which was used for more than 300 years.

Bernadino Ramazzini was the father of occupational medicine. Born in 1633 in Italy, he studied the way diseases occurred in groups of people engaged in similar work or living in a particular area. This science we now call epidemiology is an important means of associating a particular disease with exposure to a chemical. Ramazzini studied at first hand the conditions workers had to endure as part of their daily lives.

From early times working people have been exposed to toxic substances ranging from minerals and metals that were mined and then worked into products, for example asbestos, lead, and mercury, to more complex chemicals produced or used in the synthesis of the countless products we now enjoy, such as plastics and drugs. There are numerous chemicals to which workers can now be exposed, some of which may be toxic.

Occasionally, people other than the workers in the factory are exposed, as occurred during the terrible Bhopal disaster in India (see pp. 172–4) and in Seveso (see pp. 123–4). The Seveso incident, mercifully, did not result in any human deaths, causing mostly the skin disorder, chloracne. This severe acne is also a common response to other halogenated (for example, chlorinated) hydrocarbons, which is used in industry or produced as part of the manufacturing processes of pesticides and fire retardants, for example. Despite the obvious potential dangers, industrial chemicals, if used carefully in controlled conditions, often benefit us greatly by producing cheap, durable materials such as plastics, novel fabrics and dyes, and effective drugs. Unfortunately, there have been problems in the past as a result of bad industrial practices such as inadequate safety measures and poor maintenance. Some of the examples that follow will serve to illustrate the lessons that can be learnt.

Vinyl chloride

I have chosen to deal with vinyl chloride in my discussion of exposure in the workplace as it serves to demonstrate the problems that may arise from exposure to novel chemicals, how these can be prevented, and how crucial a knowledge of toxicology is to occupational health and the safe use of chemicals.

The major use of vinyl chloride (also known as vinyl chloride monomer, VCM) is in the production of the ubiquitous plastic, PVC or polyvinylchloride, which began to be used in the 1940s. This polymer is used in a wide variety of products ranging from cling film, bottles, car components, pipes, electrical insulation, and window frames. Unfortunately, vinyl chloride, which is used to make the plastic, is a hazardous chemical. For this reason and because PVC is an organochlorine compound, it has received a bad press. It has important properties, however, such as flexibility, toughness, and durability, that make it an extremely useful and versatile material.

The vinyl chloride used to make PVC is a gas at room temperature and is therefore kept under pressure during the process of polymerization. The pressure chambers used have to be cleaned out periodically, and during the early years of manufacture workers would enter the chambers to clean them after periods of use. Inside the chambers levels of vinyl chloride were high, enough at times for the workers who, presumably, were unprotected to be overcome by the fumes. Vinyl chloride is like an anaesthetic (attempts were made to use it as such in the 1940s), and hence breathing the vapour causes the victim to lose consciousness.

At first vinyl chloride was regarded as a reasonably innocuous chemical but eventually reports of toxic effects, such as numbness and tingling in the fingers, began to appear. In 1971, studies in experimental animals (rats) were carried out, which found that it was clearly carcinogenic, causing a rare type of liver tumour. Following this, in 1974 epidemiological investigations confirmed that there was a cancer risk in humans exposed in industry. There is no doubt that vinyl chloride causes a rare type of liver cancer (haemangiosarcoma), a cancer of the blood vessels in the liver, both in experimental animals and in humans. The rarity of this cancer aided the epidemiological studies, which identified vinyl chloride as a cause. Fortunately, records existed and so the studies could be undertaken, as an important adjunct to toxicity studies on this industrial chemical.

In the experimental studies animals were exposed to various doses, some of which were high. Some of the industrial workers were also exposed to high levels, in particular those who cleaned out the pressure chambers. Even though such exposure was intermittent, it was sufficient to cause cancer. There is a relationship between the level of exposure and the incidence of cancer in humans, the workers with the highest exposure having the greatest risk. Evidence suggesting a dose–response relationship, therefore, exists, as has also been shown in experimental animals. In humans the cancer takes a long time to develop, at least twenty-five years and possibly up to forty years, so cases may yet occur.

This example illustrates that had the toxicity of vinyl chloride been evaluated before workers were exposed, the 100–200 people who have had or are likely to suffer liver cancer would not have done so. Tests in experimental animals and possibly also in *in vitro* systems could have alerted manufacturers to the dangers. More recent assessments of data from animal studies reveal the risk, although they overestimate the actual risk from reported cases (see pp. 307–8).

A general awareness of the potential dangers of exposing workers to high levels of any chemical should also have prevented such working practices. In the risk averse environment in which we live in today we wonder how such things could have happened. A lack of knowledge was certainly one factor, yet in some countries such practices persist.

Vinyl chloride also causes other toxic effects as well as narcosis (putting people to sleep) and liver cancer. It causes a number of effects known collectively as 'vinyl chloride disease', including changes in the skin, blanching and numbness in the fingers (Raynaud's phenomenon), and changes in the bones of the hands—which may be due to changes in the blood supply to the extremities. There may also be tingling and pain in the extremities (that is, fingers and toes) which could be due to damage to peripheral nerves. Some workers also experienced dizziness, fatigue, headache, memory loss, and visual and hearing disturbances. Effects on the reproductive systems of both males and females have also been suggested, and decreased fertility and damage to the testes have been shown in male experimental animals. Damage to the liver apart from cancer can also occur.

These effects are generally associated with higher exposure levels than occur now. The current exposure limit set by the OSHA (Occupational Safety and Health Administration) in the USA is 1 ppm. This is the amount a worker can be exposed to for an eight-hour working day or for a forty-hour working week.[2]

So what is the answer to industrial exposure to chemicals? The principle should be that workers who are unavoidably exposed to chemicals should be exposed to the lowest level possible and practicable. Protective clothing and systems should be used, and reactions should take place in closed systems. In addition, workers should be regularly monitored and records kept. This allows populations of workers to be studied for any trends in adverse effects. Furthermore, tests should be done on chemicals in both *in vitro* systems and experimental animals to detect potential hazards. A combination of experimental data and epidemiological information is essential for the identification of hazards. These are now requirements in many countries.

As metabolism is involved, factors like the intake of alcohol are important, for example the workers who drank were more prone to develop the liver cancer. This is probably because alcohol increases the activity of enzymes involved in the metabolism of such compounds, thus changing the amount of the toxic metabolite produced.

Is vinyl chloride a hazard to the general public? PVC products may contain traces of vinyl chloride, but it is volatile and so amounts would very extremely small. In any case, the maximum amount allowed in a product is controlled. The atmosphere inside cars and drinking water may contain some vinyl chloride if it leaches from plastic components or from

How vinyl chloride causes cancer

Vinyl chloride is a simple volatile industrial chemical, which in the liver is converted by an enzyme (cytochrome P450) into a reactive product. The reactive product can interact with proteins and, more importantly, with DNA in the nucleus of liver cells. The metabolite interacts with two of the four bases in the DNA backbone, adenine and cytidine. These bases form both part of the genetic code and pairs with other bases in the DNA. The addition of the vinyl chloride alters the bases and reduces their ability to form pairs, with the result that the code in the DNA is misread which leads to mutations. When the cells of the liver divide or that part of the genetic code is expressed, mistakes, or mutations, creep in. It is now known that after liver cells are exposed to vinyl chloride some of the mutations occur in the oncogenes or cancer genes. Mutant proteins produced from these genes can be detected in the blood of workers who were exposed to vinyl chloride and suffered from the liver cancer. The level of such proteins seems to correlate with the level of exposure to vinyl chloride.

Strangely, one study found that increasing the dose of vinyl chloride did not lead to a linear increase in the incidence of tumours in rats. This is because the metabolism is saturated at higher doses, and so production of the reactive metabolite does not increase.

pipes. Ground water can also contain traces of vinyl chloride if it is contaminated with trichloroethylene, as this chemical is broken down by bacteria to produce vinyl chloride. The amounts to which we are exposed are, in general, very small.

The type of liver cancer caused by vinyl chloride is so rare that an increase in cases of it would be detected relatively easily. Studies around PVC plants in Sweden did not find any differences in the incidence of disease. Thus it would seem that vinyl chloride is not a risk for the general public, and conditions in factories making and using it ought to be such that it is no longer a hazard in the workplace either.[3]

There are many other industrial chemicals that are either known or suspected carcinogens in humans (see Table 6).

Table 6 Common occupational carcinogens

AGENT	ORGAN AFFECTED	OCCUPATION IN WHICH IT IS USED
Asbestos	Lung, pleura	Insulation workers, miners, construction and shipyards
2-naphthylamine	Bladder	Dye, rubber, paint manufacture
Benzene	Bone marrow	Shoemakers, dye users, painters
Chromium	Nose, lung, larynx	Welders, chromium production and processing
Coal tar	Lung, larynx, skin, bladder	Gas house workers, coke oven workers, coal tar and pitch workers
Nickel	Nose, lung	Nickel smelters, mixers, roasters, electrolysis workers

Methylisocyanate and the tragedy of Bhopal

Fire a blowtorch at my eyes, pour acid down my throat. Strip the tissue from my lungs. Drown me in my own blood. Choke my baby to death in front of me. Show me her struggles as she dies. Cripple my children. Let pain be their daily and only playmate. Spare me nothing. Wreck my health so I can no longer feed my family. Watch us starve. Say it's nothing to do with you. Don't ever say sorry. Poison our water. Cause monsters to be born amongst us. Make us curse God. Stunt our living children's growth. For seventeen years ignore our cries. Teach me that my rage is as useless as my tears. Prove to me beyond all doubt that there is no justice in the world. You are a wealthy American Corporation and I am a gas victim of Bhopal.[4]

This moving poem is by a survivor of the terrible industrial accident in the Indian city of Bhopal in December 1984. Just after midnight on 2 December 1984, a large cloud of vapour leaked from a factory in Bhopal, spreading out over a nearby area of shanty town, where thousands of inhabitants were sleeping. In all, between 30 and 40 tonnes of this toxic chemical were released. Many died of asphyxiation as the highly irritant gas damaged their lungs, causing fluid to accumulate, which effectively drowned them. Others survived the initial onslaught of the vapour but succumbed later to the combined effects of damage to the lung and to other organs (see box). Those who survived did not necessarily recover completely, as they continued to suffer from lung disease and blindness or poor eyesight, as well as impaired immune systems and reproductive problems. The irritant nature of the chemical meant that it attacked membranes in the lungs and eyes and the skin.

What of the death toll? The most conservative estimates were 3,000 at the time, with about 200,000 injured. More recent estimates suggest that 16,000 people may have died as a result of the accident. How did it happen? The factory, which was owned by the American Union Carbide company, was making Sevin, or carbaryl, an insecticide. Methylisocyanate was one of the components of the synthesis, of which a large amount was stored in a tank. It seems that water was able to enter the tank via a newly fitted pipe because of a faulty valve. The water caused a reaction when it mixed with the methylisocyanate, producing heat, which raised the temperature to such an extent that the liquid chemical boiled and was converted into a gas, which overwhelmed the containment system. It seems there was no trap to contain the excess, no flare system, and the scrubber system was overloaded. It was clear that the safety systems were inadequate.

The company initially claimed that sabotage was to blame, but no evidence of this has appeared; they also maintained that the Indian subsidiary was wholly responsible for the plant, but this is contested. There is little doubt that the plant design was faulty, and inadequate to contain a runaway reaction, as suggested by scientists and engineers. Even before the event the company's own engineers concluded that 'a real potential for a serious incident exists'[5] in the methylisocyanate tank at its West Virginia plant.

The story of this, the world's worst chemical disaster, is salutory: it occurred only relatively recently and demonstrates the dangers of chemical plants when management, local or distant, takes a cavalier attitude towards safety and shows a lack of respect for and knowledge of the chemicals they are using and making. It is a tragedy because such events could so easily be averted. In some respects the event in Bhopal was similar to that at Seveso, in that a process in a factory went wrong and released a chemical into the surrounding area at a time when few if any workers were around to react to the emergency. In Bhopal, however, the chemical had devastating immediate and lasting consequences for the local population. This was for two reasons: first, the quantity released was very large and, secondly, the factory was sited close to an area where a large number of people were living. One could argue that no chemical plant should be sited near urban population centres. When large quantities of toxic chemicals are stored or used there must be adequate safety systems in place to deal with them. Accidents will always happen, but their impact could at least be minimized.

A second lesson from the disaster is that toxicological knowledge of the chemical would have helped in the treatment of the victims. There was insufficient knowledge of the toxic effects and, while there is no true antidote, the lack of knowledge about the chemical meant that victims were treated inappropriately by some local medical centres who thought they had been poisoned with cyanide.[6] Data from experimental studies might have helped both in the treatment and the appreciation of the hazardous nature of this chemical. Studies in experimental animals after the event showed that even low levels of the vapour were extremely irritating to the respiratory systems in mice. Other studies in animals revealed that the chemical could stimulate the immune system, as it had done in the human victims. The long-term effects of methylisocyanate have yet to be fully determined.

How methylisocyanate kills its victims

Methylisocyanate is a volatile, reactive chemical which is highly irritant. Its reactivity means that when humans come into contact with it, it combines with components of body tissues such as the protein in eyes, skin, and lungs. It especially reacts with the sulphur atoms in proteins. These interactions disrupt the lining of the lungs and allow water to enter from surrounding tissues and blood. The result, known as **pulmonary oedema**, means the victim can drown because their lungs fill up with liquid. Studies in mice exposed to low levels have shown that the chemical is a very potent irritant to the lungs.

In the lungs there is quite a lot of a sulphur containing substance called glutathione which functions to detoxify noxious chemicals. This can react with methylisocyanate and remove it. However it seems this detoxication is but a temporary respite because the product of this interaction carries the methylisocyanate away from the lungs to distant sites in the body such as the bones and muscles where it is released and causes damage. The detoxication product has been called a 'chemical taxi'.

Studies in animals revealed that methylisocyanate, by reacting with and changing proteins in the body which would then be recognized as foreign by the immune system, caused antibodies to be produced to the altered proteins. This was also found in the human victims of the Bhopal disaster.[7]

Cadmium

This metal is widely used in industry in alloys, plating, and batteries, as well as in the pigments in paints, inks, enamel, rubber, and plastic, even though it is no longer used in pigments or paints for cooking utensils, or in plastic or rubber that may be used by children. Cadmium occurs naturally; soil contains detectable levels, and vegetables can accumulate the metal. Cigarette smoke is a significant source of exposure to cadmium for smokers, and it is more readily absorbed from the lungs than the stomach or intestine.

Dangerous levels of human exposure from diet have occurred, most notably in Fuchu, in northern Japan where rice was eaten that was contaminated with cadmium derived from an old zinc mine and present in soil and water. Low calcium intake and vitamin D deficiency may also have been factors. The victims suffered from brittle bones, and it became known as itai-itai disease from the Japanese for ouch-ouch (see box). The village of Shipham in Somerset was, similarly, found to have very high levels of cadmium in the soil, which also derived from an old zinc mine.

From the toxicological point of view, inhalation of cadmium metal fumes or cadmium oxide is particularly hazardous. Apart from causing damage to the lungs, the metal can enter the body and subsequently damage the kidneys. Inhalation of cadmium fumes over long periods of time leads to chronic bronchitis, emphysema, and possibly lung cancer. In some animals cadmium is known to damage the testes; it can cause tumours in the testes after repeated exposure.

Cadmium stays in the body for years and much of it is located in the kidneys, bound to a protein (see box). It is eliminated from the body very

Cadmium toxicity and itai-itai disease

Cadmium is chemically similar to the metal zinc, and is absorbed from the gut in a similar way using the same system. It can interfere with zinc in the body and consequently may affect the male reproductive system where zinc is important. Cadmium interferes with the metabolism of calcium too, a critical mineral in the body with which it also has similarities. This interference leads to loss of calcium from the bones which then become brittle (osteomalacia). In the case of itai-itai disease, which was the result of cadmium poisoning in Japan, the exposure to cadmium was accompanied by a deficiency of vitamin D which made the problems with the bones worse.

Cadmium, like zinc, binds to a protein, metallothionein, in the body. The presence of cadmium increases the production of this protein by many times (the body increases detoxication, when threatened, which may be a reason for the development of tolerance). Although the binding removes the cadmium and effectively detoxifies it, the cadmium-protein complex is deposited in the kidneys and cadmium is eventually released, whereupon it seriously damages the organ.

slowly, taking sixteen to thirty years for half of the amount in the body to be lost. Treatment of poisoning with metals like cadmium involves the use of chelating agents, which bind to the metal, forming a complex that is then excreted into the urine.

Solvents

There are a wide variety of solvents used in industrial processes, and exposure can also occur in other occupational settings such as laboratories and workshops. Solvents can have toxic effects ranging from narcosis, irritation, and degreasing of skin to effects on major organs such as the central nervous and reproductive systems. The trend of inhaling solvents, for their narcotic effects, which became popular in teenagers will be discussed on p. 198.

In the case of many solvents their use without protective clothing can lead to loss of fats from the skin and dermatitis. Solvents also readily pass into the central nervous system, where they cause effects ranging from loss of consciousness at high exposure levels to euphoria, giddiness, and disorientation—not desirable in a factory situation! Prolonged excessive exposure can be fatal, and there have been many deaths of workers exposed to solvents in confined spaces.

Apart from the effects of acute exposure, lower level, chronic exposure can also have toxic effects. For example, it is argued that solvent disease includes impairment of memory and coordination, and changes in personality, but this is a contentious issue. Effects on organs like the liver and kidney certainly do result from both acute and chronic exposure to certain solvents, for example carbon tetrachloride, which was used in dry cleaning but has now been replaced.

A solvent that causes problems for workers who have been exposed is hexane, used in paints, varnishes, glues, and in printing. Workers in the shoemaking, laminating, and cabinet making industries have suffered damage to the peripheral nervous system. The effects observed in humans can be reproduced experimentally in animals. The disease has a gradual onset, with first effects in the lower extremities. The effects result from the swelling and degeneration of the nerve fibres. The mechanism by which hexane causes these effects is now partly understood (see box), and so solvents that are likely to cause a similar problem can be identified. There is also evidence that exposure to hexane can lead to disturbances of the male reproductive system, by damaging the testes.

A number of solvents, such as toluene, which is widely used in glues, and

Some solvents get on your nerves

Hexane is a simple organic chemical, a volatile liquid used industrially as a solvent. Once absorbed into the body, it is changed by metabolism into other products. One of these products, in which the hexane has been oxidized, is able to react with proteins in the nerves, and this is what underlies its toxic effects. Similar solvents that do not form this product are not toxic to the nerves, while those that do, predictably cause the same type of toxic effects.

That hexane is soluble in fat allows it to distribute to most tissues in the body, including nerves. The destruction of motor nerves leads to both loss of feeling and weakness in the fingers and toes.

some chlorinated hydrocarbons, such as trichloroethane, cause adverse effects on the heart after acute exposure, leading to arrhythmias. Trichloroethane also causes toxic effects in the heart after chronic exposure.

Another solvent widely studied and known to be the cause of serious effects is benzene, which is found in petrol and is also used in the shoemaking industry. Extensive studies have been carried out in Turkey, for example, where it has been used in shoemaking. The use of benzene has now been phased out in many countries but it continues to be used in others. It has been extensively used as a starting point for the production of other chemicals and as a solvent in the manufacture of rubber, paint, and plastics, and in printing. It has been added to petrol as an alternative anti-knock additive to lead compounds.

The main hazardous effects of benzene are to the blood system. Workers chronically exposed suffer from anaemia, due to damage to and sometimes complete destruction of the cells in the bone marrow, where the cells found in the blood are formed. Benzene exposure can lead to low levels of both red and white blood cells. A more serious effect is leukaemia, which is cancer of the blood system. Because of the hazards it poses, the use of benzene has been dramatically curtailed.

The solvents most commonly used in paints are glycol ethers, which are known to cause damage to the male reproductive system by destroying the cells that produce sperm in the testes.

Finally, another halogenated solvent that causes toxic effects after acute exposure is dichloromethane or methylene chloride. It is used for degreasing engines and metalwork, and is the main constituent of paint stripper. If it is used in confined, enclosed spaces, such as rooms without adequate ventilation, it can lead to serious, unexpected toxic effects. As the solvent is breathed in by the worker it becomes localized in body fat. Later the dichloromethane emerges from the stores in the fat into the bloodstream and liver, where the solvent is metabolized. The product is carbon monoxide, which is highly poisonous and can cause death from interference with the availability of oxygen (see, further, pp. 185–7). The victim may collapse some time after the exposure has ended, not from the effects of the solvent, but from the effects of the carbon monoxide produced from it.

It should be mentioned here that pesticides are significant in terms of the risks of chemical exposure for workers both in their production and their use. There is a large variety of these, ranging from organophosphates to organochlorines, and including paraquat, for example. The organophosphates probably cause the most problems for workers after acute exposure, from the point of view of skin irritation and of more serious adverse effects on the nervous system. There has been much debate as to whether agricultural workers exposed over time suffer permanent damage to the nervous system, and this is still unclear. For more detailed discussion of pesticides, see Chapter 4.

Asbestos

CASE NOTES

A legacy of dust

Geoff, a 50-year-old builder, was ready for retirement when one day at work in 2002 he became breathless. After a series of tests, he was diagnosed with mesothelioma or cancer of the pleura, the lining of the chest, a relatively rare cancer. George was about the same age when he had a sharp pain in the left side of his chest. Tests showed that he too had the same cancer. What did these two men have in common? Both had been exposed to asbestos earlier in their working lives. Geoff had been a carpenter, and had been involved in sawing asbestos roof soffits in the 1970s. As a young man George had worked on the brakes of vans, lorries, and other vehicles, whose linings were made of asbestos and whose brakes would be full of dust when they were worn.

Although the use of asbestos has been greatly reduced and the disease asbestosis has declined, the incidence of mesothelioma has increased threefold since 1990.[8]

Asbestos is a fibrous silicate, a substance that has many uses. It is mined in places like South Africa and Canada, and miners are exposed to it during the course of their work. Asbestos is also used in the manufacture of other products and especially in the construction of warships, buildings, and installations such as power stations—although less commonly now. Asbestos has some important properties, in particular its lightness and ability to withstand heat. An excellent insulator, it has been used in both heat and electrical insulation. It can be woven and incorporated into other materials such as cement, which means that exposure to asbestos can occur when old buildings are being demolished or renovated. Because of its widespread use, there is the potential for the general public to be exposed, but for the most part those with significant levels of exposure will be workers involved in its mining or use. The families of exposed workers can also be at risk, and it is now known that those living near factories using asbestos have as a result suffered from mesothelioma. This was shown by two epidemiologists in the 1960s.[9]

There are various forms of asbestos and not all are equally hazardous. White asbestos, or chrysotile, is the form most commonly used and is the least harmful, being relatively inert in living systems. Crocidolite, or blue asbestos, is especially hazardous and may contaminate white asbestos.

Asbestos fibres may be found in water where mining takes place, and as it was once used in filters for liquids, it could also be present in drinks. Fibres were first found in drinking water in Duluth, Minnesota in 1973. The water derived from Lake Superior, into which tailings from an asbestos mine located sixty miles away were being dumped which eventually found their way into the drinking water. Evidence exists that such exposure through the gut may increase the risk of cancers.

The main route of exposure, however, is by inhalation of the fibres. Asbestos causes three types of adverse effects: asbestosis, bronchial carcinoma, and malignant mesothelioma. The latter two are types of cancer, with mesothelioma associated only with exposure to asbestos. The mechanisms by which asbestos causes cancer are not fully understood, although some features are known (see box), such as the size of fibre (only fibres of certain sizes are likely to cause disease). All forms of asbestos can cause asbestosis and bronchial carcinoma if the exposure level and length of exposure are high enough.

For abestosis and bronchial carcinoma to occur, there has to have been prolonged exposure to higher levels than with mesothelioma. After such exposure, perhaps of a miner or someone working in a factory manufacturing products from asbestos, the lungs become progressively inefficient. Breathing becomes laboured and those suffering this cannot endure much exertion.

These are now recognized industrial diseases and workers at risk can be monitored (see below). Fibres can be detected in the lung tissue and in the

Asbestos: harmful fibres

Although asbestos is an inert mineral, some forms of it can be a very hazardous material. If its fibres are incubated with human red blood cells, the cells will undergo lysis, that is, they will break open. Constant inhalation of the fibres leads to damage in the lungs, known as fibrosis, where the lungs become full of a fibrous protein called collagen which stops them functioning efficiently. The fibres appear to be engulfed by special cells in the lungs called macrophages whose job it is to remove the particles that we all inhale. Asbestos causes the destruction of these cells, leading to irritation and reactions in the surrounding tissue of the lung. This stimulates the growth of new tissue involving the fibrous protein and a scar is produced.

The length of the fibres seems to be important. Fibres shorter than 10–20 microns (μm) do not cause this fibrosis because they can be fully engulfed by the macrophages, whereas longer ones cannot.

The way in which asbestos causes cancer is not clear. It may be the constant irritation it causes plus the presence of impurities in the fibres, such as iron salts, which may lead to the production of reactive oxygen which can damage DNA. The size of the fibres again appears to be important in relation to the cancer (mesothelioma and bronchial carcinoma), with those 5 μm long and 0.3 μm in diameter being the most active. Because the fibres tend to remain in the lung tissue, a short period of high exposure *could* be sufficient to cause cancer.

sputum long after exposure, which allows a diagnosis to be made and a cause to be determined. Hence compensation for the worker is more likely, although the experiences of many of those afflicted in the UK has demonstrated what a tragically slow process it is. It is relatively difficult to find evidence of exposure to a chemical many years after the event because most chemicals are lost from the body in a fairly short time, while the damage remains. Even **DNA adducts** or protein adducts in cells last only a few months.

Bronchial carcinoma also develops in those who have prolonged heavy exposure to asbestos and occurs in approximately 50 per cent of those who have asbestosis. The risk of contracting these diseases, and their severity, are related to the exposure level, with the length of exposure being another factor.

Mesothelioma is a rare form of cancer of the pleural cavity and seems to occur only in those exposed to certain types of asbestos, especially crocidolite. The crocidolite from the north-western Cape Province of South Africa is more potent than that from the Transvaal. It does not require prolonged exposure to high levels to develop the disease, and people not occupationally exposed have developed it. The latent period is long, typically about thirty years, but once diagnosed it is rapidly fatal.

It has been estimated that cancer deaths due to asbestos will peak at several thousand per year in the UK alone, and at over 10,000 per year in the USA over the thirty-year period from 1983. The deaths from mesothelioma alone in the UK are believed to number 1,800 and will not reach a peak for another ten years. It is the most common occupational cause of death in the UK. Hence, while deaths from most other cancers are decreasing, deaths from asbestos-related cancer are increasing.

The association between lung disease and asbestos exposure was discovered as a result of the use of epidemiology (see box and pp. 287–91).

Epidemiology: how the hazards of industrial chemicals can be detected

The potential of particular chemicals to cause toxic effects can be established by carrying out experiments on animals. Alternatively, they can be detected by studies of humans. A particular group of workers can be studied specifically to look for unusual diseases. This may simply be part of the policy of an industrial company to monitor workers.

Sometimes an astute scientist or physician may notice something significant, such as the occurrence of a few cases of an unusual disease. This may be sufficient to warrant a study of the occurrence of that disease in the workers in a particular factory in comparison to those in another part of the factory or to office workers. For example, a doctor in South Africa noticed that some of his patients were suffering from an unusual tumour, mesothelioma. He investigated this by collecting detailed information about patients with the disease, including where they had previously worked. He found that thirty-two out of thirty-three cases had probably been exposed to asbestos, in particular crocidolite. His study, published in 1960, alerted the medical profession to the possibility of asbestos causing the disease. This hypothesis could then be tested by more studies either in other human populations or in animals or both.

The study of human populations is called epidemiology, and it is used to determine the agents that cause disease, including chemicals such as drugs and industrial and environmental chemicals. The potential toxicity of industrial chemicals has to be evaluated *in vitro* and in animals, but the long-term exposure that humans may experience cannot always be simulated in these experiments. Thus epidemiology is an essential tool.

Exposure permitted in the workplace has been controlled by law, having been set in 1969 at about 2 fibres per millilitre of air which workers may breathe during a working day for a working lifetime. This still represents a huge number of fibres breathed in during a normal eight-hour working day: 10 million at a resting breathing rate, while those involved in heavy work breathe in several times more air. In some countries exposure to the more hazardous forms such as crocidolite is now prohibited.

The evidence also suggests that workers who smoke and are also exposed to asbestos are much more likely to suffer cancer than those who don't smoke. The incidence of cancer in this group is much higher than the incidence in the smokers and exposed workers combined. This is an example of synergy (where the combined effect is more than additive).

Common effects of industrial chemicals on workers

The most common types of damage caused by chemicals in workers is probably to the lungs and to the skin. These are the main sites of the body that are exposed to chemicals. The most common industrial disease is dermatitis, which accounts for more working days lost than all the other industrial diseases together. If solvents or chemicals that workers handle come into contact with their hands, this can lead to irritation. More serious is contact dermatitis which results from repeated exposure over time. Sometimes serious allergic dermatitis occurs, where the chemical acts to sensitize the skin. Workers in the paper and printing industry are prone to skin diseases, and photographic chemicals, metals, resins, and coal tar derivatives can all sensitize the skin. As we have seen, certain chemicals, such as dioxins, can cause a particularly severe form of acne, chloracne.

These effects can be avoided by the use of containment, protective clothing, and barrier creams. Simply handling a metal like nickel can cause dermatitis (nickel-itch). This can also result from wearing jewellery made from nickel. Lung damage due to chemicals results from inhalation of dusts, vapours, or gases. These may primarily affect the lungs themselves, especially if they are irritant, but can also lead to absorption of the chemical into the blood and damage to other organs. As with the skin, the repeated inhalation of certain chemicals can lead to sensitization which causes occupational asthma. Other breathing disorders, such as emphysema, can result from inhaling irritant or damaging chemicals, for example cadmium fumes or asbestos. Again this can be avoided by containment of the chemical, filtration of air, and the use of protective clothing. Inhalation of cadmium fumes can also damage other organs, like the kidney.

Legislation for protection of workers

Legislation now exists to protect workers from exposure to chemicals, by specifying the acceptable exposure levels for many chemicals. Workers are also usually monitored with frequent health checks and, in some cases, specific analyses for the effects of the substances they are working with, for example organophosphate insecticides. Those working in an environment in which they may inhale particles or noxious gases or fumes will have regular tests of lung function such as vital capacity, at least in responsible companies. Monitoring workers in these ways yields data that can be analysed to detect trends and possible adverse effects of chemicals not previously known. Animal toxicity data for certain chemicals is also available and is used to determine no observed adverse effect levels (**NOAELs**), on which the maximum exposure limit (MEL) in the UK, and the threshold limit value (TLV) in the USA are based.

8

Under the Sink and in the Garden Shed

Household Poisons

CHEMICALS are used as drugs and in industry, and they have just as important a place in our homes. Some chemicals may, however, be unwanted contaminants of the home. In this chapter I shall consider a selection of chemicals used in the home in more detail to explore some of the substances we may be exposed to.

Taking a look under the sink in my own home, I found a range of products with an array of chemical constituents:

a range of cleaning products, containing non-ionic detergents and preservatives;
ant killer, containing the insecticide chlorpyrifos;
bleach (sodium hypochlorite);
dishwasher powder, containing disodium silicate;
hydrogen peroxide;
insect spray, containing the insecticide permethrin and also pyrethrum and deltamethrin;
kettle descaler (formic acid);
methylated spirits (ethyl alcohol and methyl alcohol);
oven cleaner, containing caustic soda (sodium hydroxide).

A number of these products contain chemicals that are irritant acids or alkalis, namely sodium hydroxide, disodium silicate, formic acid, and bleach. Hydrogen peroxide is a strong oxidizing agent. Oxidizing agents are reactive chemicals that cause chemical changes in other molecules. They can alter cells and damage skin and other tissues if they come into contact with the body. They are particularly hazardous if taken in by mouth, as the mucous membranes lining the mouth and oesophagus are more sensitive and delicate than the skin, which has an outer layer designed to protect the body.

The insecticides could have various effects on the nervous system but are unlikely to cause harm unless drunk in significant quantities. Chlorpyrifos is an organophosphate insecticide, and so could inhibit the enzyme acetylcholinesterase in the blood and nervous tissue. It can have a cumulative effect, and so after repeated use, say on a daily basis, the level of inhibition may be sufficient to cause unpleasant or even life-threatening effects (as described on pp. 98–103). These include difficulty in breathing, muscular weakness, and convulsions. Organophosphates vary greatly in their potency, however, and the more toxic are now less readily available. Those for home use are the least poisonous to humans and other mammals.

The methylated spirits would cause first inebriation and then drunkenness. Depending on the dose, more serious effects could occur as a result of the methyl alcohol content (see below).

A tour round my garage revealed some more chemicals:

antifreeze (**ethylene glycol**);
creosote;
herbicide (paraquat and diquat);
paint stripper, containing sodium hydroxide and methylene chloride;
paint varnish;
paraffin;
petrol;
screenwash (methyl alcohol);
white spirit (solvent).

The herbicide would be irritant to the skin but potentially lethal if taken by mouth (see pp. 104–6). Both the antifreeze and the screenwash would also be potentially lethal if drunk (see below). All the solvents (white spirit, petrol, and paraffin) would be hazardous if drunk, especially as they could be drawn into the lungs. Cases of children drinking white spirit have occurred, in which the main effect was in the lungs, where the solvent easily spread through the small air spaces and changed the ability of the lungs to function properly. Creosote is irritating to the skin and can cause acne-like eruptions with repeated exposure, like other solvents and chemicals used in industry (see pp. 122–30).

The use of paint stripper may lead to damage to the skin from both the sodium hydroxide (caustic soda) and the solvent, methylene chloride. Another, more serious hazard from the paint stripper is the danger of using it in a confined space and inhaling the solvent fumes. Apart from the possibility of being overcome by the fumes, the solvent will find its way into body fat, from where it can then be slowly released, perhaps to cause serious toxic effects later, as illustrated by the case below.

CASE NOTES

A close shave

A female student, aged 20, complained of nausea and severe headache while using a paint remover in a poorly ventilated room. An hour later, after she had left the room, she felt dizzy and shortly afterwards she lost consciousness. She regained consciousness on admittance to hospital but was disorientated. Her skin and **mucosae** were cherry red in colour. Her blood was found to contain 50 per cent carbon monoxide. She received oxygen and was discharged from hospital on the third day.[1]

What is the explanation for this? Another story illustrates the more serious effects that can result from exposure to this solvent.

CASE NOTES

Too Little Ventilation

A man, aged 25, had been exposed to the fumes of a paint stripper, Nitromors, which contained methylene chloride, for a period of three to four hours; except for mild conjunctival irritation, he had no complaints at the time. After forty-eight hours, however, he felt unwell and complained of swelling and stiffness of his ankles, knees, and wrists. A non-itchy skin rash appeared at the end of another twenty-four hours. During the following eight days he found increasing difficulty with simple intellectual tasks and was unable to concentrate. He became short of breath on minimal exertion. At ten days after exposure he had nocturia, polyuria, and loss of weight (2.7 kg). On examination he had 2 per cent glycosuria,* and a random blood sample showed a glucose level of 218 mg/100 ml.

He required diabetic treatment for six months, and even then his skin rash was not completely cleared. Slow mental improvement kept him away from his work as an accountant for six months.[2]

These effects may have been due to a combination of the solvent itself and the carbon monoxide produced from it.

* A high level of glucose in urine and blood and the production of a lot of urine are signs of diabetes.

How paint stripper can cause carbon monoxide poisoning

Methylene chloride is used in industry as a degreasing solvent and is found in most paint strippers. When used in a poorly ventilated or confined place, the volatile chemical is inhaled and rapidly absorbed from the lungs into the blood, which then distributes it to other areas of the body. Due to the way the circulatory system is organized the first organ exposed after the heart will be the brain, and this type of solvent, which is soluble in fat, can easily enter the brain. If sufficient levels are absorbed, the person exposed may suffer from the effects of the solvent on the brain: they will feel giddy or unsteady, or even lose consciousness (narcosis). Much of the solvent will be distributed by the blood to places in the body where fat is stored, while the remainder is exhaled from the lungs. After exposure to the solvent has ended, the methylene chloride will start to emerge from the fat, where it was deposited earlier, and to appear in the blood.

Some of the solvent will be metabolized in the liver, and one product of the metabolism is the poisonous gas carbon monoxide. There have been cases of severe carbon monoxide poisoning due to prolonged exposure to methylene chloride. Someone exposed to the solvent for two to three hours may achieve a level of 15 per cent carbon monoxide in the blood, which would cause only mild effects in a healthy individual but possibly more severe problems in someone with heart or lung disease.

Carbon monoxide: a common but lethal poison

As the above cases demonstrate, carbon monoxide poisoning can arise from quite unexpected sources and can be serious. Indirect poisoning from this source of the gas is probably relatively uncommon, but direct poisoning with carbon monoxide is probably the most common type of accidental poisoning that affects the general population. There are probably several hundred poisonings from carbon monoxide every year in the UK alone, some of which are fatal. Some mild cases may well go undetected as the main symptom is a headache.

Carbon monoxide is a gas that is produced in various ways in industry where it has particular uses, but it can also be easily formed in the home. When coal, wood, petrol, or natural gas burn they can all produce carbon monoxide if the level of air available is reduced. Car exhausts will always contain some of the gas (at a level of between 4 and 10 per cent), and inhaling this has long been a method of suicide. One of the methods commonly used to commit suicide is to put a tube into the exhaust from the car, feed the other end into the car, and sit in the car with the engine running and the windows closed. Alternatively, the garage is closed and the car engine allowed to run in the enclosed space. More recently, car engines have become more efficient and hence produce less carbon monoxide, and they are tested for carbon emissions.

There are many accidental poisonings from carbon monoxide in homes, and such deaths occur throughout the world. Perhaps the most famous victim was the tennis star Vitas Gerulaitis, who died from carbon monoxide poisoning when he slept in a room adjoining one containing a faulty swimming pool heater. Carbon monoxide poisoning can also often kill victims trapped in burning buildings (although it may be mixed with other poisonous gases).

The most common cause of accidental poisoning with carbon monoxide is a fire, stove, or boiler that is inadequately ventilated. When it is lit, it produces the toxic gas which accumulates in the house, especially in modern homes which are often double-glazed and have little ventilation, and especially in the winter. One reason for poor ventilation may be birds nesting in the chimney, which decreases the availability of air. In a recent case a family lit a fire in a rented holiday cottage, but unbeknown to them the chimney had become blocked with a bird's nest, and all of them died of carbon monoxide poisoning.

There have also been poisonings, some fatal, of people working on car engines in enclosed spaces while running the engine. People, in general, do not seem to be very aware of these dangers. Poisoning of drivers by carbon monoxide as a result of a faulty exhaust system can be especially dangerous. Because the effects are slow and insidious, the driver will probably not realize he or she is being affected by the gas, and judgement will be impaired, leading to accidents with potentially fatal consequences.

As illustrated by the story opposite, one of the problems caused by carbon monoxide poisoning is that the victim becomes very weak and is often unable to summon help or to escape. The reason for this is lack of oxygen in the body (see box below).

Carbon monoxide is also especially dangerous because it does not smell or taste of anything, is not irritant, and has no colour, so the victim is usually completely unaware of the fact that he or she is breathing it in. It is also very potent: a concentration in the air of only 0.1 per cent is sufficient to cause severe poisoning and death if breathed for some time. Breathing this level for one hour would give the body a lethal dose of the

CASE NOTES

Poisoned to death by a faulty boiler

■ By N. Chohan

A young accountant suffered a slow and silent death after she was gassed by a central heating boiler, a court heard yesterday. Amanda Rhys-Davies, 23, was found slumped on the bathroom floor of her flat. Her boyfriend, Wayne Court, 26, was by her side. They lay helpless, going in and out of consciousness over three days, the Old Bailey heard. Miss Rhys-Davies died of carbon monoxide poisoning while Mr Court eventually recovered. Mr Court said he came home from work after feeling ill and went to bed. Later, he felt sick and went to the bathroom, where he found his girlfriend on her hands and knees. Miss Rhys-Davies then touched him 'in the same way someone would if they were distressed', he said. Mr Court then passed out next to her. The next day he heard someone banging on the front door but didn't have the energy to answer. The following day, his mother and sister smashed the glass to the front door and shouted to him to open it, which he managed to do. Miss Rhys-Davies, who was still breathing, was taken to a specialist poisons unit in Portsmouth but she died on February 3 last year. Nigel Lithman, QC, prosecuting, said Miss Rhys-Davies died as a result of the landlord's gross negligence in not having the boiler checked and serviced. He said the gas boiler was fitted with a 'sub-standard' flue. As well as carbon monoxide being pumped back into the flat because the flue was dirty, the landlord had removed a ventilator when replacing windows.

Metro [London] (8 March 2001), 7

gas. A victim with heart or lung disease may well succumb after breathing in even lower levels than this.

The first sign of poisoning will be a severe headache, followed by nausea, vomiting, and breathlessness. After this, the victim suffers confusion, weakness in the muscles, and fainting on exertion. These symptoms of weakness and fainting mean that the victim is often unable to move very far and therefore cannot escape the poisonous atmosphere. One characteristic sign is that the victim of poisoning will be a cherry red colour as a result of the carbon monoxide becoming attached to the blood protein haemoglobin. (**Carboxyhaemoglobin** is a different colour from **oxyhaemoglobin**, which is normally present in the blood.) The final stage is unconsciousness and coma leading to death, possibly with alterations in heart rhythm and damage to both heart and brain (see Table 7). These effects are mostly due to the lack of oxygen, as the effect of carbon monoxide is to deprive the body of oxygen, not because of high

Table 7 Relation between concentration of carbon monoxide in the blood and symptoms observed

CONCENTRATION OF CARBON MONOXIDE IN THE BLOOD (% saturation of haemoglobin)	SYMPTOMS OBSERVED
Up to 20	May show no symptoms but, above 10% level, mental and physical tasks can be affected.
20–30	May have slight or throbbing headache which persists. Weariness. Exertion can cause faintness.
30–40	Throbbing and fullness in head; very severe headache which incapacitates. Nausea. Faintness without exertion. Blood pressure falls and pulse rises. Vomiting.
40–60	As above, plus weakness and incoordination. Mental confusion. Inability to walk.
60	Loss of consciousness. Convulsions. Death, unless removed from atmosphere rapidly.
80	Rapid death, unless treated immediately.

concentration, but because it excludes oxygen from specific places in the body (see box for explanation).

Coal gas contains a significant amount of carbon monoxide (5–10 per cent), and when it was widely used (before natural gas became available) poisoning was much more common, with over a thousand death each year in the UK. It was also a commonly used means of suicide, victims often putting their heads into the oven of the cooker and turning on the gas.

We know the effects of carbon monoxide and how they relate to the dose as a result of remarkable experiments carried out by the scientist J. B. S. Haldane, which were published in 1895. Using himself as the subject, he exposed himself to increasing levels of carbon monoxide and carefully recorded the symptoms. The amount of carbon monoxide in his blood was also measured to find a direct correlation between the effects observed and the concentration of the gas to which he was exposed. When the level of carbon monoxide in his blood was 27 per cent his vision was dimmed and he staggered in an attempt to exert himself. An exertion such as getting out of a chair could lead to faintness. At 40 per cent of carbon monoxide, even pricking a finger to take a blood sample was found to be very difficult by another intrepid scientist. When Haldane had 56 per cent carbon monoxide in his blood, he could not walk. His short-term memory was affected and he was unable to remember his actions during the poisoning experiment. Subsequent experiments by other courageous (or foolhardy!) scientists, as well as information from victims, confirmed

Why carbon monoxide is such a deadly killer

Carbon monoxide is a simple molecule. Chemically the molecule is written as CO, one atom of carbon combined with one atom of oxygen. Its simplicity belies its poisonous nature. It is poisonous at very low levels (0.1 per cent of the air), because it binds so avidly to haemoglobin, the red pigment in the blood. Haemoglobin, found inside red blood cells, is designed to bind the oxygen in the air we breathe in through our lungs. The blood then transports this oxygen, bound to haemoglobin, to organs and tissues like the brain, the heart, and the muscles, where it is required for the metabolic reactions that produce energy. Without oxygen, humans, like most other living organisms, die.

Carbon monoxide has some similarities to oxygen in the chemical sense, and consequently it can also bind to haemoglobin. However, carbon monoxide binds 220 times more strongly than oxygen. Thus, even though each haemoglobin molecule can bind four molecules of oxygen, at poisonous concentrations at least two molecules of carbon monoxide will be attached to the haemoglobin. Thus less oxygen is available to be transported to the places where it is needed. Furthermore, carbon monoxide also changes the haemoglobin so that it cannot so easily release the oxygen it is carrying when it is needed. The overall result is that the organs and tissues become starved of oxygen.

A level of 0.1 per cent carbon monoxide in the air breathed for 80 minutes will result in half (50 per cent) of the haemoglobin being saturated with carbon monoxide. This may sufficient to be fatal. A level of 60 per cent will certainly lead to unconsciousness and death.

Because carbon monoxide binds to the iron that is the essential part of haemoglobin, it also binds to other proteins containing the metal. This has the effect of inhibiting the metabolic processes in cells that produce energy, further incapacitating the poisoned patient.

these results. Thus we now know and understand the major effects of this toxic gas very clearly and, because measurement of the level of carbon monoxide in the blood is simple and routine, treatment is relatively straightforward.

Although carbon monoxide binds very strongly to blood, fortunately this is not permanent, and it can be replaced by oxygen when the victim breathes uncontaminated air. Thus the victim must be removed from the scene of poisoning as quickly as possible into fresh air. With severe poisoning, the patient is given oxygen and in some cases, and if available, a hyperbaric chamber may be used to increase the pressure of the oxygen.

The use of oxygen, and especially oxygen at higher pressure, will increase the speed at which the carbon monoxide is removed from the body.

The long-term effects of carbon monoxide poisoning for victims who recover are less clear. Those who recover from exposure to high levels, especially if they have been unconscious, can suffer effects on the memory and the brain and heart which may last some time or even be permanent. Some victims may suffer heart attacks some time after apparent recovery or succumb to pneumonia, especially the elderly. Similarly, despite an apparent full recovery, some weeks after the poisoning the victim may suffer from effects on the brain (for example, encephalopathy) which can cause symptoms similar to Parkinson's disease or personality changes (irritability, for example) which can persist for some time. Loss of short-term memory is common. Muscle damage sometimes occurs, which can lead to renal failure. This is because the breakdown products of the muscle are excreted into the urine and overload the kidneys. These effects are most likely in those who are victims of severe poisoning.

People whose work leads to inhalation of car exhausts (for example, traffic policemen and car mechanics) and people living in houses with a constant low level of contamination, perhaps from a slightly faulty gas fire, can suffer from low-level poisoning. The effects of such long-term exposure are not clear, but it has been suggested that heavy smokers who are exposed to carbon monoxide levels of perhaps 5–10 per cent or more in blood may suffer from effects on the heart. However, cigarette smoke does have many other toxic constituents. Oddly enough, carbon monoxide is actually produced in our bodies in tiny amounts by certain enzymes, but the levels are too low to have any significance.

In contrast to carbon monoxide, which is responsible for many deaths as well as severe poisoning cases, the caustic and irritant chemicals in our homes such as bleach, drain and oven cleaners, kettle descalers, and strong detergents/cleaners (for example, those in dishwasher powder) rarely cause death or serious injury. The most likely scenario is irritation of the skin or eyes by use of these caustic and irritant chemicals without protective measures such as the wearing of gloves. Children have occasionally drunk chemicals like bleach or caustic solutions, and can also suffer skin and eye damage after investigating the contents of the bottles found under the average household sink. This is not to say that serious and fatal poisonings don't occur as a result of the substances. Those intent on suicide may choose to end it all by drinking bleach or kettle descaler with fatal results.

CASE NOTES

Agonising death of haunted woman

■ A haunted Wendover woman died an agonising death after drinking kettle descaler

In the last year Mrs Heidi Mason, 44, of Orchard Close, Wendover, had tried to kill herself with pill overdoses, a razor blade and a plastic bag after becoming a victim of serious depression.

She was found, bleeding from the mouth, half-conscious but dying, in the grounds of St. John's psychiatric hospital, Stone, where she was a voluntary patient, on June 4.

Her stomach was almost entirely eaten away by the acid and her mouth and throat badly blistered. But despite her history of suicide attempts, Bucks coroner . . . recorded an open verdict, saying he could not be certain she had intended to kill herself at this time . . .

Pathologist Dr Andrew Tudway said that Mrs Mason's stomach was almost entirely corroded away and her mouth and throat ulcerated by the formic acid in the kettle descaler.

Aylesbury Plus (Wednesday, 16 September 1987)

Hence some suicidal people do occasionally use such unpleasant substances, but the majority of suicides involve drug overdoses.

Of chemicals in the garden shed, apart from irritants such as creosote and those that are toxic by being inhaled such as paint stripper, the most potentially hazardous are insecticides and weedkillers. One weedkiller, paraquat, has already been discussed (see pp. 104–6) and we shall come across it again in the next chapter. Paraquat has contributed to many deaths both accidental and intentional. It is a constituent of several currently available weedkillers, and when solutions of the granules are made up and left in sheds and garages, these have sometimes been drunk by inquisitive children, especially if it is stored in used cola or lemonade bottles. Fortunately, the lower strength formulations more recently available have helped to reduce the number of fatalities.

Insecticides such as chlorpyrifos, which was in one of the preparations in my kitchen (4 g per litre), would be a problem only if used carelessly (that is, allowed to come into contact with the skin) and regularly, or if large amounts of spray or liquid were inhaled or absorbed through the skin.

The more serious hazards were to be found in the garage, both liquids and both not unpleasant and therefore not difficult to drink accidentally. These are the screenwash and antifreeze.

Antifreeze: sweet tasting but lethal

Antifreeze is used in car radiators to stop the coolant water freezing in the winter. It is almost always one chemical substance, ethylene glycol, although occasionally it is mixed with methyl alcohol to make an even more toxic cocktail. Ethylene glycol would be described by a chemist as a type of alcohol, and if drunk it would at first have similar effects to a glass or two of wine. It is, apparently, sweet-tasting and indeed it was once illegally used (as was a closely related substance, diethylene glycol) for increasing the sweetness of wine in Austria. This substance is anything but sweet-natured when someone drinks it either accidentally or intentionally. My father, who was an aircraft mechanic in the Royal Air Force during the Second World War, once told me that some of the men drank the antifreeze used for aircraft engines because the label on the drums included the word 'alcohol' (this may have been because methyl alcohol was one of the constituents).

Antifreeze, and therefore ethylene glycol, is probably found in most people's homes if they own a car, and hence there is the possibility that it may be drunk accidentally by children, by adults who mistake it for alcohol, and by those who want to commit suicide. Although it looks and tastes innocuous, even pleasant, it is a wolf in sheep's clothing.

Why antifreeze is poisonous

The ethylene glycol and methyl alcohol (see below), which is also sometimes found in antifreeze, are poisonous because they are converted into more toxic products. Once inside the body, the ethylene glycol in the antifreeze is changed by metabolism into first one, and then several other chemicals. This requires the same enzyme that metabolizes the alcohol we consume in alcoholic drinks (ethyl alcohol). The ethylene glycol is converted into oxalic acid which is poisonous, and other poisonous products are also produced. Oxalic acid is also found in rhubarb leaves, which is what makes them poisonous. The result of these metabolic conversions is that the acidity of the blood increases (the pH decreases) and normal metabolic processes are inhibited. The oxalic acid formed can crystallize in the brain and the kidneys, causing damage. The oxalic acid also reacts with calcium and removes it from the body. The reduction of calcium causes other effects, like tetanus, in which muscles contract uncontrollably.

The accumulation of these products and the acidity of the tissues leads to a reduction in normal metabolic activity which will damage tissues, but it is the oxalic acid that causes the most damage, in particular to the kidneys.

Because ethylene glycol is converted to several other substances, it causes various effects and there are several stages the patient will go through. Within the first twelve to twenty-four hours the patient appears inebriated but does not smell of alcohol, then suffers nausea and vomiting. If a large amount (a lethal dose or more) of antifreeze has been drunk, the victim may become paralysed and sink into a coma. A lethal dose of this odourless, colourless liquid is about 100 ml, that is, what will fill a small wine glass. Death can occur within twenty-four hours as a result of the effects on the brain. The victim will suffer from spasms like tetanus, and possibly seizures. If they survive the first twenty-four hours then they may stay alive for several days, suffering a pain in the sides, for as many as twelve days before succumbing to renal failure. There can be effects on the heart and lungs, causing difficulty in breathing and increased blood pressure and heart rate. The heart tissue itself may be more seriously damaged, as can the nerve connecting the eyes to the brain (the optic nerve), sometimes permanently.

Fortunately, there is an antidote for poisoning with this substance, and the treatment can be very effective.

Treatment of antifreeze poisoning

Because ethylene glycol is really poisonous only when it is metabolized, it is possible to reduce its toxic effects by stopping this metabolism. Alcohol is metabolized by the same enzyme, but binds to it in preference. So in an emergency the patient can be given a double or treble whisky. The alcohol will attach to the enzyme and the ethylene glycol will thus be excluded. This competitive inhibition of the enzyme stops the formation of the toxic products. Once in the Accident and Emergency department of the hospital, more alcohol can be given to the victim by infusing it into a vein. When it is used in this way, the alcohol is acting as an antidote. Another, specific antidote can also be used, but it is very expensive and is not always available.

In addition, because one of the effects of ethylene glycol is to cause the blood and tissues to become more acidic, treatment also involves giving the patient an intravenous infusion of bicarbonate of soda (sodium bicarbonate) to reduce the acidity of the blood. If the victim is treated early enough, survival is likely.

Methyl alcohol

Methyl alcohol, or methanol, which is an alternative name, is also sometimes found in antifreeze together with ethylene glycol. It is sometimes found in other products used around the home such as screenwash, and it is a constituent of methylated spirits and industrial alcohol which are used in laboratories.

Just as ethylene glycol is toxic because it is converted into poisonous products, so methanol is converted into the poisonous substance formic acid. Again, the enzyme is the same as that which metabolizes alcohol.

Although methylated spirits, or 'meths', is a mixture of ethyl alcohol (the alcohol we normally drink in beer, wine, and spirits) and methyl alcohol, it is not normally a serious hazard in terms of accidental poisoning. This is because the presence of the ethyl alcohol acts as an antidote: it reduces the conversion of the methyl alcohol into the toxic substances that would otherwise cause poisoning. However, the regular use of meths as a substitute for alcohol, for example by 'down and outs' and alcoholics, will lead to serious damage.

There have been cases of poisoning as a result of the alcohol used in laboratories being added to punch for parties. This is usually not pure alcohol but 'industrial methylated spirits' and may contain a significant amount of methyl alcohol (4 per cent). As the lethal dose of methyl alcohol is about 70 ml, it is unlikely this could be achieved in the short space of time during a party, but drinking a strong punch made in this way would certainly lead to unpleasant toxic effects such as headache, nausea and vomiting, and blurred vision.

Some drinks naturally contain significant amounts of methyl alcohol, especially fruit brandies made from plums or apricots, for example (like slivovic, which is produced in the Balkans). The methyl alcohol content may account for some of the unpleasant effects of consuming too much of these drinks!

On a more serious note, those who make their own alcoholic drinks, in particular distil their own spirits, run the risk of methanol poisoning because, unless it is done carefully, distillation can lead to a greater amount of methyl alcohol in the final product. People who made moonshine whisky during Prohibition in the USA sometimes suffered from methanol poisoning. If it is done properly, distillation, as used in the preparation of spirits such as whisky and vodka, separates the ethyl alcohol from the water and other constituents, including methyl alcohol, so that almost pure alcohol is obtained (it is not possible to remove all the water by distillation). When it is done without the right equipment or the necessary knowledge, this can lead to an increased concentration of methanol in the final distilled spirit.

Occasionally, drinks are made illegally using methyl alcohol instead. This happened in Italy in 1986 when nineteen people died from drinking cheap wine made with methyl alcohol. Similarly, in Bombay in 1992 cheap illegal liquor drunk during the New Year celebrations caused the deaths of eighty-five people because it was made from methyl alcohol.

Drinking antifreeze or screenwash, which can contain significant amounts of methanol (the screenwash in my garage contained 20–50 per cent methanol, according to the label), would be potentially lethal. As with ethylene glycol, the first effects would be inebriation, followed by nausea and vomiting after several hours. Depending on the dose, the patient can suffer effects on the vision after about six hours, starting with dimness and blurring. In cases of severe poisoning these effects may lead to permanent blindness due to irreversible damage to the optic nerve. The product of methyl alcohol metabolism, formic acid, appears to accumulate in the eye. It damages the cells of the optic nerve which can lead to blindness. Nerves in the brain can also be damaged, leading to permanent effects. The formic acid has inhibitory effects on metabolism, which in turn affects the circulation so that less and less oxygenated blood reaches the tissues, which become more and more acidic. Severely poisoned patients suffer convulsions and descend into a coma after several hours.

The kidneys can also be damaged, leading to renal failure and, as with ethylene glycol, the acidity of the blood is increased. A lethal dose is generally about 70 ml, although there have been deaths as a result of lower doses. Damage to the eyes can occur with doses as low as 10 ml.

Treatment of methyl alcohol poisoning

Fortunately, treatment is possible, using alcohol (that is, ethyl alcohol), the same treatment as for ethylene glycol poisoning. The acidity is also treated with bicarbonate of soda. Methyl alcohol is eliminated from the body slowly, and so repeated intake, as in those who drink meths, could lead to cumulative poisoning.

Glue sniffing or solvent abuse

Other solvents that are found around the home can be hazardous if drunk and also if inhaled. The solvents found in glues, cleaning products, fire extinguishers, varnishes, paints, and spray cans are sometimes inhaled deliberately as a means of getting a pleasurable 'high'. This practice has lead to over 100 deaths in a year in the UK. Solvents commonly used for this are toluene, found in glues, and the halogen-containing solvents such as trichloroethane. Gases such as butane are also some times inhaled. The acute and long-term effects of these solvents depends on the particular solvent, for example trichloroethane will damage the heart if inhaled repeatedly. The halogenated solvents are also particularly hazardous after single exposures because they can sensitize the heart, with the result that a sudden fright, which causes a rush of adrenaline in the blood, can lead to arrythmias of the heart and sudden death from a heart attack, even in healthy young people. There are concerns that long-term exposure to toluene could affect the brain.

Alcohol: our favourite drug

One chemical that we take for granted and do not usually consider to be hazardous is the alcohol found in all alcoholic drinks. While this alcohol, known as ethyl alcohol or ethanol to chemists, is much less toxic than methyl alcohol, it is both a drug and a potential poison. Indeed, alcohol is classified as a human carcinogen, that is, capable of causing cancer, by the International Agency for Research on Cancer (IARC) because there is evidence that excessive intake increases the incidence of cancer, for example in the oesophagus.

The majority of kitchen cupboards probably contain one or more bottles of spirits (even if it is only cooking brandy). Most people would be surprised to learn that a normal-sized bottle of whisky (700 ml) contains a lethal dose of ethyl alcohol. This substance illustrates the principle of Paracelsus very well. Alcohol is a food that can supply us with energy; used in moderation, it is a handy social drug and appetite stimulant; and it has been employed in medicine as an anaesthetic and in various other roles, for example as an antibacterial solution for cleaning wounds. Yet large amounts consumed all at once or over long periods are very hazardous and cause many adverse effects on the body.

Many volumes could be written about alcohol, but there isn't the space here to give more than a relatively brief account of the chemical. The

21. *Gin Lane*, by Hogarth, depicts the ravages and evils of drinking alcohol (gin was especially popular at that time in the 18th Century).

majority of people probably don't consider alcohol to be a drug or that, like most drugs, it is a potential poison. Alcohol is a drug in the sense that it has particular desirable effects on the body: people drink it in order to experience its pleasurable effects. It can also be addictive, and has been responsible for more death, disease, and ill health than any other single drug, with the exception perhaps of nicotine/tobacco. The deaths due to alcohol number in the thousands, whereas those due to ecstasy amount to a handful. It is far more of a problem than all the drugs of abuse we tend to worry about, such as heroin and cocaine, because it is accepted by society, having been used by humans for several thousand years. Perhaps society, and politicians in particular, should be as concerned about the drug alcohol as about illegal drugs. There is now increasing concern about the problem of binge drinking among young people and its potential long-term effects as well as the immediate problems.

Alcohol was almost certainly known in prehistoric times because it is produced naturally. Wine is referred to in the Old Testament and references are made to alcohol in ancient Sumerian texts. Although alcohol for industrial purposes can be produced by a synthetic chemical process, the alcohol in drinks is normally produced by fermentation. This involves growing yeast, a micro-organism, in the presence of a carbohydrate, sugar. The yeast converts the carbohydrate into alcohol. Yeasts grow naturally on grapes, and so crushed grapes contain both the sugars and the yeast that allow fermentation to occur naturally. The result is wine. The final product usually contains about 12 per cent alcohol. The yeast usually cannot tolerate a much higher concentration of alcohol, which would poison the yeast. We ourselves produce small amounts of alcohol naturally, as the bacteria and yeasts in our gut ferment sugars in our diet into alcohol.

To make stronger drinks, like spirits, requires a process known as distillation. This involves heating the liquid that contains the alcohol produced by fermentation until the alcohol vaporizes (at above 78°C). The vapour is then condensed, and the liquid contains a higher proportion of alcohol (and some water still). In the popular imagination drinks such as vodka and whisky are generally considered to be more likely to be harmful than wine or beer, but the alcohol in them is the same, the only difference being the concentration. A measure of spirits contains about the same amount of alcohol as half a pint of beer or a glass of wine. Obviously, this is only a rough guide, as it will vary with the particular strength of beer or wine. A strong drink (that is, one in which the alcohol concentration is high), like whisky, can irritate the throat and stomach if it is drunk neat on an empty stomach, and cause gastritis.

What's all this about units?

A normal glass of wine, measure of spirits, or half a pint of beer contains approximately 11 ml of alcohol (about 9 g). One small glass of wine is approximately 100 ml; wine is normally at least 11 per cent alcohol; therefore 11/100 × 100 ml = 11 ml. This amount is about 1 unit. A similar calculation shows that a shot (25 ml) of whisky (which is 43 per cent alcohol) contains 11 ml alcohol, and half a pint (284 ml) of beer (3.8 per cent) contains 11 ml alcohol. These are only very approximate measurements as drinks vary in strength (especially beer and wine) and wine glasses vary in size. Measures of wine, for example, are often 125 ml and servings are sometimes much larger than this. So a large glass of a strong red wine (14 per cent) or a can of extra strong lager (6 per cent or higher) may contain 3 units of alcohol. Moreover, there is no international measure of a standard unit of alcohol.

After one swallows an alcoholic drink, the alcohol is absorbed from the mouth and throat to a small extent. A greater proportion (20 per cent) of the alcohol is absorbed from the stomach, but most of it enters the body through the small intestine, appearing in the blood within minutes (see pp. 13–15). The maximum concentration in the blood is reached within about forty-five minutes in most individuals. In some people, however, it may take thirty minutes longer. The maximum blood level would be about 16 mg/100 ml blood in an average man who is not accustomed to drinking, but this would also depend on various factors, such as how fast the alcohol is broken down and whether absorption is delayed (see below and boxes).

The presence of food in the stomach and intestine delay the absorption. The nature of the drink can also influence the amount of alcohol absorbed and how quickly it is absorbed. With fizzy drinks, like champagne and spirits mixed with fizzy soft drinks such as tonic water, the alcohol is absorbed more rapidly. With neat spirits, more alcohol is exhaled before it is absorbed from the mouth and throat, and the amount absorbed will often be less.

Once the alcohol is present in the bloodstream several things happen to it. It first enters the liver where it can be broken down, but as the amount of alcohol is relatively large (whereas most drugs are given in doses of milligrams, 1 unit of alcohol is about 9 grams, or 9,000 milligrams!), the processes that break down the alcohol, are saturated by the levels of alcohol in just one drink. Consequently only a proportion of the alcohol is metabolized as the blood passes through the liver, which means that effectively alcohol is removed from the blood at a constant rate of about 14 ml per hour (slightly more than 1 unit). People who drink infrequently have much slower rates, while those who drink more regularly (including alcoholics) may have much faster rates (see below). This means that it is difficult to predict the effects of a certain amount of alcohol on different individuals and it can vary even in the same individual.

Some of the remaining alcohol travels to the lungs, some to the kidneys, and a small proportion is excreted from these two organs. This is why we can detect alcohol on a person's breath (and how the breathalyser test can be used to determine the level of alcohol in a person) or in their urine. As the blood passes through the brain, some of the alcohol passes into nerve cells, where it has several effects. The alcohol tends to squash the cells and distort the receptors in the cell membrane, which leads to some confusion among the nerves and depresses their activity. Alcohol is thus called a central nervous system depressant (see box p. 208): it stops nerves talking to one another, as it were. After one drink the effects are noticeable and pleasant: we become more talkative, more relaxed and less inhibited, and so more sociable. Our blood pressure falls slightly, and for this reason it can be dangerous to give alcohol to someone who is suffering from shock, for example after an accident, as their blood pressure will already be low. Alcohol makes people less sexually inhibited and more forward, but, unfortunately for men, there is a catch. As the Porter in Shakespeare's *Macbeth* neatly put it: 'it provokes the desire, but it takes away the performance.' The effect is also known as 'brewer's droop'.

Alcohol is broken down into two products in sequence. The first causes the blood vessels in the skin to dilate and so we feel warm and may look slightly flushed. The effects will vary. The alcohol continues to circulate throughout the body and, as it passes through the liver, more is broken down. The second product is acetic acid or vinegar, which can be used by the body as part of energy generation or to synthesize fats. Alcohol is therefore a source of energy, 1 gram providing 7 kilocalories (compared to 9 kcal/g from fat).

Alcohol also affects various body systems apart from the brain. It affects the kidney by decreasing the production of a hormone, so that less water is retained in the body and more urine is produced. This is the reason we become dehydrated after drinking, especially if a large amount is consumed. Alcohol affects the ears, in particular the system of canals filled with fluid that controls our balance. This fluid is affected by the alcohol in the blood. As more alcohol is consumed, a person's balance is increasingly affected, which explains why the drunk stumbles and falls over.

As Paracelsus predicted, increasing the dose of any chemical (including

How alcohol is broken down and how it affects the liver

After we have had a drink, the alcohol it contains is absorbed into the blood and enters the liver, where it is broken down by an enzyme (**alcohol dehydrogenase**) into a chemical called acetaldehyde. This is formed slowly, fortunately (as it is responsible for some of the unpleasant effects of alcoholic drinks, flushes, headache, nausea). Thankfully, it is also removed more rapidly by another enzyme (**aldehyde dehydrogenase**), to form acetic acid, which is then incorporated into normal metabolic processes, to produce energy for the body for example.

In addition to requiring an enzyme, the breakdown of alcohol requires a coenzyme called NAD, which is converted into NADH when the alcohol is metabolized. The large increase in the level of NADH at the expense of NAD, and the lack of utilization of other sources of energy such as carbohydrate, leads to an increase in the production of fat, some of which accumulates in the liver after a heavy bout of drinking. This 'fatty liver', if maintained by regular and excessive drinking, seems to be a necessary stage in the eventual development of liver disease, like hepatitis and cirrhosis, which occurs in some alcoholics. The presence of the fat impairs the ability of the liver to function. The other effect of the change in the proportions of the coenzymes is that glucose is not produced, and this is made worse if the heavy drinker does not eat a normal diet from which glucose might be obtained. Blood glucose levels can therefore be dangerously low.

Alcoholic liver disease may have several underlying causes: there is evidence that the immune system is involved in some cases; **oxidative stress** is thought to be important, as are the processes of inflammation. It has been suggested that there may be subgroups of patients with alcoholic liver disease in which different causes and factors are important.

alcohol) increases the effects until adverse, unpleasant effects occur. There are several stages of alcohol intoxication as the dose increases, described as the four 'D signs' by a famous pharmacologist:

dizzy and delightful (blood level 100 mg/100 ml);
drunk and disorderly (blood level 200 mg/100 ml);
dead drunk (blood level 300 mg/100 ml);
danger of death (blood level 400 mg/100 ml).[3]

The legal limit for driving a car varies for different countries but is 80 mg/100 ml in the UK. Although the last stage, death from an overdose of alcohol, is relatively rare, it can and does occur.

The substance itself is responsible for the effects we enjoy—the feeling of euphoria, relaxation, and sleepiness—but by interfering with the connections between, and the functioning of, nerves, alcohol impairs our judgement and ability to do things such as drive and play musical instruments. At higher doses the control of speech and sense of balance are impaired and muscular coordination is decreased: our speech becomes slurred and we stagger and fall over. At still higher doses the anaesthetic effects become apparent, and depression of brain function leads to the victim falling asleep or, more seriously, descending into a coma. At the dangerous level of exposure the control of breathing by the brain is impaired and death can occur as a result of the collapse of breathing and the circulation.

Alcohol has many toxic effects on the body and, like methyl alcohol, it is converted into other substances. This metabolism is partly responsible for its poisonous effects. Anyone who has suffered a hangover as a result of drinking too much knows that alcoholic drinks can be 'poisonous', but is it the alcohol or the other things in the drink? The answer is: both. Most of the alcohol is broken down in the liver (and some in the stomach). The first product is a chemical called acetaldehyde, which is formed quite slowly but is more rapidly removed by further metabolism to acetic acid. This is fortunate as acetaldehyde has unpleasant effects: it causes the dilation of blood vessels which makes us feel warm after one drink, but it also causes some of the effects of a hangover, notably the throbbing headache, flushing, and sweating. The acetaldehyde also makes the hangover victim sick. Such effects are caused by pure alcohol, but the impurities in alcoholic drinks, such as methanol, other alcohols, and various impurities arising from the process of manufacture, also contribute to them. Drinking certain red wines, especially port, and certain spirits, such as brandy and rum, is more likely to result in a hangover, whereas vodka, which usually has fewer impurities and may have undergone processes of purification, is less likely to cause unpleasant effects.

One of the major effects of alcohol is to cause dehydration, which is also responsible for the unpleasant effects. Because of the effects of alcohol on the metabolism of the body, the level of glucose circulating in the blood will drop after a heavy bout of drinking (see box above).

Can a hangover be 'cured'? Unfortunately not, until the acetaldehyde has been broken down and lost from the body, but it can be minimized by:

not drinking too much or too quickly;
not drinking on an empty stomach (fatty food, especially, will slow down the absorption of alcohol);

drinking water periodically; and
drinking a pint of water before going to bed.

Eating sweet food such as honey (which is rich in fructose) the following morning may redress the metabolic imbalance, help in the removal of the acetaldehyde, and counteract the low level of glucose. While we may fall asleep easily after drinking, the quality of that sleep will not be as good as normal and we may wake up early. For this reason constant drinking before sleep can lead to excessive tiredness. When he or she sobers up and the hangover recedes the drunk may not remember what happened the night before, because alcohol blocks new memory formation. Large amounts of alcohol will also cause the liver to become fatty. These effects are temporary if the excessive drinking occurs only once or occasionally.

If a person continues to drink excessively, however, permanent changes occur. The organ most affected is the liver, because it is the main site of the metabolism and where the acetaldehyde is produced. The metabolic disruption also occurs here. The liver is most affected because it is the first organ exposed to the alcohol after the stomach and intestines. Large amounts of alcohol drunk over a few hours or repeatedly over a few days will cause the liver to accumulate fat, possibly 5–10 times the normal level. This will return to normal if the drinking stops, but if heavy drinking continues, more than a bottle of wine or a third of a bottle of whisky a day for example (about 100 g alcohol), after five or more years the liver may be more seriously damaged and the drinker may suffer from alcoholic hepatitis. Regular 'binge' drinking of large amounts (such as at weekends) can lead to this, as can a daily intake of smaller amounts.

Alcoholic hepatitis, which occurs in about 30 per cent of alcoholics, is a disease where some of the cells making up the liver are destroyed and the liver responds with inflammation. The symptoms of hepatitis include nausea and vomiting, anorexia, loss of weight, and pain in the upper part of the abdomen on the right side. The liver will be enlarged and the patient may have a fever and jaundice (they will appear to be a yellow colour). Hepatitis is reversible if the drinking stops, but in about 30 per cent of cases it progresses to the next stage. The liver is a relatively resistant organ and can regenerate, but damage to the cells that make up the liver, if continued, leads eventually to a permanent change called cirrhosis, which occurs in perhaps 10 per cent of alcoholics. The structure of the liver changes and becomes tough and more fibrous. Blood won't flow through it so easily and so it becomes less efficient. It degenerates and eventually becomes incapable of dealing with the metabolic processes in the body. Some alcoholics go on to develop liver cancer. Alcoholic men may develop breasts, their testes shrink, and they become impotent when their hormone levels change as a result of their liver failing. Ammonia levels rise in the blood because the liver cannot detoxify it, and this in turn affects the brain. The alcoholic can thus die of a hepatic coma. The effects on the brain become permanent, with memory loss, hallucinations (they may see pink elephants), and paranoia. The degeneration of nerves affects the control of muscles: alcoholics tremble, they suffer the DTs, or delirium tremens. The heart and other muscles degenerate and alcoholics can suffer congestive heart failure. Their blood pressure rises and they may suffer inflammation of the stomach (gastritis) and ulcers. Some alcoholics develop cancers in the oesophagus. Excessive alcohol intake also affects the pancreas, leading to **pancreatitis** which can be fatal.

In addition, most alcoholics suffer from a poor diet as they get most of their calories from the alcohol (see above), and the gastritis and ulcers they may suffer reduces their appetite for food. They will therefore probably have a low blood level of glucose. Pregnant women who drink significant amounts of alcohol produce smaller babies, the so-called 'foetal alcohol syndrome'.

This is quite a catalogue of harmful, toxic effects for our favourite legal drug! It has been estimated that the number of alcoholics in the USA is over 6 million. The total number worldwide must be tens of millions. The abuse of alcohol has been going on for centuries. In the early part of the eighteenth century, the drinking of gin became very popular, with devastating effects on society, as captured by Hogarth in his engraving (see Figure 21). Why do some people drink more alcohol than others, choosing to suffer these potentially fatal effects?

The reason is that alcohol is an addictive drug for some individuals. Some people are more prone to this, and there is evidence of a genetic factor as it seems to occur in families. It has been shown that alcohol can produce morphine-like chemicals in the brain. Some people become dependent on alcohol and experience withdrawal symptoms when they stop drinking. They suffer craving, headaches, trembling, insomnia, and anxiety, and become desperate for another drink, a 'fix', just like a cocaine or heroin addict. (Anyone who has seen the film *Days of Wine and Roses*, with Jack Lemmon and Lee Remick, will be familiar with the symptoms and effects.)

Tolerance to alcohol

People who drink regularly become tolerant to the alcohol, as the same amount produces less and less effect and they drink more. The reason for

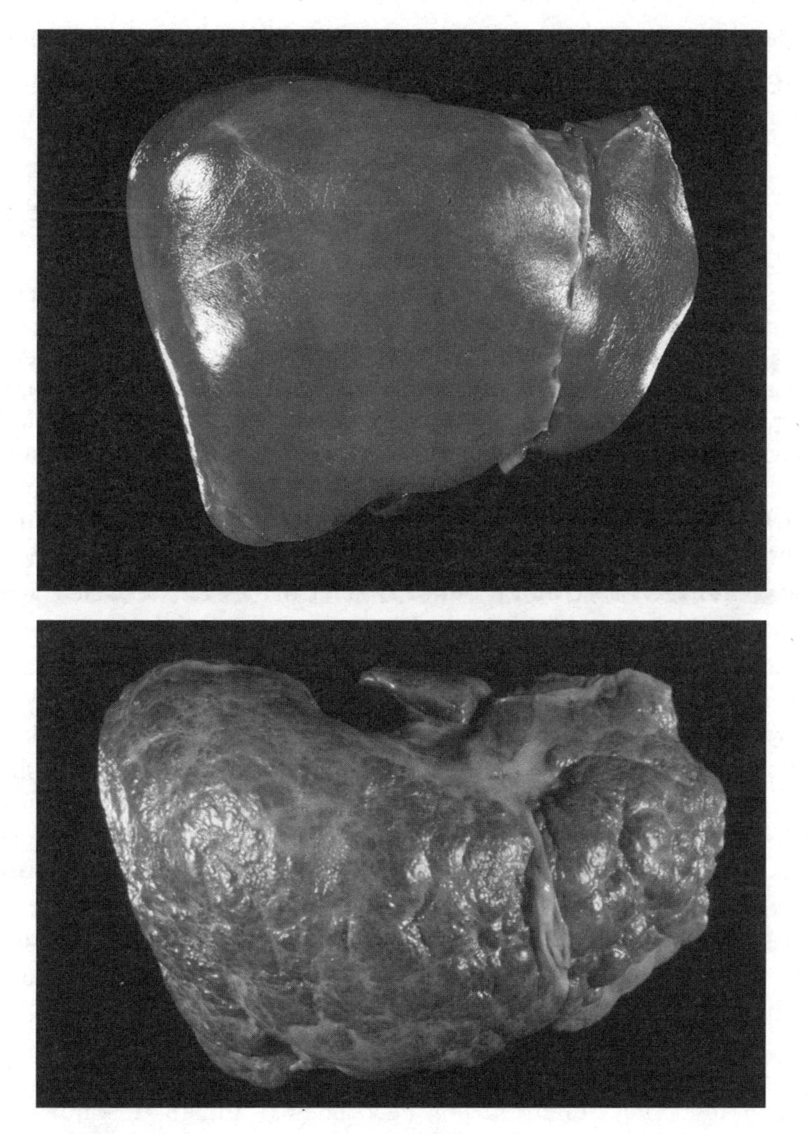

22. The effects of excess alcohol on the human liver. A liver with alcoholic cirrhosis (bottom), compared to a normal human liver (top).

this lies partly in the metabolism of the alcohol. At low doses, alcohol is mainly broken down by the enzyme alcohol dehydrogenase (as described earlier). There are two other enzymes that can break down alcohol. One of these, cytochrome P450, which is responsible for the metabolism of most drugs, becomes important at higher doses of alcohol. Levels of this enzyme can be induced by alcohol (see pp. 32–4). Constant or repeated

The effects of alcohol on the brain and addiction

Alcohol affects nerve cells in the brain, squashing them and slowing down the transmission of impulses along them. The connections between nerves are affected, and nerves stop talking to one another, in effect. Alcohol depresses the activity of nerves by interfering with receptors for a substance called GABA and by blocking receptors for another substance, glutamate.

It seems that alcohol activates the cells in the brain that produce a neurotransmitter called dopamine. It has been suggested that this interacts with alcohol to produce morphine-like chemicals (morphine is related to heroin and is addictive). Those who become addicted to alcohol have more dopamine receptors in their brains than average and this seems to be a genetically determined defect. The dopamine receptors seem to be an important factor in addiction.

Studies in rats have shown that another substance in the brain, neuropeptide Y, may be important, for when levels are low the rats drink more alcohol.

exposure to alcohol leads to an increase in the amount of the enzyme, with the result that the alcohol is broken down more rapidly and removed. The heavy drinker thus becomes tolerant to the pleasurable effects of alcohol while the adverse metabolic effects are not diminished.

Because alcoholics break down the alcohol more quickly than other people, they may appear to be sober even when they have had several drinks (that is, until they sustain liver damage, when they may be more affected than the average person). Can alcoholism be treated? Apart from simple abstinence and drug treatment to alleviate the withdrawal symptoms, there is a drug, antabuse, which helps alcoholics to stop drinking by making the effects of the alcohol unpleasant.

Treatment of alcoholism: the role of serendipity

One means of treatment for alcoholism is to give the addict a sedative/relaxant drug such as diazepam for about seven days. After this time the worst of the unpleasant withdrawal symptoms will have receded. The patient still has to be kept away from alcohol and weaned off it, but the drug helps to reduce the craving. There are other newer drugs that also do this: by increasing the substance GABA (see box above), they inhibit the activity of nerves in the brain that would otherwise increase when there is no alcohol present.

Another treatment involves interfering with the metabolism of alcohol so that the breakdown of the toxic acetaldehyde is inhibited, which then accumulates, producing the unpleasant symptoms of nausea, vomiting, and headache, even after a single drink. The drug used for this, antabuse, was discovered by chance. The chemical disulphiram was used in industry, and it was noticed that those working with it suffered unpleasant effects when they drank alcohol. In the 1940s two Danish scientists who were looking for a new drug treatment for intestinal parasites thought they had found it in disulphiram. They tried it on themselves but found that they sometimes felt ill. They eventually realized that this happened only when they had been drinking alcohol. The effect was confirmed and the drug dropped. Some time later, one of the scientists mentioned the incident at a meeting. It was reported, which led to alcoholics requesting the drug, which they hoped would be effective in helping them break their habit.

The protective effects of alcoholic drinks

Does alcohol have any beneficial effects? There is growing evidence that moderate drinking can have a beneficial effect on health. Moderate intake seems to reduce the incidence of heart disease and, apparently, people who drink have fewer colds. However, such effects may be partly due to the other constituents of drink, although alcohol itself is also believed to have some beneficial effects.

It seems clear that alcohol reduces the level of 'bad' cholesterol in the blood (known as LDL) while increasing the level of 'good' cholesterol (HDL). It dilates blood vessels and lowers blood pressure (but chronic alcohol intake increases it). A decrease in colon cancer and Alzheimer's disease have also been associated with moderate drinking. There is evidence, in particular, that one or perhaps two glasses of red wine a day may be beneficial. This is the so-called 'French paradox', for the level of heart disease in France is one of the lowest in the world despite a diet traditionally rich in animal fats. The French also drink more red wine than other countries, and evidence seems to suggest that components in the grapes, polyphenols such as resveratrol, may be partly responsible for the low level of heart disease. One theory put forward is that this is due to the antioxidant effects of the polyphenols. Antioxidants such as vitamins C and E can react with and remove reactive chemicals called free radicals which are produced often by interaction with oxygen; indeed, oxygen radicals can also be formed. These free radicals are believed to be capable of damaging tissue, and causing cancer, cardiovascular disease, and Alzheimer's disease. The polyphenols in the red wine may remove these free radicals. More recently, however, it has been found that constituents

Variability in the effects of alcohol: sex and race

It is often observed that certain people are more affected by alcoholic drinks than others, for example women are more affected than men. There are several reasons for this. Alcohol dehydrogenase is also found in the gut as well as the liver in greater amounts in men than in women. Consequently women absorb more alcohol from the stomach and intestines. They also generally have more fat and a lower body mass than men. Alcohol tends to distribute in the water in the body rather than in the fat, with the result that levels in the blood in women will be higher because they have less water and more fat. Women also have less body mass in which the alcohol can be distributed, and are therefore likely to have higher alcohol levels in the blood and therefore in the brain. As it is the alcohol itself that causes the effects on the brain, women show these effects more quickly and after fewer drinks.

It has also been shown recently that the activity / amount of alcohol dehydrogenase in a person varies throughout the day and is lower in the stomach in the middle of the day. Thus a drink at lunchtime will have a greater effect.

What about variations in effects between different races? It is well known that many Japanese men are easily affected by alcohol and, apparently, Native Americans are too. This is due to an inherited genetic deficiency which affects the alcohol dehydrogenase, with the result that it breaks down alcohol at about 70 per cent of the rate of the average Caucasian.

Some people in Asia have inherited an aldehyde dehydrogenase with reduced activity, which means that they accumulate acetaldehyde and makes drinking alcohol unpleasant. The alcohol levels in the blood are higher and stay that way for longer. Some people, both Caucasian and Japanese, also have inherited alcohol dehydrogenase which breaks down alcohol at a greater rate than average. It has been suggested that this may be a factor in alcoholism.

Studies in twins have suggested that genetic factors may be important in the development of alcoholic liver damage. While there are differences in the various enzymes involved with alcohol metabolism, some of the differences are not consistent or large enough to account for differences in susceptibility. Genetic differences in protective mechanisms and genes involved with inflammation have also been linked to susceptibility to alcohol-induced liver disease.

The removal (known as clearance) of alcohol from the blood can vary, between 8 mg per 100 ml of blood per hour in non-drinkers to 39 mg per 100 ml of blood per hour in alcoholics. This means that a non-drinker could take two to three hours to remove 1 unit of alcohol, whereas an alcoholic could do this in thirty to forty minutes.

of red wine grapes (especially Cabernet Sauvignon grapes) affect the levels in the body of a substance called endothelin-1, which constricts blood vessels. Compounds that block the production of this may reduce the fatty streaks in blood vessels and decrease heart attacks. Red wine also, apparently, leads to the dilation of blood vessels and stops red blood cells from clumping, another factor in heart disease.[4]

The principle of Paracelsus is crucial here, and the emphasis must be on the word 'moderate', as there is no doubt that higher levels of drinking lead to liver disease and a variety of other diseases. It is no accident that the level of liver cirrhosis is also relatively high in France.

Thus alcohol can be a type of food, a medicine, an antidote for poisoning (see pp. 195–7), and also a poison itself, and an apt illustration of the principle of Paracelsus: 'All substances are poisons, there is none that is not a poison. The right dose differentiates a poison from a remedy.' Is there are safe level of drinking? Is there a non-toxic dose? It is possible to give only an approximate value because individuals vary and, as we saw earlier, so do alcoholic drinks. Most people don't measure the exact amount of alcohol they consume each day or week, and there needs to be some allowance for variation.

In the 1980s the deliberations of the Royal College of Physicians and the Royal Colleges of General Practitioners and Psychiatrists concluded that drinking 21 units per week for men and 14 units for women was associated with low risk of physical harm.[5] In 1995, however, the UK government published a review of the scientific and medical evidence on the health effects of drinking which raised these limits to 28 units for men and 21 units for women. Some of those in the medical world were not happy with the new limits.

One problem is that setting a benchmark may encourage people to drink up to that level and, if their estimates are based on an incorrect assessment or understanding of the strength and volume of particular drinks, they may be well over the safe limit. There are also, as we have seen, many individual factors that affect response to alcohol including genetic factors, which may mean that some people are much more susceptible to its effects. Unfortunately, the adverse health effects can take a long time (years rather than weeks) to become apparent, and the young binge drinkers of contemporary England may not realize the risk they are taking until it is too late.

As with the use of other substances we have considered, the risks and benefits must be weighed, and individuals have to make their own choice. It is clear that there are many different factors involved in the relation between alcohol intake and adverse health effects. Everyone is different and what may be a safe intake for one person may be a risk for another, just as it is for other drugs. Clear information from manufacturers and in bars, and greater standardization of measures and strengths would certainly help drinkers make more informed choices.

9

Rasputin's Revenge

Chemicals Used to Kill

POISON has been used for execution, murder, and assassination and in warfare since ancient times. As we saw earlier, the word 'toxicology' is derived from the Greek words for the poison with which arrows were tipped. The killing of Rasputin in 1917 was a clear case of the deliberate use of poison for a politically motivated murder, but where the assassins may not have had enough knowledge of the dose of cyanide necessary (see case notes pp. 215–16).

Two thousand years earlier, the ancient Greeks were certainly aware of the potential of plant toxins, such as aconite, for murder and assassination. This is apparent from their mythology, as the goddess of witchcraft, Hecate, who lived with Circe on the island of Colchis, was aware of the poisonous nature of aconite. Aconite was in common use in ancient Rome and was a widely used arrow poison. The toxic ingredient of the plant, which was known by various evocative names, like wolfsbane, monkshood, leopard killer, brute killer, and even woman killer, is aconitine. It causes adverse effects on the heart, such as changes in the rhythm (arrythmias) and depression of breathing. A more graphic description of the effects is provided by the Elizabethan herbalist John Gerard: 'their lips and tongs swell forthwith, their eyes hang out, their thighs are stiffe, and their wits are taken from them.'[1]

Another poisonous plant that was widely used was deadly nightshade, a single berry of which contains a lethal dose. Livia, the wife of Emperor Augustus, cunningly injected some of the juice into figs on his personal tree in order to poison him without arousing suspicion. Deadly nightshade berries contain the chemicals atropine and scopolamine, which have similar actions (they are described as **anticholinergic**). A victim of poisoning with atropine or deadly nightshade will suffer from a dry mouth and find it difficult to speak and swallow. They will have a high temperature, an increased pulse rate, a rash, and possibly peeling skin. The pupils will be dilated, and they may suffer convulsions and then paralysis. Scopolamine (also known as hyoscine) may cause hallucinations and unconsciousness.

Another arrow poison, common in South America, was curare, which was extracted from a plant (*Chondrodendron*). A constituent of this extract, tubocurarine, blocks transmission of nerve impulses to muscles, leading to complete paralysis and death from cessation of breathing. The chemical has been used as a drug for the purpose of relaxing the muscles of a patient. One of the arrow poisons used in Africa was extracted from a species of the plant *Strophanthus* and contained strophanthidin and oubain (pronounced 'waabain'). These have similar effects as digitalis on the heart and, in sufficient quantities, can be lethal. The poisoned patient would suffer from erratic rhythms of the heart, a slowing of the heart, and possibly uncontrolled movements of the heart muscle (ventricular fibrillation).

Poisoning was thus very common in earlier centuries. It was so common that in the twelfth century a Jewish philosopher and physician, Moses Maimonides, wrote a book entitled *Treatise on Poisons and their Antidotes*. He may have been one of the first toxicologists. It was the Italians, however, in cities like Florence and Venice, who really developed and used poisoning as a political tool. For example, members of the Borgia family in the fifteenth and sixteenth centuries perfected the art of poisoning. It was used with great effectiveness to remove unwanted husbands, rivals, and political opponents. Some poisoners have even become household names, for example Dr Crippen who was convicted of poisoning his wife in 1910. Similarly, some of the poisons themselves, such as cyanide, arsenic, and strychnine, have become well known and associated with murder in the popular mind. The prolific writer Agatha Christie used the names of some of these poisons in her book titles, for example *Arsenic and Old Lace* and *Sparkling Cyanide*.

Poisoning has also been used as a means of official execution. The American states that use the gas chamber for executing criminals employ cyanide gas for the purpose. These days a lethal injection is the more usual method of execution. In some states in the USA and in China (which approved this method of execution only in 1997), for example, three chemicals are used. First, a barbiturate drug (sodium thiopental) is injected into a vein (this is called an **intravenous injection**), which rapidly causes unconsciousness (within thirty seconds). The victim is therefore unaware of what follows. Then pancuronium bromide, a muscle relaxant drug, is injected into the vein, which paralyses the muscles of the diaphragm within about three minutes and so stops respiration. Finally, a large amount of the chemical potassium chloride is injected, which rapidly stops the heart (that is, causes cardiac arrest) by interfering with the electrical activity of the heart muscle. Large doses of all these chemicals are given so that the effects are rapid and final. The victim is likely to be connected to a machine called an electrocardiogram which monitors the activity of the heart. After the drugs have been administered, a physician will determine that the victim's heart has stopped and that he or she is no longer breathing.

An older example of execution using poison is that of the philosopher Socrates in ancient Greece, who was made to swallow hemlock (as described on pp. 152–3).

Cyanide: bitter almonds, Herman Goering, and Rasputin

CASE NOTES

Was Rasputin poisoned, shot, or drowned?

A self-proclaimed holy man, who held great sway over the Tsarina of Russia in the early part of the twentieth century, Gregory Efimovich Rasputin owed his power to his apparent ability to heal her son, the Tsarevich. The Tsarevich suffered from haemophilia, a hereditary disease afflicting males, which was passed on through Queen Victoria to various members of the royal families of Europe, in which the blood fails to clot. The Tsarina became dependent on Rasputin, often sending for him to tend her son when he was bleeding uncontrollably upon the slightest injury. During the First World War the Tsar left Petrograd (as St Petersburg had been renamed) to take control of the army personally, and left the Tsarina in charge of affairs of government. Rasputin seized this chance to strengthen his hold on power.

Members of the Russian aristocracy became so concerned about his influence on the Tsarina that they plotted to kill him. The question was how and where, as Rasputin was a suspicious man. They decided to invite him to a party at the palace of Prince Felix Yusopov in Petrograd to poison him. He was lured there on 16 December 1917 by the promise of an orgy, wine, and chocolate cake (of which he was reputed to be very fond). The conspirators prepared a chocolate cake incorporating potassium cyanide and dissolved more cyanide in wine. It was said that Rasputin ate the cake and drank copious quantities of wine but without apparent effect. In desperation, Prince Yusopov shot him in the back, but after some time he regained consciousness and attacked the Prince. Rasputin was a giant of a man but those present managed to restrain him and hit him over the head several times. His body was wrapped in a carpet and dumped into

23. Rasputin, who was poisoned with cyanide, shot, and beaten, before he succumbed to death through drowning.

the River Neva. It was later found and, apparently, showed evidence that he had regained consciousness after being thrown in the water and had tried to struggle free. This served only to fuel the mystery surrounding Rasputin's alleged powers. The conspirators were arrested, but the murder of Rasputin came too late to save the country. Within a few months the Tsar was deposed and the imperial family imprisoned and eventually shot.[2]

Why did cyanide not kill Rasputin immediately? It has been suggested that he had a deficiency in his stomach acid, a condition known as achlorhydric gastritis, which might have decreased the production of hydrogen cyanide from the potassium cyanide. There is no evidence for this, however, and it has been disputed: the potassium cyanide should still have killed him. The cyanide had been taken with food and wine, which would have delayed the absorption and therefore slowed down the lethal effects.

A more likely explanation for Rasputin not having been killed by the cyanide was not any superhuman power but that the potassium cyanide may have been allowed to get damp and had degraded into relatively harmless products. There have been other well-documented cases where what should have been lethal doses have been taken without fatal consequences. A combination of food in the stomach and degraded cyanide salts would be enough to account for this.

The other well-known event in which cyanide (as the volatile hydrogen cyanide, HCN) featured was the killing of millions of Jews, gypsies, and political prisoners in gas chambers by the Nazis during the Second World War. Cyanide has also been used in gas chambers in some states of the USA to carry out the death penalty. Cyanide is thus another chemical inextricably linked in the popular imagination with poison, even though it is not commonly associated with individual suicides, homicides, and accidental poisonings. Most individual deaths from cyanide poisoning are suicidal, although there is the occasional accident. Apart from the mass 'suicide' (or murder) of the People's Temple (see case notes below), the most high-profile individual suicide involving cyanide was probably that of the Nazi Hermann Goering, following the Nuremberg war trials. Had he managed to keep a cyanide suicide capsule in his possession in prison or managed to acquire it from one of his captors? Whatever the source, he took his own life with cyanide after the verdict was delivered.

Cyanide comes in many forms and not all are equally poisonous. The most potent and most likely to be lethal is hydrogen cyanide, also known as prussic acid when in solution in water. This is a volatile liquid but the gas can be generated by the action of an acid on cyanide salts (for example, a mixture of potassium cyanide and sulphuric acid). The Nazis generated the hydrogen cyanide for the gas chambers in places like

CASE NOTES

Mass suicide

In 1978 about 900 people, members of a religious sect called the People's Temple, died in Guyana, South America. Founded in California by Jim Jones, a faith-healing preacher, the sect comprised a motley group including drug addicts, the maladjusted, mentally afflicted, and some ex-convicts. When a group of relatives arrived determined to investigate the activities of the sect, the founder managed to convince his followers to drink a potion prepared by the medical officer, which contained potassium cyanide and was highly effective. Jones then shot himself.

Auschwitz using a commercial insecticide called Zyklon B, which liberates hydrogen cyanide. There are a number of cyanide salts, the most commonly known being potassium cyanide. There are other chemicals that contain the cyanide group, for example sodium cyanide, potassium ferricyanide, and acetonitrile. Some of these chemicals are used in industry for case-hardening steel, in photographic processes, in electroplating, and in the extraction of gold and silver from ores. The potential for accidental poisoning therefore exists in these industries.

Cyanides are also used in laboratories. At one time it was used to destroy wasps' nests and to fumigate buildings, but when this led to accidental poisonings other chemicals were used instead. Volatile cyanides can be released when certain materials burn, notably polyurethane foam, which means that people trapped in fires may suffer from cyanide poisoning by inhaling the fumes.

There are also natural products that contain cyanide, for example the plant cassava (see pp. 255–6) and various fruit kernels such as almonds, apricots, and apples. Oil of bitter almonds, which is used in cooking, contains enough cyanide to be lethal, and people have committed suicide by drinking it.

Two cases of accidental poisoning with cyanide reported in 1981 involved groups of children in Israel who ate apricot kernels, which are, apparently, 'sweet and tasty'. In the first, involving a group of thirteen children, three died about half an hour after eating the 'sweets'. In the other case, which involved eight children, one died and the others were ill within two hours of eating the kernels. Apricot and almond kernels contain a substance called amygdalin,* a sugar-like molecule which has a cyanide group within it. The cyanide is released when the kernel is chewed because an enzyme in the kernel is released and becomes active when it is chewed in the presence of saliva. The enzyme breaks down the amygdalin to release cyanide. The exact number of apricot kernels that need to be eaten before adverse effects appear depends on the individual and on the type of kernel. Fifty or more kernels will certainly have adverse effects, but as few as twelve bitter almond kernels have been reported to cause serious toxic effects in an individual.

The cassava plant has also been responsible for a number of poisoning cases and can be a major problem in parts of the world where the crop is

* Amygdalin was sold to cancer patients as an alternative medicine called Laeotrile in the 1960s and 1970s. In the USA the FDA legislated against its use and the perpetrator was charged with fraud. Demand persisted from some cancer patients, and the US National Cancer Institute and the FDA conducted a clinical trial. It was not found to be effective and was labelled 'a toxic drug that is not effective in cancer treatment'.

part of the staple diet. Laurel leaves, when crushed, will also release small amounts of hydrogen cyanide and these were once used by entomologists for killing their insect specimens. Occasionally, humans have suffered poisoning from crushed laurel leaves, for example Dr Price, described as the 'last alchemist', who in 1783 poisoned himself with a solution made from crushed laurel leaves. In 1781 Captain Donallan murdered Sir Theodosius Boughton by giving him cherry laurel water in place of his normal medicine.

Cyanides have even been used in legitimate drugs, such as sodium nitroprusside, a cyanide-containing chemical used as a drug for lowering blood pressure. The chemical releases cyanide during its metabolism, and when large doses have been given this has caused the poisoning of patients.

Thankfully, the body has a detoxication system for cyanide, presumably because of the presence of naturally occurring cyanides in plants. This defence mechanism, however, can be easily overwhelmed and fatalities do occur (see p. 220). If someone working with a cyanide is found frothing a little around the mouth, an unusually brighter red colour than normal, and smelling of bitter almonds (a bit like marzipan), all is not lost, however. If they are still alive, there is hope, for there are several antidotes and treatments for cyanide poisoning. Unfortunately, not everyone can smell cyanide as there appears to be a genetic deficiency in some people which means that they cannot detect it by its smell.

Most laboratories and industries where cyanide is used should have an antidote available.

Detection and treatment of cyanide poisoning

There are several sensitive methods for detecting cyanide in blood samples, so murder and suicide cannot easily be disguised. One simple method uses a special indicator paper which changes colour in the presence of cyanide.

The key to the treatment and the antidotes is to remove the cyanide from the blood and to promote the normal detoxication process. The body detoxifies cyanide by combining it with a substance called thiosulphate, which is found in the diet. Giving a victim this substance as an injection speeds up the detoxication, but it won't always be enough and so the cyanide is usually also removed. As the binding of the cyanide to the crucial enzymes in the body is reversible, removal is possible and effective. This is done by giving the victim a chemical that binds cyanide, called a chelating agent. These are often used in the treatment of poisoning. The chelating agent, with the cyanide bound to it, is excreted into the urine.

How cyanide is poisonous

The important part of cyanide is the same, whether it is derived from salts, prussic acid, hydrogen cyanide, or released from kernels. It is written by chemists as CN^-, which means that it is the cyanide ion. When cyanide is absorbed into the body from the stomach, some will exist in this form in the blood and be transported to the liver, where it can be detoxified by an enzyme that converts it into something less toxic, thiocyanate. This is then excreted into the urine. Humans and other animals have probably evolved this mechanism to protect them from the various cyanide-containing plants that occur naturally. If the amount taken is too much for the system, the victim suffers poisoning which can be lethal.

The cyanide in the blood is a small molecule and can easily get inside cells, from which it finds its way to a part of the cell where the energy is generated. This is a structure called the mitochondrion, of which there are many in a cell. Inside this powerhouse, the cyanide binds strongly, but reversibly, to a particular enzyme and blocks the system that produces energy and uses oxygen. Sufficient energy cannot be produced and the tissue or organ, for example the heart, can no longer function properly. The metabolism of the body changes in order to try to produce more energy but excess acids are produced.

At first there is sufficient oxygen in the blood, but eventually breathing becomes laboured as respiration starts to fail, and the victim may begin to look blue (cyanosis) because oxygen is not being absorbed from the lungs. The normal causes of death are failure of respiration or the heart, or the malfunction of the brain. The heart is particularly vulnerable to a decrease in the supply of energy and the victim will start to suffer arrythmias. The brain is also vulnerable and eventually the lack of energy will lead to dysfunction of the brain, giving rise to convulsions. The victim will fall into a coma before death.

Alternatively, if this agent is not available, the cyanide can be made to bind to the haemoglobin in the patient's blood, which will bind cyanide if it is in a particular (known as oxidised) form. Some of the patient's haemoglobin can be converted to the oxidised form by giving him/her a substance called **nitrite**. The cyanide in the blood then binds to the oxidised haemoglobin. Oxygen can also be given with these treatments and seems to help.

Contrary to popular belief, cyanide poisoning is not necessarily very rapid, unless hydrogen cyanide is inhaled when death can occur within a few seconds to a few minutes. The presence of food will delay absorption of cyanide salts taken by mouth and so delay the onset of poisoning, so there may be time to give an antidote. Some victims may survive thirty minutes or more (even two to three hours has been recorded) after a fatal dose. If the patient survives for several hours, they are likely to recover. The symptoms include headache, vertigo, anxiety, and confusion. Victims who have survived have described strange experiences, such as the feeling of being able to walk through walls.

Arsenic

Arsenic has long been recognized as a poison, and was reputed to have been used by Agrippina to assassinate the Roman emperor Claudius and by the Roman emperor Nero to kill Claudius' son Britannicus. In the latter case, after the first attempt had failed and had instead aroused suspicion, the arsenic was put into the water used to cool his soup rather than into the soup itself which was tested by a taster. In the seventeenth century a woman by the name of Tofana produced arsenical powders which became known as 'les poudres de succession', as they were used to remove obstacles like rivals, husbands, and so on. The powders contained arsenic sulphide, aconite, box, caustic lime, powdered glass, and honey. She is reputed to have committed some 600 murders. Her most well-known poison was Aqua Tofana, probably a solution containing arsenic and lead.

Since the time of Tofana arsenic has featured in many cases of poisoning. The availability of various forms of the poison, such as arsenious oxide used in rat poison and weedkiller, and potassium arsenite in flypapers, was probably one reason for its popularity. In England and Wales, for example, of 541 fatal poisonings between 1837 and 1838, 185 were due to arsenic.

Later in the nineteenth century, the number of cases of arsenic poisoning decreased dramatically. This was primarily for two reasons. First, restrictions were placed on the sale of arsenic and preparations containing it. Secondly, with the introduction of the Marsh test in 1836 it became possible to detect arsenic in bodies reliably, at levels that were likely after poisoning. However, cases of homicidal poisoning have continued to occur, for example in 1968 and 1970. Suicidal poisoning by arsenic has been less common, although between 1918 and 1951 in New York, for example, there were 145 such cases. Arsenic has also been used for mass killing by poisoning, and was a key constituent of one of the poison gases used in the First World War (see p. 235).

In his novel *Madame Bovary*, Gustave Flaubert describes the characteristics of arsenic poisoning in the suicide of Emma Bovary who has taken

CASE NOTES

Was Napoleon poisoned?

Napoleon Bonaparte died on the island of St Helena where he was imprisoned. He suspected that he had been poisoned, asserting that 'I die at the hands of my enemies'. Arsenic was a poison commonly used at the time and it is known that arsenic will localize in hair, where it can be detected. Locks of Napoleon's hair exist and have been analysed for arsenic. Sure enough, arsenic was detected in some of the specimens, in some cases at apparently high levels, but more recent evidence has not shown such high levels. Furthermore, Napoleon did not show the classic signs of arsenic poisoning in the skin, and the post-mortem revealed that he had a large tumour in his stomach.

The levels of arsenic found, up to 11 ppm, are more consistent with therapeutic use of arsenic, such as in Fowler's solution (see p. 224). An alternative explanation is that the wallpaper and draperies in the house where he lived were coloured with pigments, such as Scheele's green dye (copper arsenite), which contained arsenic. Samples of the wallpaper have been found which do indeed contain arsenic. It is known that in damp conditions, in the presence of mould, this dye will release a volatile organic form of arsenic, methylarsine, which is very toxic. Such a case of poisoning occurred more recently in the US Embassy in Rome in the 1950s.

Thus it is unlikely that Napoleon was deliberately poisoned but he may have been chronically exposed to arsenic from one or more sources.

arsenic (**arsenic trioxide**), 'a white powder which she crams into her mouth': 'An acrid taste in her mouth awoke her'; ' "I'm thirsty . . . I'm so thirsty!" she whispered.' 'At eight o'clock the vomiting began again. Charles noticed a sort of white sediment clinging to the bottom of the basin'; 'he passed his hand over her stomach. She gave a shriek'; 'Her wavering pulse could hardly be felt at all now'; 'Drops of sweat stood out on her blue veined face'; 'Not long afterwards she started vomiting blood'; 'A convulsion flung her down upon the mattress . . . She was no more.' The description includes symptoms such as intense thirst and abdominal pain, sweating, a weak pulse and low blood pressure, vomiting, irritation of the stomach and intestines, and hence blood in the vomit. The eventual death is usually due to circulatory collapse. Failure of the circulatory system means that the blood is not efficiently circulated through the lungs and is therefore not well oxygenated. This results in the veins having a blue appearance.

Arsenic trioxide or white arsenic has some ideal properties for the would-be poisoner. It is odourless and almost tasteless, and its solutions are colourless. It does have the disadvantage of being rather insoluble in water and therefore difficult to administer. Porridge or soup can, however, hold large amounts of it, a feature that was put to effective use by Mary Blandy (see case notes).

CASE NOTES

The case of Mary Blandy

Mary Blandy was an unlikely poisoner. She was so incompetent that she left a trail of evidence that led quickly to her downfall. It may be that she was at first duped into poisoning her father by Captain Cranstoun, the man who wished to marry her. Mr Blandy had advertised the fact that his daughter was relatively rich (it transpired that he had been exaggerating), but was apparently not amenable to Mary's choice, or indeed her previous choices, for marriage. The son of a Scottish peer, Captain Cranstoun was not rich and was probably penniless. Furthermore, it appears that he was already married. After failing to persuade Mary to marry him secretly, he convinced her to administer a potion to her father that would make him more amenable to their marriage. At first it seemed to work, but when Mary discovered that Cranstoun also had a mistress in London he tried to persuade her to poison her father. He sent her a present of Scottish pebbles and some 'white powder' to clean them with.

First she tried putting the 'powder' (which was arsenic oxide) into his tea. He drank some of it, but disliked the taste and left it. Later he became ill with pains in his stomach and vomiting. Mary did not think to remove the tea some of which was then drunk by a maid who also became violently sick. She wrote to Cranstoun, who advised her: 'You must make use of the powder . . . by putting it in anything of substance wherein it will not swim atop of the water.' Mary then tried giving it to her father in oatmeal soup and he was subsequently ill. Again he left some, and Mary allowed the remainder to be drunk by another servant who also became ill. The housemaid, Susan, was suspicious and tasted the remains of the soup. She also fell ill. She took what was left in the pan, a white sediment, to the doctor for analysis, and told Mr Blandy that his daughter was poisoning him and that she had seen Mary tampering with the soup. Mr Blandy confronted Mary with this whereupon she panicked and threw a pile of letters and a package of powder into the fire. Both were retrieved by the cook.

As Mr Blandy's health was deteriorating, a doctor was summoned. He was of the opinion that the patient was being poisoned and that Mary's position would be serious if he were to die. The contents of the soup pan and the white powder were examined by another doctor, Dr Addington, who carried out a series of tests and compared the results with those of a known sample of white arsenic. These tests were very thorough and included adding the powder to water, sprinkling it on red-hot iron, and adding various other substances such as syrup of violets and spirit of vitriol to a solution of the sediment. The doctor observed an exact 'similitude' between the results with the sediment and the known sample of arsenic oxide.

Mary again wrote to Cranstoun but, incredibly, gave the letter to another servant to post. This servant, aware of the situation, handed the letter to the chemist who had examined the powder. Mr Blandy duly died, and on the same day Mary dug a deeper hole for herself by offering first one servant and then the cook money to go abroad with her. They both refused. She was arrested, tried, convicted of murder, and executed. Despite appearing to be a pleasant young woman, 'an emblem of chastity and virtue', she was certainly aware of her part in the crime, which was clear from letters found in Cranstoun's possession when he died.[3]

Arsenic: useful drug or deadly poison?

Arsenic is an enigmatic substance. Recommended by Hippocrates and Pliny, it has been used in one form or another as a medicinal agent. Many compounds containing arsenic are also highly toxic, and for centuries have featured regularly in intentional and accidental fatal poisoning cases. At the same time, it has been used for the treatment of a number of diseases and for other beneficial purposes. For example, it is the main ingredient in Fowler's solution (Liquor Arsenicalis), which was invented in 1780. Supposedly a tonic, reputedly an aphrodisiac, an appetite stimulant, as well as a treatment for fevers, it contained white arsenic and tincture of lavender. Arsenic certainly does affect cellular metabolism and has been used to improve racehorse performance and to fatten pigs.

Fowler's solution was recommended for other ailments, such as asthma and common neuroses. It was also found to be successful in the treatment of malaria, and was used to destroy nerves in teeth in dentistry. It was still being described for such purposes in a textbook of *Materia Medica and Therapeutics* published in 1921.[4] Arsenic was also used in cosmetics because it was believed to improve the complexion, imparting to it a 'milk rose' appearance.

Arsenic was, without doubt, important in the successful treatment of syphilis. The drug Salvarsan (arsphenamine) was discovered by Paul Erhlich during a systematic study of arsenic compounds for a potential cure for syphilis. Salvarsan is a synthetic organic chemical that contains arsenic; he had tried more than 600 chemicals when he happened on this one. His quest was based on the belief that he could find a substance that was selectively toxic to the organism that causes syphilis (a spirochaete) but less toxic to the patient, a 'magic bullet'. Only recently, arsenic trioxide (the form of arsenic commonly used for homicide) was licensed by the

FDA in the USA as a treatment for a type of leukaemia. It causes an increase in the production of red blood cells.

There is evidence that tolerance to this lethal poison can develop. According to a physician, the 'arsenic eaters' of the Styrian Alps in the seventeenth century thrived on a twice-weekly dose of arsenic. It was supposed to improve both their ability to work at high altitude and the women's complexions. The idea was initially ridiculed by scientists, until a peasant demonstrated at a conference that it was possible to eat 400 mg of arsenic trioxide and survive.

How is arsenic toxic?

Arsenic exists in many forms, and its toxicity and other effects will depend to an extent on the particular form. Most forms of arsenic bind strongly to the sulphur atoms found in proteins, which are known as **sulphydryl groups**. These groups are often crucial to both the structure and the function of the protein. Enzymes, which are proteins, will not function correctly when arsenic is bound to these sulphur groups. Arsenic thus interferes with the production of energy in the body by interfering with its metabolism. Furthermore, a similarity between arsenate and phosphate, a crucial component for metabolism in cells, leads to inference with energy production by metabolism. Proteins involved in the structure of cells may be weakened by the binding of the arsenic, resulting in many non-specific effects. When taken by mouth arsenic has an extremely damaging effect on the stomach and intestines, leading to diarrhoea, in which blood and parts of the lining of the intestine will be passed out (this is known as 'rice-water stools'). Once absorbed, arsenic will damage many internal organs, especially the liver and kidney to which it is distributed after an acute dose. The cause of death is usually due to circulatory collapse. The lack of energy production causes the heart to fail, and so blood pressure and the pulse rate fall and blood is not effectively pumped around the body to provide oxygen. This affects the brain and nervous system, leading to convulsions and coma.

Like mercury, arsenic is a double-edged sword which also illustrates the Paracelsus principle. Salvarsan was an effective treatment for syphilis, and it seems that arsenic trioxide is currently useful in the treatment of certain forms of leukaemia. Fowler's solution could have been effective in the treatment of fevers, although probably not for all the ailments for which it was recommended (see above). But for most people, the word 'arsenic' is synonymous with poison.

CASE NOTES

Was James Maybrick poisoned?

The trial for murder of Florence Maybrick is famous. The Maybricks were an American couple who had moved to Liverpool in the 1880s. They were reasonably affluent, had two children, and employed several servants. James Maybrick, a cotton broker, was a hypochondriac and regularly treated himself with arsenic, declaring that 'I take this arsenic because I find it strengthens me'. He also took other patent medicines including strychnine. In 1889 Florence Maybrick took a lover, a Mr Brierly, and spent three nights with him in a London hotel, probably unbeknown to her husband. However, he probably came to suspect a liaison between them, for at the Grand National race at Aintree he met and had an argument with Mr Brierly. The same evening he hit his wife, Florence, giving her a black eye.

A month after this, on 27 April, James Maybrick's health started to deteriorate. He seemed genuinely ill, with vomiting and diarrhoea, both of which were symptoms of arsenic poisoning. He died on 11 May and his wife came under suspicion. Mr Maybrick's relatives said that they had noticed various strange events in relation to his medicines. One of the servants claimed to have seen fly-papers soaking in the basin in Florence Maybrick's room. Fly-papers were known to contain arsenic (as potassium arsenite and arsenious oxide). It was found that soaking one of these papers for only an hour yielded a solution with 50 mg of arsenic. Florence Maybrick claimed to have used this for cosmetic purposes.

The post-mortem revealed that the body of James Maybrick did indeed contain arsenic. Hyoscine, strychnine, morphine, and prussic acid were also found in his stomach. Florence Maybrick was charged with the murder of her husband. The discovery of love letters, one of which was intercepted by a servant just three days before her husband's death, did not help her case. The letter described James being 'sick unto death': 'I now know he is perfectly ignorant of everything.'

Florence Maybrick was found guilty of murder and sentenced to death, but this was later commuted to life imprisonment. She was freed in 1904 but always maintained her innocence of the crime of murder, though not of adultery. Did James Maybrick eventually succumb to the toxic effects of self-administered arsenic and the other dubious medicines he was taking?[5]

Symptoms and detection of arsenic poisoning

Some of the symptoms of arsenic poisoning may easily be confused with other diseases, but the ease with which the presence of arsenic in a corpse can be detected means that a would-be murderer rarely if ever gets away with it. The symptoms, which have already been alluded to, include severe gastrointestinal upset with intense pain and bloody diarrhoea, which are sometimes mistaken for food poisoning, for example ***Salmonella*** or *Campylobacter* infections. The presence of tissue from the lining of the intestine in the diarrhoea ('rice-water stools') is more specific, however. Stomach tissue would also be present in the vomit (the white sediment in the basin described in *Madame Bovary*), which is due to the corrosive action of the arsenic trioxide. A very weak pulse and convulsions reflect the effect on the heart and circulation and on the central nervous system. Chronic poisoning would cause different symptoms, with loss of hair, a blush to and increased pigmentation of the skin.

Arsenic is readily detected: the Marsh test was devised in 1836 and was used for over a hundred years. More sophisticated and very sensitive methods are now available, using X-ray analysis or a technique known as mass spectrometry which detects the arsenic atoms. Even before the Marsh test a careful scientific process of experiment would sometimes be successful, as demonstrated by the doctor in the case of Mary Blandy in 1752 (see case notes p. 223). Arsenic remains in tissues of the body for a long time, hence analysis of Napoleon's hair was possible long after his death (see case notes p. 222). The sensitive analysis of hair and nails from the body of American president Zachary Taylor, exhumed in 1991, also established that he was not poisoned in 1850.

Treatment of arsenic poisoning relies on removing the chemical from the body with a chelating agent. The one commonly used is dimercaprol or British anti-lewisite (see pp. 236–7).

Strychnine

The word 'strychnine', like 'arsenic', is in many people's minds, synonymous with poisoning. This natural toxin, an alkaloid, became known as strychnine only in the nineteenth century, after it was isolated from the plant nux vomica in 1817. It had been used as a rat poison, and sometimes as a medicine, since the sixteenth century, and had thus been readily available for hundreds of years. It is also a constituent of the upas tree, the sap of which was used for executions in Malaya, as described earlier. (See also p. 155.)

In some respects, strychnine is an unlikely poison for a murderer to use as it has an extremely bitter taste. It has been described as the most bitter substance known, but this does not appear to have deterred poisoners who have resorted to ingenious ways of dosing their victims. One murderer convinced the intended victim that a strychnine pill was a particular medicinal treatment. Two such homicides occurred in Romania where two old men were murdered by a relative who wanted to inherit their property. In another case a blind woman was murdered by her daughter who gave her mother a paper capsule containing strychnine. Unusually for strychnine poisoning, the old woman vomited and her cat proceeded to devour some of her vomit. The cat died of convulsions, as did the mother.

Thus getting strychnine past the taste buds bypasses the problem of bitterness. Another way is to add the toxin to strong-tasting food or drink. This was the method used by William Palmer.

CASE NOTES

William Palmer: the Staffordshire mass murderer

William Palmer was born in the English town of Rugeley in 1824. His father owned a saw-mill and was relatively prosperous, leaving £70,000 when he died. When Palmer was 21 he inherited about £9,000. By this time he had been apprenticed to a firm of druggists in Liverpool. He was associating with criminals in the world of racing, and eventually had to leave the firm after stealing money from them. He was next apprenticed to a doctor in Cheshire, during which time he ran a private practice in abortion and fathered fourteen illegitimate children. It seems that he followed the family tradition of drunkenness and irresponsibility.

He next worked at Stafford Infirmary, and it was here that he first poisoned someone, an acquaintance named Abbey. He used strychnine added to a glass of brandy, apparently just to see how the poison worked. It was successful and no one was suspicious of Abbey's death. Palmer then murdered one of his illegitimate children when she visited him.

At the age of 22 he went to London and qualified as a doctor, which allowed him to practise in Rugeley. Despite his inheritance, Palmer had debts and one of his creditors, a man named Bladon, was poisoned while he stayed in Palmer's house. The man's widow did not go to the police despite being urged to by friends. Another creditor, a Mr Bly, suffered the same fate and Palmer even asked the man's widow for the money, claiming that Bly had been the debtor.

The other doctor in Rugeley, who attended the deaths, was an elderly man in his eighties. Palmer proceeded to murder his uncle, attempted to murder the wife of another uncle, and probably murdered four of his children who died of convulsions. He may also have murdered his mother-in-law and several more of his illegitimate children.

He had married the daughter of a colonel in the Indian Army who brought with her a large dowry. His profligate ways continued and, having lost £10,000 on a horse, he took out insurance on his wife. When the creditors became insistent, she followed his other victims to the grave, apparently having suffered from 'English Cholera', as certified by two elderly doctors.

The insurance proved insufficient and another victim was set up, this time Palmer's brother, Walter, who was insured for £82,000. Walter was invited to stay, during which time Palmer tried to persuade him to drink himself to death. This having failed, he poisoned Walter during a drinking bout. The insurance company was by now suspicious, and refused to pay up. The poisoning, it seems, had been observed by someone at the hotel where it took place, a man named Myatt. Palmer attempted to poison him too, but was not successful: Myatt suffered but survived.

The final victim was a young gambling friend, also in debt, called Cook. Cook put all the money he had left on his own horse, in a race in Shrewsbury, and as it won he and Palmer celebrated. During their celebrations Palmer tried to poison the brandy Cook was drinking. Cook was suspicious, but Palmer demonstrated its safety by drinking some himself. That evening Cook was ill but he still lent Palmer money the next day, which Palmer promptly lost by backing another horse which turned out to be a loser. Back in Rugeley Palmer systematically poisoned Cook, who took to his bed, while his friend went off to collect Cook's winnings from London. Cook continued to get worse despite the attention of another doctor. Palmer had dosed Cook with prussic acid (a solution of hydrogen cyanide) and strychnine, and he finally succumbed with convulsions and contortions, which are characteristic of strychnine poisoning.

Palmer tried unsuccessfully to forge a cheque with his victim's signature and a document showing that Cook had negotiated a considerable sum of money in Palmer's favour. Suspicions having finally been aroused, Cook's stepfather demanded a post-mortem. Palmer was arrested and the bodies of his wife and brother were exhumed. Resourceful to the end, Palmer was not only present at the post-mortem but tried to escape with the stomach of his last victim. He also tried to bribe the driver taking the stomach to London for forensic examination to break the jar and so lose the specimen. Although no strychnine was found in the stomach, which may have been lost while it was in the jar, there was sufficient circumstantial evidence to convict him of murder and he was condemned to death.[6]

Symptoms and detection of strychnine poisoning

The effects of strychnine usually appear rapidly, within ten to fifteen minutes. As with most poisoning cases, the victim appears to be well and in good health, and then suddenly falls ill following the eating or drinking of something or the taking of a preparation or medication. Within minutes the victim will complain of stiffness, often in the back of the neck. Tremor and twitching start, followed by convulsions. Occasionally there is only one massive convulsion before death, but usually there are five or more. They are painful and last about a minute, and the contortions of the body are extreme. The body is arched as the muscles contract excessively, so much so that the head may touch the heels (as in the case of one of Palmer's victims). The muscles that control breathing are also affected and may stop, leading to temporary asphyxia. After each convulsion there is a relaxation period of about fifteen minutes during which the victim is not only exhausted but terrified, as strychnine causes heightened awareness. Stimulation of the senses can precipitate a convulsion and so treatment involves keeping the patient in a darkened room, sedated, and sometimes even under anaesthesia. This may not be available quickly enough, however, as death may occur after five or so convulsions. The effects can be reduced by removing some of the strychnine from the stomach by washing it out, if the patient is lightly anaesthetized.

Detection of poisoning relies on toxicological analysis—detection of strychnine in the body—which is possible for a long time after death, for which there are methods available. There are often no signs at post-mortem other than possible signs of asphyxia. The jaw is sometimes twisted by muscular contraction to give the victim's mouth what appears to be a sardonic grin at the time of death. For further details and the way in which strychnine works, see pp. 155–6.

Thallium: almost a perfect poison

Thallium, a metal, is probably largely unknown to the general public, but it has featured in a number of cases of homicidal, suicidal, and accidental poisoning. It has been used as a pesticide for killing insects and rats and has various uses in industry. At one time it was even used for removing unwanted hair and in the treatment of ringworm, for which a dose of 8 mg of thallium acetate per kilogram of body weight was given to children. This is dangerously close to the lethal dose of 12 mg per kilogram in adults. Needless to say, there have been a number of fatalities due to dosages having been incorrectly calculated, and some patients have suffered from its poisonous effects. The worst occurrence was in Granada, Spain in the 1930s when fourteen out of sixteen children who had been given the treatment died. The children had been given thallium acetate at a dose of 8 mg per kilogram of body weight, but the scales may have been inaccurate with the result that the dose was higher than intended.

The main cases of poisoning have been as a result of its use in preparations for killing pests such as cockroaches and rats. In the first half of the twentieth century several hundred accidental poisonings, suicides, and homicides were due to the use of Zelio paste (containing thallium

CASE NOTES

Graham Young: the Bovingdon poisoner

In 1971 Graham Young started work at John Hadlands, a photographic instruments company in Bovingdon, Hertfordshire, north of London. Within a few weeks a number of his workmates fell ill with diarrhoea and sickness. So many of them suffered from the symptoms that they called it the 'Bovingdon bug'. The management were mystified and more than a little worried. Bob Egle, head storeman, had been the first to be taken ill, in July 1971. As well as sickness and diarrhoea, he had numbness in his fingers. He died eight days after the first symptoms appeared, apparently of **polyneuritis** and broncho-pneumonia, on 19 July. In October another storeman, Fred Biggs, became ill with similar symptoms. The illness continued and he was transferred, first to the local hospital, and then to a specialist hospital in London where he died in November. Other men at the company were also feeling ill, suffering from stomach pains and pins and needles in their feet. Two of them were hospitalized, and by the time they came out of hospital both had lost all their hair. Several of the men appeared to have become ill after drinking tea.

One victim, Jethro Batt, who survived, had been told by Young that 'it was easy to poison someone and make it look like natural causes'. Young had then proceeded to lace his coffee with thallium when Batt was out of the room. When Batt discarded the coffee, complaining that it tasted bitter, Young had asked, 'Do you think I am trying to poison you?'

Eventually the company called a meeting at which a doctor spoke about the illness. Young asked him if he thought the symptoms were consistent with thallium poisoning, demonstrating a considerable knowledge of toxicology in discussion with the doctor after the meeting. The police were alerted and an inquiry into Young found that he had previously been imprisoned in a secure psychiatric institution, Broadmoor Asylum, for poisoning members of his family. He had administered antimony tartrate (known to pharmacists as tartar emetic) to his sister, stepmother, and father. He had been arrested, found to possess bottles of antimony tartrate, and was committed to Broadmoor in 1962 (where he may have poisoned another inmate). He was released nine years later and his previous history was never made fully known to his employers.

In 1972 he was convicted of the murders of his two colleagues and of several attempted murders. The police found thallium in the room where he lived and a notebook with damning entries such as 'I have administered a fatal dose of the special compound to F'. The final proof was the presence of thallium in the body of Mr Biggs, which was exhumed, and in the ashes of Mr Egle, who had been cremated. This is particularly significant as it was the first time in the UK that a poison has been detected in the ashes of a cremated victim.[7]

sulphate), which was widely available in the USA and continental Europe. There is a case of a murderer attempting to kill her husband by putting thallium in his coffee, but it tasted so bad that he did not drink much of it and suffered only diarrhoea and sickness. Later she spread Zelio paste on his sandwiches, using the contents of two tubes, but he survived and she was convicted of attempted murder. Another murderer also spread the paste on bread and butter, and committed seven successful murders and thirteen attempted murders before thallium poisoning was recognized.[8]

Thallium salts have some of the characteristics of the ideal poison, being tasteless and colourless, but detection of thallium, for example in urine, is straightforward and can occur even two months after a single dose. The metal remains in the body after death and is not destroyed or lost by burial or cremation.

Thallium is a cumulative poison, which means that the substance remains in the body and repeated doses become poisonous as the total amount in the body increases. It takes some time to kill the victim, who first suffers from severe pains in the stomach, nausea, vomiting, and severe constipation or, sometimes, diarrhoea. These symptoms do not occur immediately, but only after twenty-four to forty-eight hours. Depending on the dose, the victim may survive for weeks. If they do, one important symptom that helps in the detection of the poisoning is loss of hair, which happens after about two weeks. The other symptom is polyneuritis, that is, irritation or inflammation of the nerves leading to tingling in the hands and feet and then pain and weakness in various parts of the body. There can be degeneration of the optic nerve and consequently blindness.

Fortunately, thallium is no longer used as a pesticide or a drug today.

How is thallium toxic?

Thallium is a metal that is very similar to the metal potassium, which is essential in the body. Thallium interferes with systems in which potassium is important, such as in certain enzymes and also in the nerves.

Treatment of thallium poisoning

Fortunately there are ways to treat thallium poisoning once it has been diagnosed. Its similarity to potassium gives a clue to the treatment of thallium poisoning. Apart from washing out the stomach and giving charcoal to absorb the poison, techniques often used in the treatment of poisoning, there is an antidote. This is the dye Prussian blue, which contains potassium and is given by mouth. It exchanges its potassium for thallium. Like potassium, thallium is excreted through the kidneys into the urine but it is also secreted into the gut. Therefore any thallium remaining or secreted into the gut is bound by the Prussian blue and eliminated. The thallium in the blood is also lost much more rapidly after the Prussian blue is given.

Murder by carbon monoxide

There are many other poisons that have been used for murder, suicide, and execution, both natural and man-made substances. As we saw earlier, carbon monoxide has been widely used for suicide, when coal gas was used in homes and also by way of car exhausts. Both coal gas and, more recently, car exhausts have also been used for murder. Even pure carbon monoxide has been used, for example, as in a case where a college lecturer killed his wife by using the gas from a cylinder he had acquired from the laboratory. Having convinced her to sleep alone in their caravan, he fed a tube from the cylinder in through the window of the caravan and released the gas into the caravan when she was asleep. When carbon monoxide poisoning was diagnosed following her death, he maintained that she had been poisoned by the faulty stove in the caravan. However, the level of carbon monoxide in her blood detected at the post-mortem was too high for it to have come from a faulty stove. He was convicted of her murder and eventually confessed to the crime. Carbon monoxide remains in and is detectable in the blood of a victim as long as there is haemoglobin present. Consequently, the cause of death of the victim is easily established long after death and burial.

Murder by weedkiller

CASE NOTES

The case of Susan Barber

As well as being used in many suicides, paraquat has also featured in a murder case. The murder went almost undetected but for the persistence of a pathologist.

The Barbers lived in Westcliff-on-Sea in Essex and had been married for over 10 years by 1981. Michael, who worked in a local factory, was unskilled and had been in trouble with the police several times. His wife, Susan, had married him at 17, already with child although, unbeknown to Michael, the child was not his.

Susan continued her infidelity, and had a regular lover who lived a few yards away. Michael's job required him to start early, at 5 am every morning, and once he left Susan's lover, Richard, would be round to share her bed. One morning in March Michael returned sooner than expected from an early morning fishing trip and discovered Susan and her lover. He attacked both of them.

Some time after this, on Thursday 4 June 1981, Michael complained of feeling unwell. First it was a headache, and then stomach pains and nausea. By the Saturday the doctor was called and antibiotics prescribed. The following Monday he was having difficulty breathing and had to be taken to the local hospital. His condition worsened and he was transferred to Hammersmith Hospital with severe kidney dysfunction. The medical staff were at a loss to explain his condition. Paraquat poisoning was considered a possibility, and junior staff were asked to collect urine and blood samples for analysis by the National Poisons Unit. Michael eventually died twenty-three days after the first symptoms had been detected. It had been anything but a quick death. A post-mortem was carried out by the pathologist, Professor Evans. He suspected paraquat poisoning and samples of the body tissues were sent for preparation for **histology**. Samples of the organs were also preserved. Professor Evans was told that there was no evidence of poisoning with paraquat but was unconvinced. Meanwhile Susan Barber took up residence with her lover and collected £15,000 in death benefit and pension plus regular payments for each child from her husband's employers.

In September the histological slides were returned to Professor Evans, and again he saw evidence suggestive of paraquat poisoning. He called a conference of staff involved and, while preparing for it noticed that the notes made no mention of a test for the presence of a poison such as paraquat having been carried out. A check soon found that the National Poisons Unit had never received the samples. Fortunately the serum and other tissue samples from Michael Barber had been preserved and were still available. Analysis by both the National Poisons Unit and the manufacturers of paraquat, ICI Ltd, revealed that paraquat was indeed present.

The police were alerted and Susan Barber and her former lover (she had by now acquired a new one) were arrested, nine months after her husband had died. She confessed that she had found the weedkiller Grammoxone in the garden shed and had added some to the steak and kidney pie she prepared for her husband. She claimed that she had only wanted to make him ill, and repeated the poisoning twice when there seemed to be no effect. She had not known that paraquat takes some time to work its unpleasant effects. Susan Barber was convicted of murder on 1 November 1982. An astute and persistent pathologist, together with sensitive and specific chemical analysis, had been her eventual undoing.[9]

This recent murder case featured a chemical, paraquat, which is more often involved in accidental poisonings and suicides. The toxicity of paraquat has already been described (see pp. 104–6). What this story shows, from a forensic toxicology point of view, is that, even when the murderer thinks they have got away with the crime, an inquisitive scientist or doctor can be their undoing. It also shows that when normally healthy people suddenly fall ill and die, especially with unusual symptoms (as in this case), it is cause for suspicion. In this case even an early correct diagnosis would not have helped Michael Barber, as paraquat poisoning cannot be treated except at an extremely early stage. Once a fatal dose has been taken, the outcome is a foregone conclusion and the prognosis extremely unpleasant. However, as paraquat is reasonably easy to detect in the body of a victim, it is not an ideal poison for homicide.

Poisons used in time of war

Chemicals have, unfortunately, been enlisted in warfare with the design of lethal and unpleasant agents with which to attack enemies during battle. This is not a new technique. The ancient Chinese were familiar with 'arsenical smokes', and during the Peloponnesian War in 429 BC, according to Greek historians, the Spartans set fire to wood soaked with pitch and sulphur which burnt furiously and produced poisonous, choking fumes. In 1456 the Christians of Belgrade were saved from their Turkish attackers by a toxic cloud generated by burning rags dipped into a chemical prepared by an alchemist.

The large-scale use of chemicals in warfare, especially the use of gases, began in the First World War. Some of the gases used in warfare, for example chlorine and phosgene, were already known but had not been designed for the purpose. Others, like lewisite, were specifically designed for warfare. During the First World War chlorine gas was first used in Flanders, where it was released by the Germans in April 1915, killing 5,000 Allied soldiers and injuring a further 15,000. Other poisons used were phosgene, which replaced chlorine, and later in the war mustard gas and lewisite. The use of gas masks reduced the effectiveness of gas after the first few attacks. The Geneva Conference in 1925 outlawed the use of such gases, but it has not stopped them from being used altogether. In 1936 the Italians used mustard gas against the Ethiopians in Abyssinia, and more recently Saddam Hussein used it in the war against Iran in the 1980s, and in the massacre of thousands of Kurds in northern Iraq in 1988.

Chlorine and phosgene are both reactive gases which destroy lung

24. Victims of poison gas in the First World War. In this picture, the soldiers are believed to be affected by a tear gas, the effects of which were not permanent. Other gases used were much more toxic and lethal. Fortunately this terrible weapon has now been outlawed.

tissue. They cause the lungs of victims to become filled with fluid, a condition known as pulmonary oedema, and the victims asphyxiate. The junctions between the lung cells and the blood vessels are damaged, allowing fluid into the air spaces in the lung. Phosgene was the most effective and lethal agent used in the First World War.

Mustard gas, which smells like garlic, causes blistering of the skin, severe burns to eyes and lungs, and damage to internal organs such as the bone marrow and gut. The symptoms are delayed for some hours. Mustard gas or sulphur mustard was the most effective incapacitating agent used in the First World War.

Lewisite, dichloro(2-dichlorovinyl)arsine is a chemical that contains arsenic, which though a liquid is sufficiently volatile to be dispersed among enemy troops. The arsenic atom in the lewisite reacts with proteins and causes terrible blisters on the skin and damage to the eyes and lungs if inhaled. Fortunately, an antidote was devised as a result of the work of the British biochemist Rudolf Peters. The antidote was appropriately named British anti-lewisite (dimercaprol), and abbreviated to BAL.

The BAL molecule contains two atoms of sulphur which are able to bind with the arsenic in the lewisite and remove it. This removes the chemical from the skin of the victim and from the body. The antidote was based on an understanding of the way in which arsenic is toxic. BAL has also been found to bind metals and, as the first metal complexing agent (chelator), has been used clinically for the treatment of metal poisoning. Although it has now largely been superseded, it has been widely used as an antidote for other forms of arsenic poisoning and for poisoning by heavy metals such as mercury, gold, and sometimes lead poisoning in children.

During the Second World War far more lethal and potent poisons were developed, the nerve gases. In 1936 Dr Gerhard Schrader, working in the German chemical company I. G. Farbenindustrie synthesized a compound based around phosphorus for possible use as an insecticide. It was one of the first organophosphorus compounds to be made and was found to be highly toxic to mammals. It was manufactured under the name tabun. Further development led to another similar chemical, sarin, which was far more toxic. A third, soman, was developed in 1944. These chemicals were highly potent, lethal nerve gases. The mode of action of the nerve gases was similar to that of the insecticides that were developed from them, but they did not have the selective toxicity only for insects (see pp. 98–101 and box below). By the end of the war the Germans had 12,000 tons of tabun. About 1 mg is lethal to a human, which means that 1.2×10^{13} lethal doses were present in the stockpile, or enough to kill 10 billion people!

How nerve gases work

The effect of the nerve gases tabun and sarin is to block the enzyme acetylcholinesterase, just as the organophosphate insecticides do. This enzyme is found in many tissues but it has a particularly important job in the nerves where it removes the substance acetylcholine at the ends of the nerves. The blockade of the enzyme at nerve endings means that the acetylcholine, a neurotransmitter, is not removed. This results in the receptors responsive to the acetylcholine being over-stimulated. The result is constriction of the intestine, bladder, and air passages of the lungs. The muscles that enable us to breathe fail and the victim dies of asphyxiation. As the nerve gases are so potent, this effect is very rapid. Although there are antidotes (see pp. 101–102), the action of the nerve gas is so rapid that they would be ineffective unless used immediately.

The USA adopted sarin as its nerve agent after the war and then developed a more sophisticated gas, VX. This is not volatile and is described as persistent, remaining in a sprayed area, whereas sarin is non-persistent, that is, it evaporates and therefore a sprayed area becomes less hazardous with time.

10

Ginger Jake and Spanish Oil

Toxic Food Constituents and Contaminants

On 1 May 1981 an 8-year-old boy died in Madrid, apparently from a respiratory disease. Later six more members of the family fell ill with similar symptoms. Within a week at least 150 cases a day were being recorded and by June there were 2,000 cases in hospitals in Madrid. Over 20,000 cases were eventually recorded and there were 351 deaths.[1] The epidemic became known as the toxic oil syndrome and attracted world-wide attention and scientific interest. It was the result of the illicit sale and use of contaminated oil for cooking. This relatively recent example shows how vulnerable we can be to contaminants in our food (see below pp. 263–5).

Contamination can be natural as well as man-made. We all eat food regularly and only rarely do we think about what it may contain. Our complacency can easily be shattered by a bout of food poisoning, which is most commonly the result of bacterial contamination, although similar symptoms can occur as a result of contamination by other agents, such as metals, as happened in Camelford (see pp. 143–4). If the bacteria are *Salmonella* or *Campylobacter*, the consequences may be extremely unpleasant and can be serious. These bacteria infect the gut of the unfortunate victim, multiply, and affect the lining of the intestine, causing the symptoms of diarrhoea and sickness.

Food and drink may already contain toxins produced by bacteria, fungi, or other micro-organisms. This chapter deals with the toxins and toxicants present in our food and drink as a result of contamination by an organism such as a bacterium or fungus, or a toxin found naturally in the food. Alternatively, toxic chemicals may be produced by the process of cooking, or deliberately or inadvertently added to the food by humans.

Natural contaminants of food

Aflatoxin and the mouldy peanut

In 1960 some turkey farmers in the United Kingdom lost over 100,000 birds to a mysterious disease. At first it was thought to have been caused by a virus and was named Turkey X syndrome. Eventually it was traced to the use of a particular feed produced by the Oil Cake Mills company, mostly from groundnuts imported from South America. Veterinary pathologists found that the internal organs of the dead birds, for example the liver, showed a number of changes including cancer. What could have caused this sudden outbreak of cancer among turkeys? It was particularly puzzling as the feed had been given to turkeys before with no untoward effects. It was then discovered that the peanuts had become contaminated with what was eventually found to be a fungus, *Aspergillus flavus*. Scientists working on this discovered that contamination of feed with this fungus produced the same effects when it was given to animals under experimental conditions. The fungus produced a toxin, which was named aflatoxin after the fungus (*Aspergillus flavus*). The discovery gave rise to great concern, as nuts, particularly peanuts and peanut products, were of economic importance.

Turkey X syndrome, together with earlier episodes which had not been understood at the time, revealed that aflatoxins are highly toxic chemicals—potent carcinogens, which caused liver cancer in several species of laboratory animals including monkeys. Not all strains of the fungus produce the toxin, and the food on which the fungus grows, as well as the weather conditions prevailing, determine the extent of contamination. The toxin-producing strain of the fungus grows well on peanuts and the harvesting process and conditions of storage in South America, where the peanuts for the turkey feed were grown, favoured its growth.

Animals vary in their sensitivity to aflatoxins but it is certain that many different species of mammals succumb to liver cancer. There is now good evidence that humans are similarly susceptible to these potent toxins and that liver cancer may follow dietary exposure (see box). Agricultural workers may also be exposed by breathing in the dust from infected groundnuts.

The incidence of primary liver cancer is particularly high in certain parts of China (for example, the Guanxgi region), where there is exposure to aflatoxin, mainly through contaminated corn. Other areas of the world where the incidence is high are in Africa—the Gambia, Senegal, and Kenya, where exposure is through contamination of groundnuts and maize, staple foods for these regions and highly susceptible to *Aspergillus* infections. In some parts of the world local customs regarding the handling of foodstuffs make contamination more likely. For example, the Bantu tribes of Africa, apparently, prefer the taste of corn contaminated with fungus. It may not be coincidental that they too have a high level of primary liver cancer.

A connection has now been established between the occurrence of liver cancer and exposure to aflatoxin in the diet.[2] Other factors are also known to be important (see box).[3] High levels of exposure may also lead to acute hepatitis.

As peanuts are a popular snack, and many people enjoy peanut butter, this is a concern for many. Can contamination with aflatoxin be avoided? Exposure probably cannot be completely avoided, as aflatoxin appears to

Aflatoxins and how they cause cancer

The aflatoxins are a group of related **mycotoxins** produced by the mould *Aspergillus flavus*. There are four toxins, B_1, B_2, G_1, and G_2. The mould typically grows on crops such as grain and peanuts in hot, humid climates. There is evidence from epidemiology of an association between exposure to aflatoxin B_1 in the diet and liver cancer in humans. Aflatoxin B_1 is metabolized by the enzyme system cytochrome P450 in the liver to a chemically reactive metabolite (see pp. 19–23 and fig. 25), which reacts with molecules such as DNA and protein in liver cells.

At high doses the interaction with protein can cause immediate damage to the liver (acute hepatitis). At lower doses interaction with DNA will lead to mutations in the genetic code which can cause cancer. Laboratory studies in experimental animals like rats have revealed that aflatoxin B_1 can cause liver cancer, and the products of the reactive metabolites of aflatoxin can be detected in the blood of these animals. Furthermore the same metabolites have also been detected in the blood and urine of humans eating fungus-contaminated food in China, for example. Most recently a correlation has been shown between exposure to these fungal toxins, as indicated by metabolites of aflatoxins bound to protein in blood samples, and liver cancer in humans in China.

A specific mutation in the genetic material has been detected in people exposed to aflatoxin. An important factor in determining susceptibility may be individual variation in the metabolism of aflatoxin. A significant risk factor in liver cancer is infection by the virus that causes hepatitis B, and it appears that a combination of aflatoxin exposure and the viral infection makes individuals especially susceptible.

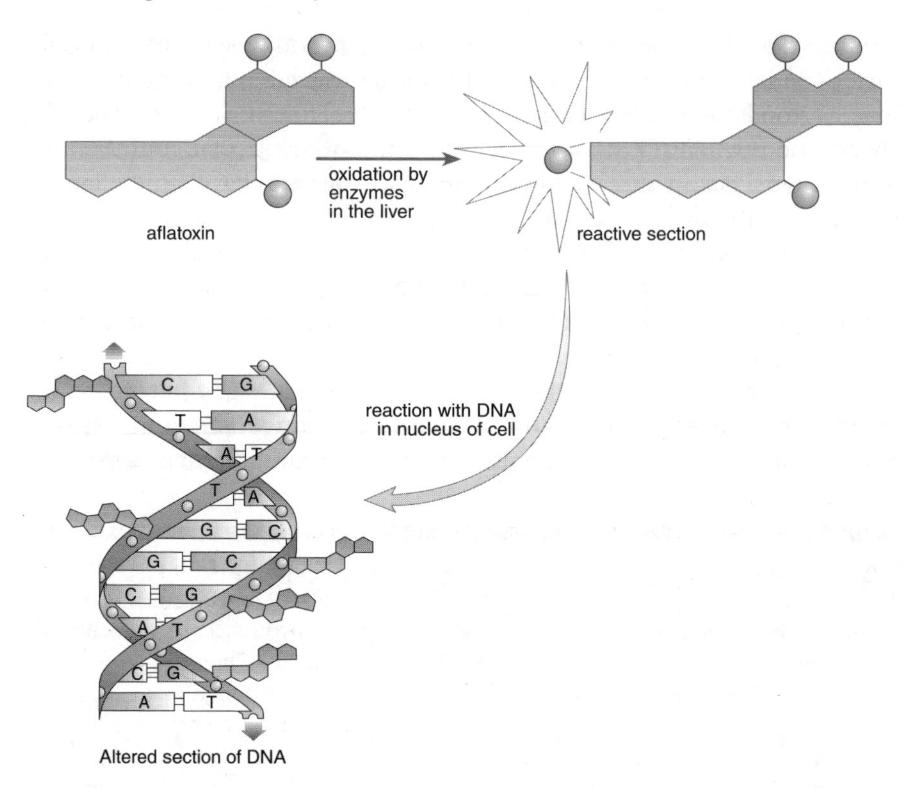

25. How the natural toxin aflatoxin, produced by a fungus, damages DNA. The aflatoxin is converted into a reactive product in the liver that interacts with components of DNA (the bases A, T, C, G). This changes the base code, leading to mutations and cancer.

have been detected in many foods such as rice, beans, sunflower seeds, spices, and figs as well as peanuts and corn. Nuts and other crops can be stored in better conditions and chemicals (insecticides and fungicides) employed to help reduce infestation with the mould, but this of course creates a dilemma. The use of chemicals such as pesticides on food crops may leave a residue in the food product. People are increasingly unhappy with the use of chemicals in food production and the possible contamination of food, and so-called 'organic' foods (produced without the use of chemical fertilizers, pesticides, and so on) have become very popular in the developed countries. However, significant levels of aflatoxins have been found in samples of organic peanut butter, whereas little or no aflatoxin was detected in standard (non-organic) peanut butter, made from peanuts in whose harvesting, storage, or processing chemicals had been used.

Fungicides, such as those used on seeds or food crops, are tested for toxicity in laboratory animals, and the levels allowed in food products are strictly controlled through government expert committees. There is always the possibility that some residue will remain in food, but the level is unlikely to pose a threat to human health. This is because safety factors are used to calculate the allowable limits for residues in food (see Chapter 12). Furthermore chemicals found to cause cancer in animals are not used as pesticides in human food production. Therefore the risk from fungicide is very low, whereas the risk of suffering a toxic effect and cancer from aflatoxin, a known and very potent carcinogen, in contaminated peanuts and other foods may be significant. The sensible course is to balance the relative risks, taking into account the likely exposure levels and the known potency of the contaminating chemicals, both natural and synthetic.

Fungal contamination of food crops can produce a range of other toxins, known collectively as mycotoxins.[4] These include ergot, tricothecenes like T2 toxin and nivalenol, ochratoxin, sterigmatocystin, fumagillin, and the delightfully named vomitoxin. It has been estimated that 25 per cent of the world's food supplies are contaminated with mycotoxins and in the past large numbers of people in many countries suffered ill health, infertility, and even death as a result of the contamination of food with these toxins. Even within the last century or so outbreaks of poisoning have occurred, for example ergotism was reported in Ethiopia in 1978 and tricothecene (T2 toxin) poisoning was known in the Soviet Union in the period 1931–47 (see below pp. 246–7).

Another mycotoxin is patulin, produced by certain species of the moulds *Penicillium* and *Aspergillus*, which may grow on apples. There have been occasions when apple juice has been contaminated with patulin but, provided producers remove infected apples and levels are monitored, there is not normally a danger from exposure (see Chapter 12).

The decline in the incidence of mycotoxin poisoning is probably a result of changes in climate, greater awareness of the problem of fungal contamination, improvements in agriculture and in the storage and processing of crops, and changes in diet. It is difficult for those of us who live in the developed world, and eat high-quality food bought in supermarkets, to appreciate how poor the quality of some food was, before modern farming and processing techniques, fertilizers, antibiotics, pesticides, and detailed scientific knowledge became widely available.

Ergot

Ergot is a fungus that grows on crops like rye, wheat, oats, and barley and can be recognized as a black growth ('ergot' is French for rooster spur). The fungus, of which *Claviceps purpurea* is a major variety, produces a mixture of chemicals, perhaps as many as twelve different related substances, known as alkaloids. The proportions of alkaloids will vary depending on the climatic conditions, the type of plant, and other factors. The different alkaloids have various effects, hence the symptoms vary between different outbreaks of poisoning. The chemicals produced by ergot include ergometrine (sometimes called ergonovine) and ergotamine. Ergometrine constricts blood vessels such as small arteries and can cause gangrene and loss of fingers or even limbs, abortion, and hallucinations.

The Salem 'bewitchings'

In Salem county in the north-eastern United States an outbreak of 'bewitchment' started in December 1691 and continued into 1692, reaching a peak in April and then again in July. Of the thirty people affected, twenty-four complained of fits, of being pinched, pricked, or bitten, all of which were considered to be symptoms of 'bewitchment'. These 'bewitchment' cases led to the infamous Salem witch trials, of which there are court records. The trials continued until October 1692 when they were prohibited by the governor of Massachusetts, by which time twenty people had been executed. The events have been chillingly portrayed in Arthur Miller's play *The Crucible*.

There have been various explanations for the strange occurrences, including hysteria, the influence of particular individuals in the local society, and rivalry between factions pro and anti a particular minister of the church. An alternative explanation is that the 'bewitched' individuals, mostly young people, were suffering from ergotism, poisoning with toxins produced by ergot. What is the evidence for this? The settlers in New England had started growing rye as a crop only after the wheat crops failed in previous years due to a fungal infection known as rust. The information available from records such as tree rings and contemporary diaries indicate that in 1690–1 the weather was unusually cool, and that there was a wet spring in 1692. These are ideal conditions for the growth of ergot, and rye is particularly susceptible to ergot infection. More important perhaps are the recorded symptoms. As well as fits, other symptoms mentioned in the court records include burning sensations; visions and out-of-body experiences; and temporary blindness, deafness, and speechlessness.[5] Some of these symptoms along with fits, are characteristic of ergotism.

According to the Salem court and other records, a number of people, as well as cattle which also exhibited strange symptoms, had died. These symptoms were not confined to Salem county, according to records for other counties in New England in the same year. There may have been other factors operating in Salem that led to the witch trials but ergotism is a plausible explanation for the underlying cause of the strange goings-on.

The Salem 'bewitchings' were by no means the only case that may have been due to ergotism but it is probably the most well known. Another example was 'La grande peur' in northern France in 1789, when civil disobedience and rioting coincided with a particularly serious infestation of the rye crop with ergot. The peasants were described as having lost their heads and an association was drawn between the behaviour, poor health, and 'bad flour'.[6] The chemicals produced by ergot cause the flour and the bread made from infected grain to be coloured red. (Indeed a reference to 'red bread' was made in a report from Salem in 1692.) The association between ergot infection of grain and adverse effects were described as early as 600 BC by the Assyrians who spoke of 'a noxious pustule in the ear of grain'.

One of the features of ergotism that has created some confusion is the variation in symptoms. This is due to the ergot fungus producing a number of toxins that have different effects on the body. For example, there is a reference in the writings of the Parsees around 350 BC to 'noxious grasses that cause pregnant women to drop the womb and die in childbed', a description of one of the toxic effects of substances produced by the ergot fungus. In contrast, reports from the tenth century describe victims 'tortured and twisted by a contraction of the nerves', or having their 'limbs eaten up by the holy fire that blacked like charcoal'. A seventeenth-century record describes slightly different symptoms: 'It seized upon men with a twitching and kind of benumbedness in the hands and feet. Terrible pains accompanied this evil, and great clamours and screechings did the sick make . . . and some had epilepsies, after which fits some lay as it were dead six or eight hours.'[7]

Ergotism

There are two forms of ergotism—convulsive and gangrenous—which reflects the mixture of alkaloids that the fungus can produce, with different effects on the human body. The convulsive form is characterized by writhing and shaking and was often reported as fits and convulsions. It may also be accompanied by muscle spasms, hallucinations, and delusions depending on the particular alkaloid content. The gangrenous form of ergotism involves constriction of the blood vessels in the limbs, fingers, and toes, which leads to a reduction of the blood supply and can result in the loss of fingers or toes or, if prolonged, even a whole limb.

Thus a confusing array of signs and symptoms have been reported as a result of ergot poisoning, including gangrene and convulsions, heart attack, vomiting and diarrhoea, blindness and deafness, delusions, hallucinations, writhing, and burning, itching, and even blistering of the skin. Fertility is reduced and pregnant women exposed experience stillbirth and abortion. Milk production is reduced in nursing mothers and the alkaloids pass into the mother's milk so exposing breast-fed infants.

The confusing range and the variety of symptoms led many people, during the Middle Ages and later, to regard this affliction as divine retribution for their sins. The disease was called 'holy fire' because of the burning sensations in the skin and the blackening of the hands and feet in cases with gangrene. Prayers were offered to St Anthony, a saint believed to have special powers against fire, and the affliction also became known as 'St Anthony's fire'. Thousands died in some outbreaks, for example in France in 994, where a well-documented report of ergot poisoning indicates that as many as 40,000 might have died. It is probably Russia that has suffered most from ergot poisoning epidemics where it was called zlaia korcha (evil writhing) and the mortality was at times as high as 40 per cent. Russia, with its continuing use of rye and rye bread, has been particularly prone to ergotism. The weather conditions also favoured the growth of the fungus and the combination of this with periods of famine increased the likelihood of poisoning. Outbreaks have occurred even in the twentieth century, for example in 1932–3.

As mentioned above, a more recent outbreak was reported in Ethiopia in 1978 following two years of drought. The barley crop was allowed to become dominated with wild oats and heavily contaminated with ergot. There were ninety-three cases in this outbreak, forty-four of which had symptoms of gangrene, as a result of which some lost limbs. Other symptoms reported during this outbreak were burning and crawling sensations on the skin, nausea, and vomiting and diarrhoea. The milk of nursing mothers dried up, leading to the deaths of fifty to sixty infants.

Ergot is not all bad, however, as some of the constituent chemicals (alkaloids) have important medical uses. As long ago as 1582 ergot was used to quicken labour in childbirth, while more recently ergometrine has been used to control bleeding after childbirth as it constricts blood vessels in the uterus. Ergotamine has been successfully used for the treatment of migraine where the blood vessels of the brain become relaxed: the alkaloid partially contracts these vessels. Ergot is yet another illustration of the principle of Paracelsus, that at the right dose a toxic chemical can have a beneficial therapeutic use.

Lucy in the sky with diamonds: hallucinogenic compounds

In 1943 a chemist named Albert Hoffman experienced and later recorded strange effects while working on the ergot alkaloids and similar chemicals in the laboratory: 'I sank into a not unpleasant intoxicated-like condition, characterized by an extremely stimulated imagination. In a dreamlike state . . . I perceived an uninterrupted stream of fantastic pictures, extraordinary shapes with intense, kaleidoscopic play of colours.' He had been working on lysergic acid diethylamide, otherwise known as the **hallucinogenic** drug LSD, a substance he had synthesized previously. As LSD is very similar to the ergot alkaloids in chemical structure these were used as a starting point to make LSD.

Hoffman's strange experience persuaded him to take what he thought was a small amount of LSD (a quarter of a milligram) but he was then to experience more alarming effects: 'Everything in my field of vision wavered and was distorted . . . Pieces of furniture assumed grotesque, threatening forms . . . The lady next door . . . was no longer Mrs R. but rather a malevolent, insidious witch with a coloured mask . . . A demon had invaded me, had taken possession of my body, mind and soul. I jumped up and screamed, trying to free myself from him, but then sank down again and lay helpless on the sofa . . . I was seized by a dreadful fear of going insane.'[8]

These effects would have been similar to those produced by the ergot alkaloids—imagine what people in medieval times would have felt when they experienced such effects during ergot poisoning. Lysergic acid derivatives are also found in plants such as Morning Glory, which was used by the Aztecs in their magical brews.

Tricothecenes and other mycotoxins

In 1942–3, during the Second World War, thousands died in the Soviet Union after eating bread made from flour contaminated with mycotoxins. The flour was made from grain that had been left in the fields over the winter and had become wet and mouldy before it was harvested. The mould was of the *Fusarium* type which can infect a variety of crops from rye to rice, and produces tricothecene mycotoxins (see below, this page). The disease caused by these toxins, alimentary toxic aleikiia, has a high mortality rate and a confusing variety of signs and symptoms, as described below.[9] Symptoms such as a persistent sore throat and fever often led to the belief that victims were suffering from infectious diseases like diphtheria or scarlatina. The effect of the toxins on the immune system may indeed have increased susceptibility to infection.

Fusarium species produce a number of toxic tricothecene alkaloids: T2 toxin, nivalenol, and deoxynivalenol can all be readily detected and measured. Contamination with deoxynivalenol, particularly of wheat and wheat products, seems to be the most common. The symptoms vary, depending on the extent of contamination and therefore the level of exposure to the toxin. In relatively mild cases the victim suffers from headaches, nausea and vomiting, painfully inflamed tonsils, sore throat, and skin eruptions. In severe cases diarrhoea, sweating, swelling and necrosis of the mouth, swelling of glands such as those in the armpits, blood in the urine, fever, delirium, convulsions, and brain haemorrhage can all occur. The toxins can also reduce the production of white blood cells in the bone marrow, which results in increased susceptibility to infections as white blood cells are involved in our defence against bacteria.

In addition to the former Soviet Union, contamination has been reported in other countries, including the UK, Germany, Finland, and New Zealand. In the UK in 1982 tricothecenes were detected in over 50 per cent of a sample of breakfast cereals. In an outbreak in India in 1987, several hundred people were affected with severe symptoms following the eating of contaminated bread, and several different mycotoxins were detected in the flour used to make the bread.

These toxins are stable and can survive cooking and processing. Deaths have even resulted from drinking beer made from contaminated grain.

Ochratoxin and Balkan nephropathy

Another important and widespread fungal toxin is ochratoxin, which is also found in cereals and, to a lesser extent, in coffee and cocoa beans. The toxin Ochratoxin A is the most commonly found and is produced by the *Aspergillus* type of fungus. Exposure occurs in many countries in Europe and affects farm animals as well as humans. The major toxic effect in both humans and animals is kidney damage and cancer of the kidney. The available epidemiological evidence indicates that the disease called Balkan nephropathy is associated with consumption of food contaminated with ochratoxin, and the toxin has been detected in the blood of people living in areas with a high incidence of the disease. Balkan nephropathy is a chronic (persistent) disease involving kidney dysfunction and failure which occurs particularly in the Balkan countries. The disease is also associated with a high level of kidney cancer.

Botulinum toxin and botulism

In addition to toxins from fungi, food may be contaminated with toxins produced by bacteria, such as botulinum toxin. Produced by the bacterium *Clostridium botulinum*, it is one of the two most potent toxins known to man, the other being ricin (see pp. 150–1). As little as one-hundred-millionth of a gram (1×10^{-8} g) of the toxin—about 0.00000000035 oz—is lethal to a human. Fortunately the toxin is destroyed by heat, and so cooked food is unlikely to be contaminated (although the bacterial spores are quite resistant). The bacteria grow in the absence of air (they are anaerobic), therefore the foods most likely to be contaminated are those that are bottled or canned and eaten without cooking, for example raw or lightly cooked fish.

CASE NOTES

Fish with botox

In October and November 1987 eight cases of botulism occurred, two in New York and six in Israel. All the victims had eaten Kapchunka, air-dried, salted whitefish, which had been prepared in New York and taken by individuals to Israel. All the patients developed the symptoms of botulism within thirty-six hours and one died. Some were treated with anti-toxin and two received breathing assistance.[10]

The first cases of poisoning associated with the toxin were probably described in eighteenth-century Germany where there was an outbreak, during which many people died from eating contaminated sausage. The syndrome caused by the toxin, known as botulism (derived from the Latin *botulus*, a sausage), has a high mortality rate. As the effect is essentially irreversible, victims who do survive can suffer paralysis for months. The symptoms can appear within a few hours of eating the contaminated food or may be delayed for several days.

The toxin attacks the function of the nerves, which cause contraction of the muscles, and this leads to weakness of the muscles, resulting in blurred vision and difficulty in swallowing and breathing. Breathing may stop completely and the victim may die if sufficient toxin has been absorbed into the body (see box).

Cases of botulism are rare and there is an antidote available, known as an anti-toxin. Refrigeration of food and the use of preservatives such as sodium nitrite have reduced the likelihood of contamination, but cases still occur.

Botox targets nerves

Botulinum toxin is a mixture of six large molecules, each of which consists of two components. One binds to the walls of nerve cells which then allows the whole toxin molecule to be transported inside the cell (rather like a Trojan horse). Once inside the nerve cell, the second component destroys a protein, synaptobrevin. By destroying the protein, the toxin prevents the release of the substance acetylcholine from small packets at the ends of nerves. These nerves, attached to voluntary muscles, need acetylcholine to allow the flow of signals (or impulses) between the nerve and the muscle (see Figure 26). By preventing the release of acetylcholine, botulinum toxin causes paralysis of the muscle. If the muscles affected are vital to life, such as those involved with breathing, the outcome is fatal.

Surprisingly perhaps, given its extreme toxicity, botulinum toxin was introduced into medical practice in 1983 to treat patients with squint. Since then its use has been expanded to include other disorders of muscle control, including those suffered by patients with cerebral palsy, or after a severe stroke where the brain cannot control the muscles, which may remain permanently contracted. Tiny amounts of the toxin are injected into the affected muscle which then becomes paralysed and so relaxes.

There are several forms of the toxin which are now marketed as Botox (type A toxin) and Myobloc (type B toxin). Botox has recently been used in cosmetic medicine as a way of reducing lines and wrinkles in the face due to ageing. There have, however, been cases of unwanted long-term effects after its repeated cosmetic use. The toxin must be used with great care. Here again we have an example of the principle of Paracelsus and the poison paradox: all substances are poisons, and it is the right dose that differentiates a poison from a remedy.

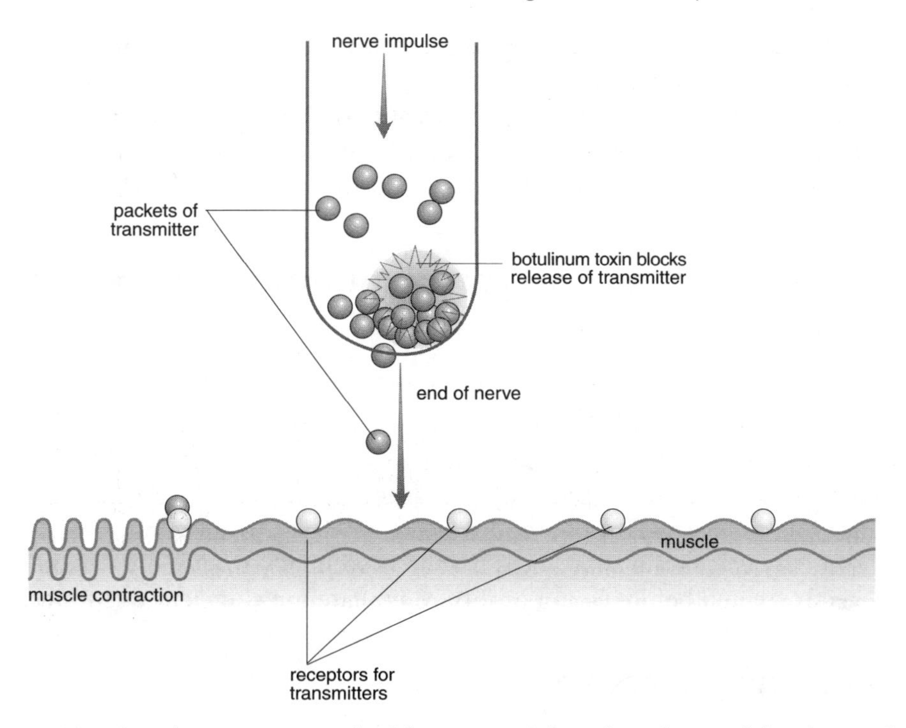

26. How botulinum toxin works. The toxin inhibits the release of the chemical acetylcholine, a neurotransmitter present in the ends of nerves. The nerve impulses are inhibited, so the muscles relax and the victim is paralysed. For more detail see the explanatory box.

Red tides and shellfish poisoning

. . . and all the waters that were in the river turned to blood. And the fish that were in the river died; and the river stank and the Egyptians could not drink of the water of the river.

Exodus 7: 20–1

A probable explanation of this is that the water was full of microscopic organisms known as dinoflagellates (phytoplankton), which impart a red colour to the sea or river and are the cause of the so-called 'red tides'. A more recent example was the red tide that occurred in the Gulf of Mexico in 1946 which resulted in the death of large numbers of fish. The dinoflagellates, which may be eaten by shellfish, produce highly toxic brevetoxins. If the shellfish are then eaten by humans serious poisoning may result.

A recent episode of serious poisoning in Canada, when poisoned seafood was eaten, highlights another toxin associated with shellfish, domoic acid, which is also produced by phytoplankton. A similar toxin is saxitoxin, produced by cyanobacteria, which may also infect shellfish such as mussels, and is responsible for cases of paralytic shellfish poisoning. All three toxins are known as neurotoxins because they act on nerves and cause neurological toxicity such as paralysis.

Foods that contain toxic substances

We have seen how micro-organisms such as fungi, bacteria, and plankton can cause toxic contamination of food. Plants themselves may also have naturally occurring toxic constituents. For example, as we saw in the discussion on herbal remedies in Chapter 6, the pyrrolizidine alkaloids are found in many plants which may contaminate crops such as wheat and hence find their way into the food we consume.

Some foods contain substances that are potentially toxic if they are not correctly treated before eating or if they are eaten in sufficient quantities.

Fugu: the puffer fish

About three to four o'clock in the morning we were seized with most extraordinary weakness in all our limbs attended with numbness of sensation like that caused by exposing one's hands or feet to the fire after having been pinched much by the frost. I had almost lost the sense of feeling nor could I distinguish between light and heavy objects, a quart pot full of water and a feather was the same in my hand. We each took a vomit and after that a sweat which gave great relief.[11]

So wrote Captain Cook in 1774 during a sea voyage. A fish unknown to him or his naturalist had been caught and was prepared for dinner. Thankfully Cook did not eat very much of it, otherwise it is unlikely he would have survived to recount his experience.

The poisonous properties of the puffer fish have been known for centuries, and were described as early as AD 220 in China. A Chinese treatise written in AD 600 by Chaun Yanfang (*Studies on the Origin of Diseases*) described the liver, ovaries, and roe as the most poisonous. Despite the risks, puffer fish is still eaten as a delicacy in Japan, where it is known as fugu. Fish that contain the toxin are also accidentally eaten sometimes. Consequently poisoning cases do occur. Between 1974 and 1983 there were 646 cases and 179 fatalities. It has been estimated that as many as 200 cases a year may occur, with a mortality rate of up to 50 per cent. A trained chef should know which parts of the fish to avoid.

27. Puffer fish (*Fugu rubripes*). The picture shows one type of puffer fish which, although eaten as a delicacy, have been responsible for a number of fatal poisonings.

The fish contains a very potent toxin, tetrodotoxin, which, it has been suggested, is produced by bacteria but stored by the fish, presumably to discourage predators such as fishermen. The toxin is lethal at a dose of about half a milligram for an average man. The toxin imparts an unusual flavour to the fish, and the pleasures and risks have been described thus: 'The pleasures of fugu are ambiguous, being both gustatory and tactile. The presence of traces of tetrodotoxin in the food apparently leads to a delicious feeling of warmth and euphoria. The first sign that something is going very, very wrong is the appearance of the same sensation in areas not extreme and the disappearance of the euphoria.'[12] Initial effects of the toxin are a tingling in the mouth followed within ten to forty-five minutes by muscular incoordination, salivation, skin numbness, vomiting, diarrhoea, and convulsions. Death occurs as a result of paralysis of muscles which makes breathing impossible (see Figure 28).

Polar explorers and vitamin A

In 1911 the geologist Sir Douglas Mawson led an expedition to explore the unknown region of Antarctica west of Cape Adare on the Ross Sea. Mawson set off east of Cape Denison with fellow explorers Xavier Mertz and Belgrave Ninnis. Having completed only 320 miles, Belgrave Ninnis

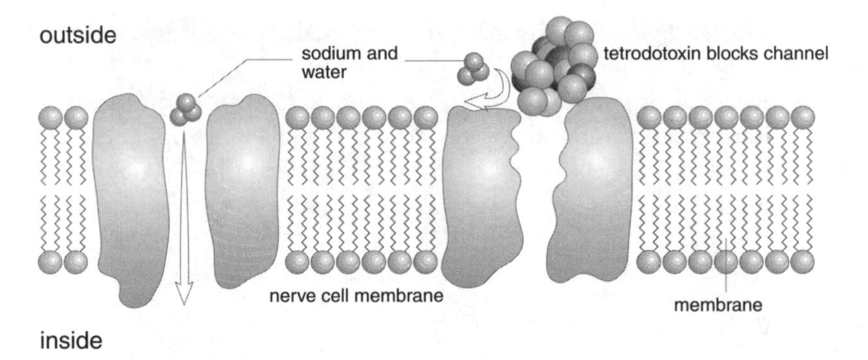

28. How *tetrodotoxin*, the toxin found in the puffer fish, is poisonous. This toxin blocks channels in the membrane of nerve cells. This stops nerves working and so the victim is paralysed and stops breathing.

fell into a crevasse and disappeared with most of the food and tools. Mawson and Mertz, in danger of starvation, killed the huskies for food. They fell ill with bizarre symptoms such as wasting, cracking and loss of skin, and dementia, and Mertz died. Mawson struggled almost 100 miles back to his base alone, only to see the ship *Aurora* sailing away. When he was eventually rescued he was so ravaged by disease that a horrified colleague cried out 'My God! Which one are you?' Mawson and Mertz suffered from vitamin A poisoning as a result of eating the dogs' liver.[13] This was unknown at the time but subsequent cases that occurred and research have confirmed it. The liver is an organ that stores and accumulates certain substances, for example vitamin A (a fat-soluble vitamin). The liver of certain animals, like huskies and also seals, polar bears, and some fish contain relatively high concentrations of vitamin A. It is hazardous and potentially fatal to eat these livers.

Vitamin A is an essential part of a healthy diet and a deficiency of it, which would be rare in the developed world, leads to defects in the eye, sterility, and increased susceptibility to infections and stunted growth in children. However, it is toxic if eaten in greater amounts than necessary. There are occasional cases of people eating meat such as seal liver, as well as cases where people become fanatical about taking dietary supplements and take too many vitamin pills, assuming that if one vitamin pill is good for them, ten must be even better.[14] As we have seen, the principle of Paracelsus warns us of this. Some vitamins do have a very low toxicity, which means that taking more than the recommended dose, within reasonable limits, may not be hazardous. Vitamin A, however, causes a number of toxic effects, as indicated above, but it is also a teratogen, that is, it can cause birth defects in the baby if too much is taken by a pregnant woman.

This reveals another interesting principle, that with some chemicals, such as vitamins and minerals (for example, metals like copper and zinc, which are essential for the human body to function), there is an essential dose 'window', above which there is toxicity and below which there is deficiency.

Cassava, apples, and almonds

These three edible plants all contain something in common, cyanide. This is not normally a problem with apples, as the cyanide is found in the pips which are not eaten, but almond kernels are eaten occasionally (see p. 218). By contrast cassava is eaten and is the staple crop for 40 per cent of people in sub-Saharan regions of Africa. Cyanide is found in the roots of cassava in the form of a compound called linamarin. After the plant is eaten the digestive juices act on the linamarin to release the cyanide (see box). Fortunately the cassava can be prepared for eating in such a way that the cyanide is removed. This is done by crushing the plant and steeping it in water. After soaking for some time, the remains of the plant are thoroughly washed and are safe to eat. If the cassava is not prepared in this way the cyanide remains in the plant and is poisonous. Cooking the cassava does not destroy the linamarin.

The cyanide released can be fatal if sufficient is eaten all at once (acute toxicity)—for example, a family of two parents and three children died in Nigeria after eating unwashed cassava—but this is relatively uncommon. The cyanide is more likely to cause a persistent type of toxicity (chronic toxicity) if it is not properly prepared. This is because cyanide is detoxified by the body quite efficiently using sulphur which is normally gained from protein in the diet (see box). Those who eat cassava tend to have a poor diet which is deficient in protein, and are therefore less able to detoxify the cyanide. With repeated exposure to cyanide in cassava the protein-deficient person suffers paralysis of the legs, which may result in their being crippled for life. It has been estimated that several thousand people in the African countries of Tanzania, Mozambique, and Zaire suffered from this in the 1980s, with up to 20 per cent of the population affected in some villages in Zaire. It may be that other nutritional factors and dietary deficiencies are involved in addition to the cyanide content of the cassava. An additional twist is that the detoxification of cyanide can exaggerate goitre and cretinism (mental retardation) in the offspring of people who also have low iodine in the diet (see box).[15]

How cassava can poison people

Cyanide in linamarin is in a form where it is attached to a sugar, the combination being known as a glycoside. The cyanide content can be 100 mg or more per kilogram of cassava. In some areas of Africa 1.5 kg of cassava may be eaten every day. A safe level in properly prepared cassava would be less than 20 mg cyanide per kilogram of cassava.

The cyanide in cassava that is not properly prepared can be released by the action of enzymes in our gut and can poison an individual. It does this by binding to an enzyme in the mitochondria inside the cell which blocks the production of energy. In vulnerable tissues, like the nerves of the spinal cord, this causes damage. The detoxication of cyanide with sulphur, derived from sulphur-containing amino acids in protein, produces thiocyanate, which is excreted into the urine, but this system is easily overwhelmed especially in those deficient in sulphur. Drought and poor soil seem to increase the production of linamarin in cassava, and it is possible that other dietary deficiencies may contribute to the nerve damage.

In areas where there is also low iodine in the diet a further complication may arise. The thiocyanate that is formed from the cyanide inhibits the uptake of iodine into the thyroid gland. Iodine is used to make hormones that regulate the metabolism of the body, and its deficiency leads to goitre (enlarged thyroid) and cretinism (mental retardation) in the offspring of women with the disorder.

Some other food plants contain linamarin, for example beans like lima beans. Provided these are soaked and boiled before eating, there is no hazard.

Fava beans and haemolytic anaemia

There are many other foods that contain toxic chemicals from a particular part of a plant or at some stage of their processing, which it is beyond the scope of this book to cover. It is well known, for example, that rhubarb leaves are poisonous, as they contain oxalic acid which is toxic, especially to the kidney. Unripe, green potatoes are also poisonous because they produce a poison called solanine. Some foods, for example fava beans, are poisonous only to certain individuals.

Fava beans cause a severe illness, favism, in individuals who have a particular genetic deficiency. It occurs only in men, and its incidence is very high in some communities. As many as 50 per cent of male Sephardic Jews from Kurdistan suffer the deficiency, which reduces the level of natural protection (glutathione) of the red blood cell. When a susceptible individual eats fava beans, the chemical divicine in the beans damages his haemoglobin, the essential oxygen-carrying protein in the red blood cell. The damaged red blood cells are removed from the blood by an organ in the body called the spleen and destroyed. The victim suffers from a deficiency of red blood cells (anaemia) known as haemolytic anaemia (see also p. 32).

Unnatural food contaminants

In addition to toxic substances of natural origin occurring in our food, man-made toxic contaminants can also accidentally get into, or even be intentionally added to, our food, sometimes with disastrous consequences. Potentially toxic contaminants can also be produced in our food by cooking.

Dioxin and the Belgian poultry scandal

In many developed countries random samples of food are regularly tested for substances such as dioxin, to see whether levels are within the acceptable range. In May 1999 it was reported by the Belgian government that relatively high levels of dioxin had been detected in chickens, dairy products, and eggs and in some pork and beef products.[16] This had occurred as a result of 2–4 kg of a substance, probably arochlor 1260, having been added to a batch of animal fat (80,000 kg) in a storage tank at a processing company near Ghent in Belgium. Arochlor 1260 has been used as a transformer oil and is a mixture of chemicals called polychlorinated biphenyls (PCBs). The arochlor had been heated at some stage, a process that produces dioxins. The company had sold the contaminated fat for use in animal feed and it had then been mixed with over 1,000,000 kg of feed. By this route the dioxin had found its way into chickens, as well as cows and pigs. There were nine producers of animal feed involved, and feed from several of them was found to be contaminated.[17] Several hundred farms were supplied with the feed, some of which was contaminated. Dioxin is not readily eliminated from the body in humans or other animals, and is absorbed into fat tissue. It therefore remains in contaminated animals and is transferred to humans when they eat the meat. If sufficient contaminated food is eaten over a period of time, the dioxin or similar contaminant level in a human body could increase (see pp. 128–31).

The scandal lead to the arrest of the owner of the processing company and the resignation of two Belgian government ministers, and cost the Belgian food industry billions of Belgian francs. But was there a risk to human health from the contamination?

The analyses showed wide variation in the dioxin levels of a small group of contaminated chickens but, taking the mean level, a 4 oz (112 g) portion of meat would have contained about 18,000 picograms of dioxin. The **tolerable daily intake** (TDI, see pp. 300–1) set for dioxin and similar compounds in the UK is 2 picograms per kg body weight per day or about 120 picograms for an adult per day. Clearly a portion of chicken contaminated in the Belgian incident would exceed the TDI by more than 100 times, but if the contaminated meat was eaten only once the consequences would probably not have been serious, as there are large safety factors employed in setting the TDI (see pp. 300–1).[18] It appears that there were no initial reports of illness following this incident, unlike the exposure at Seveso following which a number of children suffered from chloracne (see pp. 123–4).

The Michigan farm disaster

A similar incident occurred in 1973 in Michigan, where the chain of events was again clear but the health effects of greater concern. A company accidentally substituted a batch of fire retardant chemical, firemaster BP-6, for an animal feed supplement, with the result that the fire retardant chemical was added to animal feed. Contaminated animal feed was made and dispersed to affect hundreds of farms across the state of Michigan. The fire retardant chemical was a mixture of polybrominated biphenyls (PBBs), which are highly fat-soluble, persistent chemicals similar to PCBs and dioxin, although not as toxic as the latter. Sufficient of the contaminating PBBs was absorbed by the different farm animals for them to suffer toxic effects, some of which were bizarre, such as the development of long curved hooves in cattle. Thousands of animals were destroyed, but not before some of the chemical had spread through the food chain and many humans, both dairy farmers and local residents, were contaminated. Adverse effects were observed in many of these individuals, in particular suppression of the immune system which left them more susceptible to infection. The localization of PBBs in body fat and their persistence means that such effects can last for many years. It has recently been suggested that girls born to mothers who were exposed to the PBBs have entered puberty earlier than usual.

Other examples of food contamination have been discussed on pp. 128–31.

The ginger jake tragedy, or the jake walk blues

Then he would eat of some craved food until he was sick; or he would drink Jake or whiskey until he was shaken paralytic with red wet eyes.

John Steinbeck, *The Grapes of Wrath*

In early 1930 a new illness began to appear in the south and mid-west of the United States. First described as the '1930-type of polyneuritis', it was reported in the newspapers and seemed to affect a growing number of people. The victims experienced aching calf muscles and numbness in the legs, followed by loss of sensation, which progressed to weakness and then paralysis, which in some cases was permanent. This would result in the syndrome of 'foot drop'. For some victims a similar process would later occur in the arms, giving rise to 'wrist drop'. These symptoms were significantly different from any known disease and puzzled physicians. When the number of known victims reached the hundreds, there was even greater concern. Over a six-month period in 1930, 400 cases of the new illness were admitted to Cincinnati General Hospital. The areas affected were mostly in the southern and mid-western states of Oklahoma, Missouri, Arkansas, Texas, Kansas, and Ohio.

Eventually it became clear that the syndrome was associated with the use of an extract of Jamaica ginger, normally sold for medicinal purposes.[19] Remedies like this were popular at the time and some of them were well known, for example the Vegetable Compound of Lydia Pinkham ('Lily the Pink'). The ginger was dissolved in up to 90 per cent alcohol to produce a syrup or tincture (for example, Tinctura Zingiberis was 90 per cent alcohol). A few drops in water would usually be taken, and the extract was sold in pharmacies in 2 oz bottles. Popularly known as 'jake', the remedy had been in use since the nineteenth century for the treatment of minor ailments such as colds, period pains, headaches, and flatulence and to aid digestion.[20]

This episode happened shortly after the stock market crash of 1929, during the Depression, and in a time of Prohibition, when the sale and consumption of alcohol were forbidden. Many people who wanted a drink did not have the money to buy liquor at the inflated prices charged in speakeasies in Chicago or New York by the illicit liquor pedlars and bootleggers like Al Capone. Ginger jake, which contained at least 70 per cent alcohol, was seen as a solution to the dilemma, and at 35 cents a bottle it was affordable. The pharmaceutical preparation was allowed for sale as it contained sufficient irritant ginger to be described as 'undrinkable'. The ginger flavour could be diluted by mixing the jake with a soft drink like Coca-Cola which gave those who were determined to drink at least as much alcohol as they would have had in a legal drink before Prohibition. In 'dry' parts of the United States, jake had already been used as a source of alcohol by some individuals and this had been reported by newspapers since the turn of the twentieth century. During Prohibition there was every incentive to produce more acceptable versions of jake, containing little if any ginger and adulterated with other substances such as castor oil or molasses. These illegal brands were for the most part consumed without any apparent ill effects, apart from those caused by the alcohol.

The first case of the new syndrome associated with the use of extract of Jamaica ginger was seen by Dr Goldfain at an Oklahoma hospital on 27 February 1930. By the end of the day he had treated four patients, all apparently suffering from the same syndrome, a multiple neuritis (inflammation of the nerves). It so happened that one of the patients was a pharmacist who had sold the jake extract and had himself taken some ten days earlier. Another sixty-five patients with the same symptoms presented themselves, all of whom had taken Jamaica ginger extract seven to sixteen days prior to the onset of their symptoms. While the association between the use of the Jamaica ginger extract and the illness was made, this was puzzling because the remedy had been used since the middle of the nineteenth century without any apparent problem.

When the federal authorities became aware of the problem, which was growing to epidemic proportions, samples of the jake used by affected victims were collected and analysed by the Treasury Department's Bureau of Industrial Alcohol. The chemists soon found that a substance identified as tri-orthocresyl phosphate (TOCP) was present to the extent of about 2 per cent in the adulterated samples. In experiments the adulterated samples were found to produce similar effects when given to rabbits. Calves that had been fed the adulterated jake for medicinal reasons also developed the syndrome but, curiously, monkeys and dogs did not. Eventually it was discovered that monkeys and dogs were not susceptible, unlike rabbits, due to differences in the absorption of the TOCP from the digestive system. It was suggested that differences in the way TOCP is broken down in the different species may also account for the differences in susceptibility between the types of animals (see box).

Many shipments of Jamaica ginger extracts were seized by the Food and Drug Administration and the primary source of the adulterated samples was eventually traced to a company in Boston, Hub Products. The president of the company, Harry Gross, and a part-owner of the firm were charged with conspiracy to violate the Prohibition Act and the Food and Drug Act. They pleaded guilty and were convicted.[21]

How tri-orthocresyl phosphate caused jake leg

Tri-orthocresyl phosphate (TOCP) is an organophosphate used as a solvent in industry. It has also been used as an additive for aero engine oil. It causes degeneration of the peripheral nerves (the nerves serving the limbs and hands and feet), a disorder called peripheral neuropathy. This toxic effect is due primarily to a specific interaction between the organophosphate and a particular protein in the nerves, which is an enzyme. TOCP becomes bound to it and then undergoes a change, known as ageing. The reaction with the protein occurs relatively rapidly, probably within an hour of exposure, but the toxic effect, delayed neuropathy, may not be apparent for ten days or more. It seems that the protein, which is attached to the nerve cell, is critical to the function of the nerve, which starts to die after the TOCP binds to it. Thus in the long nerves that serve the legs and arms a process of dying back occurs, which involves degeneration of the nerve and the myelin sheath that surrounds it. The nerve dies progressively, starting at the end and working upwards towards the spinal cord. This degeneration of the nerves means that the muscles of the legs and arms are not stimulated and consequently the poisoned victim suffers paralysis. Post-mortem examination of the victims of the ginger jake episode who died showed inflammation and degeneration of the nerves in the legs.

Why had TOCP been added to the ginger extract? Tri-orthocresyl phosphate was readily available as it was a constituent of lacquers and varnishes, and was extensively used in the leather industry. It was seen as an ideal solvent to use in the preparation of jake because it was odourless, tasteless, colourless, and cheap. It was also miscible with the resinous extract of ginger and soluble in alcohol. Unfortunately, it was very toxic to humans, a fact that appears not to have been known. At the time there was no law requiring food additives or medicinal products to be tested for safety, and thus the suppliers had broken the law, not by supplying a contaminated and unsafe product, but by selling a product that was not as described in the US pharmacopoeia. In June 1930 twenty-one men and six New York corporations were indicted for conspiracy to violate federal laws.

There were at least 35,000 victims in the ginger jake poisoning incident, and some estimates claim that as many as 50,000 people were affected. There seems to have been little sympathy or help for the victims, perhaps because they were thought to be alcoholics or down-and-outs. While the majority of victims were relatively poor, many were ordinary people who were using the preparation for a legitimate purpose. For example, a woman in Boston became a victim after drinking five 2 oz bottles of jake over several days to help her recover from influenza.

There appeared to be considerable variability in the response to the TOCP, with some victims suffering paralysis and even death after only one drink of the contaminated jake. Most of the victims suffered paralysis for a long time, some permanently, and required crutches or sticks to walk, which they did with a shuffling gait, in some cases dragging their feet ('foot drop'). This became popularly known as 'jake leg', and the way victims walked became known as 'the jake walk'.

Although it may not have received much official recognition, the epidemic of ginger jake poisoning passed into popular legend in songs and novels, hence the quotation from the John Steinbeck's novel *The Grapes of Wrath* at the beginning of this section. At least eleven blues songs referring to the poisoning episode were recorded between 1930 and 1934, for example:

> I went to bed last night, feeling mighty fine
> Two o'clock this morning, the Jake Leg went down my spine.
> I had the Jake Leg too.
>
> I woke up this morning, I couldn't get out of my bed,
> This stuff they call Jake had me nearly dead.
> I had the Jake Leg too.
>
> Ray Brothers, 'Got the Jake Leg Too' (1930)

Songs like this were all that remained for the victims, some of whom continued to live for forty more years without compensation or official recognition.[22]

Since then other occurrences of tri-orthocresyl phosphate poisoning have been documented.[23] An incident in the early 1930s in Europe involved the use of apiol (an alcoholic extract derived from parsley seeds), which was adulterated with TOCP for use in inducing abortions, and led to paralysis in several hundred women. Another episode, which occurred in Morocco in 1959, has disturbing similarities with the Spanish oil disaster which occurred twenty years later (see below). As in the USA, there was an outbreak of a 'paralysing disease', which was believed to be the result of either a viral infection or poisoning. The symptoms were similar to those described in the ginger jake episode and resulted in an ungainly, high-stepping gait and a weakness in the hands of some victims. The epidemic centred on the city of Meknes in Morocco and affected only the Muslim population, sparing both Europeans and Jews. Among the Muslims, it was generally the poorer people who were affected, and there was a greater incidence in women than in men, and in adults than in children.

These observations seemed to rule out an infectious disease as the cause.

Some victims associated their symptoms with the use of a cooking oil they had bought in local shops and markets. One family were so suspicious of the 'dark oil' they had bought that they fed some of the food cooked in the oil to their dog first. When the dog showed no immediate ill-effects they proceeded to eat the meal themselves. Within a few days, however, both the family and their dog suffered the typical symptoms of aching muscles and paralysis. There was thus suspicion about the oil, and the similarities with symptoms of tri-orthocresyl phosphate poisoning suggested that the oil may have been contaminated. Analysis of the suspect cooking oil and comparison with other cooking oils showed that the 'dark oil' was vegetable oil containing about 3 per cent cresyl phosphates.

The oil was in fact for industrial use, produced to withstand the high temperatures in turbo jet engines, the tri-orthocresyl phosphate being a specific additive. Due to a change in engine design the oil had been discarded, and someone had taken the fateful decision to sell it as cooking oil. At least 2,000 people suffered toxic effects from the contaminated oil, but thankfully there were no deaths. In this case the majority suffered only relatively mild and probably reversible paralysis.

The Spanish toxic oil syndrome

On 1 May 1981, in a district of Madrid in Spain, an 8-year-old boy died, apparently from a disease affecting the respiratory system. Later six more members of the family fell ill with similar symptoms. Other cases were reported at around the same time and within a few days the Directorate of Public Health became aware of the problem. This marked the beginning of an unusual outbreak of respiratory disease in and around Madrid. Within a week 150 or more cases per day were being recorded in various areas of Spain, and by June there were 2,000 cases in hospitals in Madrid. The outbreak rapidly became an epidemic, with over 10,000 cases being seen in hospitals in two months, and more than 20,000 cases were recorded by December 1982. There were a total of 351 fatalities. The illness was widely reported and was so serious that the World Health Organization convened a conference in 1983.[24]

Initially an infection was thought to be the cause of the pneumonia-like symptoms, but patients did not respond to treatment and the spread of the disease and the location of victims suggested otherwise. Poisoning was considered a possible cause by a Spanish physician, Dr Tabuenca, who tested his hypothesis by studying a group of children affected by the syndrome. He found a strong association between the disease and the use of cooking oil bought from door-to-door salesmen. The distribution of some cases was directly related to the route taken by an itinerant door-to-door cooking oil salesman. This was a clever piece of epidemiological detective work.

It transpired that the toxic oil was rapeseed oil that had been adulterated by the addition of 2 per cent of a chemical called aniline, as required by law in Spain for imported rapeseed oil so that it cannot be used for cooking. This oil had been refined and was sold for human consumption, as had been done before without adverse effects. It would appear that a batch of oil may have been refined differently or somehow became contaminated. The toxic oil was found to come from only one source within a particular refinery. The Spanish government agreed to replace the suspect oil with pure olive oil, after which the number of new cases subsided. Collection of contaminated oils known to be associated with the syndrome for analysis was a difficult task due to the government exchange programme. The results showed that the oil associated with the syndrome had a similar composition to rapeseed oil and contained contaminants as a result of the added aniline and the subsequent refining process.

The disease appeared after a latent period of at least one to two weeks, longer in some cases, and an apparent relationship between the extent of use of the oil and the effect (a dose–response relationship) was noted in one report. The syndrome had an initial phase lasting one to two months, with effects mainly on the respiratory system and the accumulation of fluid in the lungs. There were many deaths at this early stage from respiratory failure. In the next phase (two to four months) there was muscle pain and liver damage. In the final phase there was muscle wasting and weight loss and the skin was affected (see box).

There have been suggestions in the press that the agent responsible for the toxic oil syndrome was an organophosphate pesticide which contaminated the food eaten by the victims. However, with the exception of peripheral neuropathy, an effect seen in only some victims, which may be caused by some organophosphates (for example, tri-orthocresyl phosphate, TOCP), the effects observed in the toxic oil syndrome are very different from those caused by organophosphates. The weight of current scientific evidence points to the cause as contamination resulting from the aniline-adulterated, refined rapeseed oil. While the association between the oil and the syndrome is not proven, information has recently emerged that demonstrates a strong association between one substance in the contaminated oils used by victims and the disease. The mechanism by which it causes the syndrome is not yet understood, however. The most recent

The pathology of toxic oil syndrome

It appears that the primary targets of the 'toxic oil' were the cells lining the lungs which became damaged, leading to problems in the respiratory system such as the accumulation of fluid in the lungs. The initial phase lasted for one to two months and accounted for many of the deaths.

Similar cells in blood vessels were also damaged, leading to inflammation of the vessels, and to blocked or narrowed small blood vessels (capillaries) and the narrowing of larger vessels (for example, arteries). This resulted in a reduction in the blood supply to organs such as the lungs and the liver, as well as to the nervous system and muscles, and accounts for the dysfunction seen in these organs and tissues. This probably occurred during the second phase which lasted up to four months following exposure. There were also features of the syndrome, such as a marked rise in the number of certain white blood cells (eosinophilia), that suggested that the immune system was involved. It is not clear, however, whether this was an early or a later development.

The final phase, which included weight loss and muscle wasting, may have been a result of the changes in both the blood supply and the **immune response**.

results suggest that there may be a genetic factor at work, with some individuals being more susceptible than others (see box below).

This episode of poisoning highlights the difficulties of studying problems associated with food contaminants, which are often factors beyond the control of the scientist. In this case the difficulty of obtaining samples of oil reliably associated with the syndrome, and the absence of an animal species that shows similar effects, have greatly hampered the research (but see box).

The episodes of poisoning we have seen show how a large number of people may be affected by a toxic contaminant in a foodstuff. A more subtle toxic reaction to a food contaminant, however, could affect many more people before it was detected.

Food contaminants produced by cooking

When food is cooked in certain ways it can become contaminated with potentially toxic chemicals. For example, it has long been known that cooking meat over charcoal will result in chemicals called polycyclic aromatic hydrocarbons (PAH) contaminating the meat. This is because when

A possible mechanism underlying the toxic oil syndrome

It is known that the rapeseed oil had been adulterated with aniline, and that the aniline reacted with **fatty acids** present in the oil to form anilides. These were detected in the oil collected from victims. A good correlation between the amount of a particular anilide and the occurrence of the syndrome was found, which suggests a relationship between consumption of contaminated oil and the syndrome. However, there were individuals who had apparently consumed the contaminated oil but had not shown symptoms, and vice versa.

Very recently a substance has been detected in the oil used by the victims that shows a much higher correlation with the occurrence of the syndrome in particular households and that was found only in oils from the one suspect refinery. This substance, which was derived from the aniline adulterant, was found to be toxic when administered to mice. The correlation between the presence of the aniline derivative and the occurrence of the syndrome was found to be very high (the odds ratio was 26; see p. 289 for discussion of the odds ratio).

Research on the victims of the poisoning also found that they were more likely to have a deficiency in an enzyme, N-acetyltransferase, than those who did not suffer the effects. This enzyme detoxifies aniline and similar chemicals. There is a known genetic variation in the ability to acetylate in the human population, with about half the European population showing a relative deficiency in this enzyme (see pp. 29–31, 70, for more on this). Such individuals, known as slow acetylators, would be less able to detoxify the aniline. This could account for the variability in susceptibility, whereby some individuals consumed contaminated oil but did not suffer toxic effects.[25]

juice from the meat drips onto the hot charcoal some of the substances in the juice are broken down and altered. The heat will also change these substances in the meat itself. One of the products of this process, a PAH called benzo(a)pyrene, is volatile and mixes with the smoke that envelopes the meat on the grill. This substance, a known carcinogen that is also found in tobacco smoke, has been measured in meat cooked in this way. Although it is not known whether the substance has been responsible for any cases of cancer in humans eating food cooked in this way, research has shown that sufficient of this and similar substances is present in barbecued meat to cause biochemical changes which can be measured in the human liver (the change detected was induction of certain enzymes; see p. 34).

Another group of chemicals formed in meat by cooking are

heterocyclic amines, for example the chemicals abbreviated PhIP and MeIQX, two of at least twenty that have been identified. The two heterocyclic amines PhIP and MeIQX have been found to cause mutations (changes and/or damage to the genetic code) in the DNA in bacteria. These compounds are therefore **genotoxic**, a feature shared by many, though not all, carcinogens. The two compounds also caused cancers in various tissues, including the colon and mammary glands, in animals such as rats and mice. Heterocyclic amines are produced by the effect of heat on components of lean meat, in particular the amino acids in proteins and the substance creatine.

Research has shown that there is a relationship between certain cancers, for example cancers of the large intestine, breast, and pancreas, and the intake of well-done, fried, or barbecued meats. By studying patients who suffered stomach cancer, it was also found that those who ate beef that was well done had three times the risk of developing the cancer.

Dangerous French fries and a Swedish tunnel

In 2002 new concern was expressed about the effect of cooking, in particular baking and frying, on foods rich in starch and other carbohydrates. The story starts, improbably, in a railway tunnel being constructed in southern Sweden in 1997. Owing to the level of groundwater, the work on the tunnel was not progressing well, and a cheap sealant was used on the inside walls to stop the seepage of water. The sealant contained acrylamide, a toxic substance suspected of being a carcinogen. Unfortunately, the acrylamide contaminated groundwater that fed local wells, potentially exposing the population to a toxic substance. Cows grazing in the vicinity showed symptoms of paralysis and salmon in nearby rivers were found to be contaminated with acrylamide. Worse still, a significant number of the tunnel workers reported numbness in their fingers, legs, and arms, and headaches and dizziness. These are recognized symptoms of damage to nerves, especially those in the hands and feet (an effect known as peripheral neuropathy), an effect of acrylamide previously observed in both animals in the laboratory and humans exposed to the chemical. A number of the workers with the highest exposure had evidence of damage to the nervous system.

When the workers were investigated at the university in Stockholm and the level of acrylamide in their blood measured, the results were very surprising. As is normal in such a study, a group of control subjects, volunteers who had not been exposed to the sealant or any other source of acrylamide, was investigated at the same time as the workers. The control subjects were also found to have high levels of acrylamide in their blood but at the time the source was unknown. After initial scepticism, the data in the control subjects was replicated in other countries and eventually the source of the acrylamide was found to be food, especially foods such as French fries and crisps. It was found that a large range of foods were contaminated with acrylamide, especially foods that had been fried. The higher the frying temperature and the longer food was fried, the greater the contamination, hence overdone chips had the highest concentrations measured (12,800 ppb).

In the Swedish studies estimates were made of the likely daily intake of acrylamide and the doses were found to be about a thousandfold lower than were likely to cause effects on the nervous or reproductive systems. However, the effect of repeated exposure over many, many years in those whose regular diet consists of fried foods that contain relatively high levels of acrylamide needs to be assessed.

Whether any one of the contaminants or mixture of contaminants of food introduced by cooking, if any, is responsible for some cancers is not currently known and awaits further research. What is clear is that cooking food in certain ways will introduce known carcinogens, such as polycyclic aromatic hydrocarbons, heterocyclic amines, and acrylamide into our diet. By avoiding this style of cooking any risk can be minimized.

In addition to the food contaminants that have been discussed, there are also man-made chemicals that can inadvertently enter the food chain or water, for example pesticides, industrial chemicals, and substances fed to animals like antibiotics and anabolic steroids. Some of these have been discussed in Chapter 5. In addition, there are also contaminants that are derived from the packaging or processing of the food. An important example is the **phthalates**, chemicals that are added to plastics, some of which are used in packaging food or drink. These substances may leach out of the packaging and into the food. As they are believed to have effects on the reproductive system (endocrime disruption) there is clearly concern about their potential appearance in food. This type of substance has also been discussed in Chapter 5.

Finally, our food may contain specific additives, chemicals intentionally added to food in order to improve it in some way. As we shall see in Chapter 12, these are tested for safety and cannot be added to food unless they are deemed to be safe.

11

Butter Yellow and Scheele's Green

Food Additives

THE intentional addition of substances to food is not a recent practice, as might be supposed. For centuries salt has been commonly used as a preservative and spices to flavour, and perhaps disguise, poor-quality food. In Ancient Rome sapa was used to sweeten food and wine, and the ancient Greeks also added it to wine. It eventually became associated with adverse effects, such as constipation, tiredness, colic, infertility, and anaemia.

What was in the sapa that caused these effects? We have come across these symptoms before but in a different context. The sweetening agent sapa was made by boiling grape juice or wine, often after it had started to turn to vinegar. The boiling was done ideally in lead pots, most of the liquid evaporating to produce a syrupy solution. In describing the process, the Roman writer Pliny (AD 23–79) said that the use of a lead pot was essential for good sapa. In grape juice or wine there are acids like tartaric acid and citric acid, and in vinegar the alcohol in the wine has been turned into acetic acid by bacteria. When the grape juice was boiled in lead pots, these acids reacted with the lead-producing salts, for example lead acetate, which were soluble in water. The lead acetate was very sweet—it was called 'sugar of lead' in later times when its deadly composition was known. The concentrated solution, or the crystals that formed from it, were used to sweeten food and wine. It also preserved wine as the lead inhibited the growth of any bacteria or fungus. As recently as the nineteenth century lead shot was added to bottles of port for this reason.

The use of sapa became associated with adverse effects on health, and it became known that wine to which it had been added would induce abortions, for example. For this reason prostitutes used it and at the same time acquired a pale complexion as a result of anaemia caused by the lead. Pliny knew of the adverse effects of wines to which sapa had been added, writing that 'from the excessive use of such wines arise dangling paralytic hands', no doubt a reference to the effects of lead on the nerves (see p. 142). Dioscorides, a Greek physician writing before Pliny, stated with prescience that 'wines so treated are most hurtful to the nerves'.

While the use of 'sugar of lead' as a sweetener died out after the fall of the Roman Empire, the use of lead salts in food did not. Much more recently, for example, lead chromate, or 'chrome yellow', has been used to colour sweets and custard powder. As recently as the nineteenth century toxic metal salts were used as food colourants, for example the dye Scheele's green was used to colour blancmanges green. This dye is copper arsenite, containing both copper and, more importantly, arsenic.

Early in the twentieth century a very hazardous chemical was used as a food colourant: 4-dimethylaminoazobenzene, so-called 'butter yellow', was a yellow **azo dye** used in some countries to colour butter, before extensive testing was required. When it was studied in 1947, the dye was shown to be a potent carcinogen capable of causing liver tumours in experimental animals, and it was rapidly withdrawn. Fortunately, the treatment of food with additives, which has now become more extensive, is now safe.

Substances that are intentionally added to food and do not contain any nutritional value are termed food additives. These are added to, or used in the preparation of, food as preservatives or to change its texture, consistency, taste, colour, alkalinity, or acidity. There are several types of food additives, details of which are given below. In Europe food additives are given an E number[1] (see Table 8 for examples), while different code numbers are used in the USA. These numbers appear on the packaging. It has been estimated that there are well over 3,000 different additives, most of which are used purely for cosmetic purposes, and relatively few of which are employed as preservatives or in processing.

Almost all of us are exposed to these additives. A cursory look around my kitchen revealed:

orange squash, containing **beta carotene** (colour);
cola drink, containing caramel (colour), aspartame and acesulfame (sweeteners), and sodium benzoate (preservative);
tonic water, containing saccharin and aspartame (sweeteners), as well as quinine (flavour);
a packet of soup, containing caramel (colour);
a jar of pickle, containing caramel (colour);

glacé cherries, containing erythrosine (colour) and potassium sorbate and sulphur dioxide (preservative);
custard powder, containing annatto (colour); and
a packet of prawn crackers, which while it boldly claimed to contain 'no artificial preservatives or colouring agents' admitted the addition of monosodium glutamate (flavour enhancer).

On venturing into the bathroom and inside the medicine cabinet, I discovered to my surprise a cold cure containing sunset yellow (colour). However, pride of place in the list of additive-laden products went to the vitamin pills which contained the following:

colourants: titanium dioxide, iron oxide;
anticaking agents: magnesium stearate, silicon dioxide;
sweeteners: aspartame, acesulfame K;
glazing agents: shellac, carnauba wax;
stabilizers: sodium carboxymethyl cellulose, acetylated monoglyceride;
tableting aid: polyvinyl pyrrolidone;
and the vitamins, of course!

Types of food additives

Preservatives are substances added to food to prolong their shelf life. They prevent or reduce bacterial or fungal growth. Examples of preservatives are salt, nitrites, sulphur dioxide, propionate, and benzoate.

Antioxidants are added to oils and fats to stop them going rancid because of oxidative damage. They may also be added to fruits and vegetables. Examples of antioxidants are butylated hydroxytoluene, **ascorbate**, and **α-tocopherol**.

Emulsifying, stabilizing, and thickening agents are added to food to improve consistency, stability, and homogeneity of the product. Examples are **monoglycerides and diglycerides**, agar, and various vegetable gums.

Colouring agents are added to make food and drink look more appealing. Some are synthetic, for example tartrazine and erythrosine, while others are natural, for example carotene and annatto.

Flavouring agents and enhancers are used to create or improve the flavour of a particular foodstuff. This is the biggest group of additives. Some flavours are natural, for example plant extracts and resins, while others are synthetic, for example esters and ketones. The most widely used flavour enhancer is monosodium glutamate.

Artificial sweeteners are use to sweeten food or drink without adding calories. Examples of artificial sweeteners are saccharin, aspartame, and acesulfame K.

Nutrients are considered as food additives only in the USA, and not in other countries. The group includes vitamins, minerals, and essential amino acids.

Miscellaneous additives, including acidity regulators used to adjust the pH of fruit juices, for example; anticaking agents (for example, magnesium stearate, silicon dioxide) which are added to sugar or salt so that they flow freely; anti-foaming agents to stop liquids foaming; flour treatment agents which improve baking properties; glazing agents, like shellac and carnauba wax; propellants; and raising agents.

The use of food additives has increased enormously in the past thirty years, especially as the Western diet now comprises many processed foods. This means that we are exposed to a variety of food additives. Are all of these additives necessary? And what are their effects on our bodies?

Some additives clearly serve an important function. Preservatives help to prevent food from spoiling and enable processed food to be stored for much longer. They reduce the likelihood of bacterial contamination in the food we eat. Sodium nitrite is added to cured meat, for example, to prevent the growth of organisms like *Clostridium botulinum*, which causes severe toxicity, botulism (see pp. 249–51). Preservatives also reduce chemical degradation and so allow food to have a longer shelf life. Other additives may also have a beneficial function, for example artificial sweeteners reduce the sugar intake of people who suffer from diabetes or obesity.

Table 8 Examples of food additives

ADDITIVE TYPE	NAME	E NUMBER
Preservative	Sodium nitrite	E250
Antioxidant	Butylated hydroxytoluene	E321
Emulsifier	Pectin	E440a
Colourant	Caramel	E150
Flavour enhancer	Monosodium glutamate	E621
Anti-foaming agent	Dimethylpolysiloxane	E900
Stabilizer	Acetylated monoglyceride	E472a

Colours and some of the other agents added to food, however, have less obvious benefit for the consumer and are probably more important for the manufacturer. Enhancing the attractiveness of food, for example, increases its saleability, which would appear to be the main reason for the use of some additives such as colouring agents. Often we as consumers prefer to have attractively coloured food, but recently many consumers have begun to question the unnecessary use of additives in food and to demand additive-free food or the use of 'natural' additives. While this may satisfy those who believe that natural substances are intrinsically safe, the fact is that natural products can be just as toxic as synthetic ones (see pp. 145–58, 240–57). Therefore each 'natural' food additive also needs to be assessed for safety. For example, a tin of custard powder may claim to contain 'no artificial colours' but contains annatto as colouring, which while it may be of natural origin has been associated with adverse effects (see below).

Are food additives harmful?

The use of food additives on a wide scale is beginning to be questioned by some toxicologists, especially as the long-term effects of the substances are often not known. With the increase in the use of food additives, there has been data suggesting that some people have an intolerance to certain food additives and linking various physical and mental disorders, including childhood hyperactivity, with their use. In response to public pressure, food manufacturers have begun to supply foods that are 'organic', that is, additive-free or, for example, containing only 'natural' colouring agents. The term 'organic' applied to food also means food that has not been exposed to pesticides at the growing or harvesting stage.

At present there is insufficient reliable scientific data on the adverse effects of food additives in humans but there is much public concern and many anecdotal reports of problems relating to food additives, particularly allergic reactions. The extent of intolerance to food additives in the population at large is, however, only around 3 in 10,000 or, to be more precise, 0.026 per cent).[2] Most of the data available refers to patients who have symptoms such as skin rashes (**urticaria**). Some of the studies carried out have, however, highlighted certain effects and the same food additives appear to feature in the reports: tartrazine and benzoic acid are the most common offenders, especially in relation to hyperactivity in children.[3] There may also be **cross-reactivity**, where an additive causes an effect in a person who has been sensitized by another additive. This can occur between additives and naturally occurring food constituents, for example between salicylates, which occur naturally in some foodstuffs, and tartrazine (see below).

Tartrazine

The food colour tartrazine is a well-known food additive which is currently still in use, although it was removed from many products as a result of consumer pressure. Known also as E102 in Europe and FD&C yellow no. 5 in the USA, tartrazine is an azo dye with an intense yellow/orange colour. Its similarity in colour to natural orange juice has meant that it is used extensively in the soft drink industry, and it has been also used in products as wide-ranging as breadcrumbs and medicines. It is one of the colours most frequently implicated in food intolerance studies and also in reactions to pharmaceutical preparations to which it was sometimes added. Effects occur most commonly in children, and it is thought that 1 in 10,000 children are sensitive to tartrazine. According to Feingold and his team, adverse reactions to tartrazine seem to occur most commonly in subjects who are also sensitive to aspirin and salicylic acid, a finding that has been confirmed by other studies.[4] Depending on the test protocol followed, it has been found that between 10 and 40 per cent of aspirin-sensitive patients are also affected by tartrazine.[5] The reactions include asthma, urticaria, rhinitis, and childhood hyperactivity (see box). The most extreme reaction to tartrazine reported has been anaphylactic shock which occurred in a person who was known to be allergic to tartrazine. The use of tartrazine has declined in the UK and is no longer permitted in the USA, as it is now a recognized sensitizer.

Hyperactivity is difficult to diagnose and distinguish from restlessness, which may be due to other factors, like hunger, boredom, or inappropriate treatment by adults. Whether hyperactivity is caused by food additives is controversial:[6] some studies have shown an improvement in behaviour after children switch to a diet, such as the Feingold diet, that is free from artificial colours and flavours, while other studies have shown no improvement. One double-blind crossover study of fifteen hyperactive children found some improvement when the Feingold additive-free diet was used,[7] but it may be argued that a major change in dietary habits would be expected to lead to behavioural changes. Another double-blind crossover study using objective laboratory and classroom observation failed to find any effect from the Feingold diet.[8] Another trial on twenty-two hyperactive children found a statistically significant improvement in the mothers' ratings of their children's behaviour but not in objective tests.[9] According to Juhlin, the one study carried out to the most rigorous scientific standards, where objective, non-involved observers were used, a change in diet showed no effect on behaviour.[10]

Urticaria due to intake of tartrazine is more widely accepted as an adverse effect and has been demonstrated in a number of studies. During this reaction substances such as histamine are released into the blood which cause the symptoms of red weals on the skin and itching. A number of other food colours and other types of food additives can also cause urticaria and there may be cross-reactivity between colours such as erythrosine and sunset yellow. A challenge of patients whose urticaria had improved on a colour-free diet with 0.15 mg of tartrazine, resulted in three out of thirteen developing urticaria within three hours of exposure.[11]

Asthma may also be a symptom of hypersensitivity to tartrazine and one study showed that 11 per cent of asthmatics reacted to an orange drink containing colouring agents.[12]

Tartrazine sensitivity

The mechanisms underlying sensitivity to tartrazine are unknown but some observations may suggest a possible cause. Tartrazine is broken down by the bacteria that normally live in the digestive system, giving rise to several products. It has been shown that the urine of animals that have been fed tartrazine, and presumably containing these products, is mutagenic. This means, for example, that when bacteria were exposed to the urine mutations occurred in the genetic material. This suggests that a chemically reactive substance (or substances) is present that reacts with DNA. This reactive substance might be expected to interact with proteins also and (although this is hypothetical) hence produce altered proteins. Altered proteins may be recognized by the body as foreign. The presence of foreign proteins (antigens) can stimulate the immune system to produce antibodies directed against these antigens. This may lead to allergic reactions and other responses of the immune system, that is, symptoms such as asthma, runny nose, and itching skin.

However, a range of foreign proteins, which are potential antigens, are present in the diet and are absorbed from the gastrointestinal tract. The immune systems of most individuals become tolerant via a regulatory system which prevents adverse reactions to food constituents and additives. However *some* individuals seem predisposed to allergic diseases and do not become tolerant and so may develop adverse reactions to dietary constituents. It is possible that the interaction of a breakdown product of tartrazine with a protein in the body could be sufficient to cause an allergic reaction in some sensitive individuals.

In conclusion, although there is much seemingly contradictory information, the weight of evidence indicates that tartrazine is responsible for a variety of adverse effects, including allergic reactions (that is, involving the immune system) and hyperactivity in children. However, the incidence of these reactions is low and additive-induced hyperactivity in children has been probably overdiagnosed.[13] Moreover, some studies have used unreasonably large doses to elicit adverse effects in subjects.[14]

Other food colours

While tartrazine seems to be the food colouring most frequently associated with adverse reactions, other colouring agents are also known to cause mental and/or physical ill effects, for example cochineal, a natural agent; Ponceau 4R (E124); and erythrosine (E127). Both erythrosine and cochineal have been associated with hyperactivity in children, while Ponceau 4R should be avoided by those with asthma and those who are sensitive to aspirin, a salicylate.

Other colourants in use are sunset yellow (E110), another synthetic azo dye, and caramel. Caramel colours are manufactured from sugars and are very widely used. This food colour has caused some concern due to effects in experimental animals, for example the reduction in the number of white blood cells in rats. This may be due to the effects of contaminants at the high doses given which may be enhanced by a reduced intake of vitamin B6 in the diet. Annatto and beta carotene are naturally occurring colourants, but in one study 26 per cent of patients with chronic urticaria were shown to react to annatto.[15]

Benzoic acid and benzoates

This group of chemicals, which are found naturally in many fruits, are used as preservatives, a function that they also perform within fruit. Benzoic acid itself may be used, or more commonly sodium benzoate (a salt) or ethyl or methyl para-hydroxybenzoate, also known as parabens. They have sometimes associated with adverse effects. Anaphylaxis-like reactions have been reported as well as urticaria. About 4 per cent of people who suffer from asthma may experience breathlessness and wheezing when exposed to benzoates. Parabens are often used as preservatives in cosmetics.

Antioxidants

A variety of antioxidants are added to food including vitamin C and vitamin E. Two antioxidants that may occasionally cause problems are butylated hydroxytoluene (BHT) and butylated hydroxyanisole (BHA), which have been associated with symptoms such as eczema, skin rash, runny nose, wheezing, headache, chest pain, flushing of the skin, and red eyes. At the same time, it has been reported that these antioxidants may be helpful in preventing cancer, possibly by removing damaging free radicals such as reactive oxygen and its by-products from fatty acids.

Saccharin

Saccharin[16] has been for a long time a widely used and very successful sweetener. It was discovered in 1879 by chance, by Constantin Fahlberg, a graduate student at Johns Hopkins University in the USA. Fahlberg was a student of Professor Ira Remsen who had instructed him to synthesize derivatives of the substance toluene. He made one derivative, and then noticed at lunch that his bread tasted sweet. He realized that he had not washed his hands since working in the laboratory and the sweet taste was due to the chemical he had made. He called it saccharine, left the university department, applied for a patent and, despite opposition and annoyance from the university and Professor Remsen, eventually succeeded in his application for a patent in 1885.

Saccharin, as it became, was first commercially manufactured in Germany in 1894, and then in the USA in 1901. Under the US Food and Drug Act of 1906 it was allowed to be used as a sweetener, and in the early years of the twentieth century it became increasingly popular as a sweetener. Its purpose was to provide people with sweet food and drinks without their gaining weight as they would have done with sugar, and it has continued to be widely used by diabetics. Although safety tests were not required at the time, evidence of its safety was eventually demanded and human volunteers took relatively large amounts (5 g per day for six months) without apparent harm. Despite this, the recommended intake was later restricted to 0.3 g per day. By the time of the Second World War it was used by millions of people and continued to be popular up to the 1970s. It was used in sweeteners such as Sweet 'n Low, in processed foods, and in diet drinks. In 1959, owing to its apparently safe use for fifty years, the US regulatory agency, the Food and Drug Administration (FDA), designated saccharin as Generally Recognized As Safe (GRAS). As expected of a food additive, saccharin has low toxicity, which is indicated by studies in animals.

In the years that followed a number of further animal studies were carried out in the USA and Canada which involved giving rats or mice large daily amounts of the sweetener. Human populations were also studied using epidemiological techniques. While most animal studies continued to show that saccharin was not toxic, studies in which rats were treated from before birth revealed some tumours in the bladders of male rats. The doses given in these studies, however, were huge, roughly equivalent to an average man drinking about 1,000 cans of diet drink every day. Even salt would be toxic at this level!

These studies persuaded the Canadian and American authorities to suspend use of the sweetener in 1977. In the USA it was banned under the Delaney Clause of the Food, Drug, and Cosmetic Act which prohibits the use of any food additive that has been shown to produce cancer in laboratory animals. There was a public outcry against the ban, as saccharin was the only general purpose artificial sweetener approved for use and available to diabetics and those with an obesity problem as well as to people who wished to reduce their consumption of sugar. The result was a moratorium on the ban in the USA to allow the evidence to be examined. Further studies were carried out and, while they confirmed the findings, it proved possible to rationalize the results in relation to human safety, as we shall see in the next chapter. In 1991 the FDA eventually withdrew its proposal to ban the sweetener.[17] The cloud hanging over it was finally removed in 2000 when official concerns were dropped.

Saccharin is an example of a food additive that was unnecessarily banned as a result of inappropriate testing, inflexible and conservative interpretation of the results of tests, and the failure to appreciate other evidence. For more details of the safety assessment of food additives and further discussion of this case see pp. 299–308 and below pp. 280–1.

Monosodium glutamate and the Chinese restaurant syndrome

Monosodium glutamate (MSG, or E621) is the flavour enhancer most commonly added to food. It is used especially in Chinese and Japanese cooking and foods. It is not a synthetic additive, as it is found naturally in many foodstuffs, especially seaweed and parmesan cheese which have some of the highest levels, and also tomatoes, mushrooms, and soy sauce. E621 is the sodium salt of glutamic acid, a naturally occurring amino acid which has important functions in the body, especially in the brain. Glutamic acid is found both free and as part of protein, and when the latter is broken down and digested in the gut glutamic acid is released (see box). The glutamic acid is released relatively slowly, however, and while excess glutamate can be removed by metabolism it is possible that large amounts may accumulate. Too much free glutamic acid in the diet, which can occur if food that contains naturally high levels is consumed with food to which large amounts have been added, can be toxic, leading to unpleasant symptoms. A variety of studies into the effects of MSG have been carried out, but they have shown conflicting results. In one of the most recent properly conducted, **double-blind trials**, where the subjects did not know what was being tested, the authors concluded that there was no effect of MSG. It seems that some individuals suffer an intolerance reaction when exposed to glutamate, especially when the free form is present in food in large amounts. In asthmatics and those who suffer from allergies, excess glutamate may cause asthma, runny nose, and sneezing. Individuals who are particularly sensitive to monosodium glutamate suffer persistent headaches, thirst, palpitations, pains in the neck, and dizziness. It has been suggested that one reason for the increased susceptibility in some individuals may be a deficiency in vitamin B6, which is involved in the removal of glutamate (see box).

This is another illustration of the truth of the principle of Paracelsus, 'All substances are poisons . . . The right dose differentiates a poison from a remedy.' Even natural substances, like glutamate, which is found in the human body and which has important functions, can have adverse effects if the level of intake is higher than necessary and normal.

Glutamate and other amino acids

The body contains many amino acids, some of which can be synthesized (non-essential amino acids), while others must be absorbed from the diet (the essential amino acids). Amino acids are the building blocks of protein, being attached together in groups to form the proteins that the body needs as part of its cellular structure or as enzymes, which are biological catalysts. Glutamate can be synthesized in the body, and it is also absorbed from food, in which it occurs as the free amino acid, or it can be released from proteins by digestion.

Glutamate is utilized in metabolic interactions with other amino acids and is formed into new proteins. One route by which glutamate can be removed is through reactions that require vitamin B6 (pyridoxine), and individuals who are deficient in this vitamin may suffer greater effects from excess MSG. Glutamate also has an important role in the brain where it acts as an excitatory neurotransmitter, increasing the activity of nerve cells. It is responsible for most of the interactions between nerve cells in the brain, for example those involved in memory.

Safety evaluation of food additives and contaminants

As the historical examples at the beginning of the chapter demonstrated, it is essential that food additives be carefully tested for the potential to cause adverse or untoward effects before humans are exposed to them. The testing is normally carried out in experimental animals *in vivo* (in the living organism) but information from *in vitro* (in the test tube) experiments may also be taken into account.

The tests usually consist of lifetime exposure of experimental animals to the substance at different doses, but with the maximum level several times greater than that expected to be consumed by humans. Such testing, however, may not always be predictive, as experimental animals may not show the same type of behavioural or immunological effects as humans and the fate of the chemical can also be different (see p. 27). Moreover, although the quantities of food additives consumed by humans are very small, their consumption generally occurs over a lifetime. This chronic exposure is often sporadic rather than continuous, which is difficult to simulate in laboratory animals.

When sufficient information has been accumulated this is assessed by the FDA in the USA or the Committee on Toxicity (COT) and other committees in the UK. The committees, consisting of various scientifically and medically qualified individuals, determine whether the substance is safe to be used as an additive as well as the level at which it can be added to food.[18] This level is known as the **acceptable daily intake**, or ADI. How this is determined will be discussed in the next chapter (see pp. 299–301). In the USA an amendment to the Food and Drug regulations, known as the Delaney Clause, was introduced which specified that 'no additive shall be deemed safe if it is found to induce cancer when ingested by man or animal or it is found, after tests which are appropriate for the evaluation of the safety of food additives, to induce cancer in man or animals'. This allowed little flexibility for the scientific judgement of the expert panels of scientific experts and has led to much controversy, compared to the situation in other countries where regulations are more flexible and room for judgement is allowed.

Contaminants also have to be evaluated for safety. As we saw earlier, there are many substances in food that occur naturally or as a result of contamination from human activities. Thus shellfish may be contaminated with heavy metals and oily fish with substances such as dioxin. The levels of these contaminants in our food have to be regularly monitored and if the level rises above that permitted, known as the tolerable daily intake (TDI), the sale of that foodstuff is restricted.

Using information gained from the literature, both from whole animals and *in vitro* studies and from studies that may have been carried out specifically in relation to a particular food contaminant, the tolerable daily intake level is determined.[19] How this level is set will be discussed in the next chapter.

CASE NOTES

When the acceptable daily intake was too high

At one time, cobalt chloride, a salt of the metal cobalt, was added to beer in the USA as a foam stabilizer. Such salts had been used to treat people with anaemia at doses of 300 mg per day without any apparent problems. However, heavy beer drinkers, who drank about 10 litres per day, suffered effects on the heart, known as cardiomyopathy, a degeneration of the heart muscle, which was eventually ascribed to the cobalt. It transpired that, while cobalt alone was apparently not toxic to the heart, even at single doses many times higher than the exposure from the additive in beer, in the exceptionally heavy drinkers the toxic effects of cobalt were greatly increased. It turned out that the victims were malnourished individuals and deficient in particular amino acids, which was an important factor in increasing the toxicity of the cobalt. Furthermore, the excessive alcohol intake was itself an additional factor. The potentiation of the toxicity of cobalt was unexpected and so had not been taken into account in the safety assessment. This illustrates the difficulty of anticipating all possible circumstances in safety evaluation.

In addition to natural food contaminants that may be present in food, there are also man-made contaminants that can inadvertently enter food, such as pesticides, industrial chemicals like dioxin, or substances fed to animals like antibiotics or anabolic steroids. Some of these are discussed in Chapter 5.

12

A Risky Business

Assessment of Chemical Hazards and Risks

CHEMICALS, as we have seen, can pose a risk to humans and other animals. There are at least 65,000 chemicals in use in the USA and several hundred are added each year. In the past some of these were used without due care, for example DDT, a case Rachel Carson highlighted in *Silent Spring*. The general public is now rather suspicious of all chemicals, and often has an exaggerated fear of the harm they may cause. There have been cases where the effects have been devastating, for example those due to the drug thalidomide and the industrial chemical methylisocyanate. In other incidents, for example the exposure to dioxin at Seveso, the effects have (at least so far) turned out to be much less serious than expected: apart from chloracne, which was fortunately a reversible toxic effect on the skin, no extra cancers, birth defects, or other effects have been observed. Many of the incidents involving serious adverse effects from chemicals that have been discussed in this book have resulted from a lack of respect and care when using them, a lack of knowledge about their effects, or sometimes deliberate misuse.

How do scientists, and toxicologists in particular, assess the risk from chemicals and what does their assessment mean? Sometimes we can assess the risk of a particular activity from past experience and knowledge. With poisoning cases we know that if someone eats or drinks more than a particular amount of potassium cyanide or Weedol (paraquat) they are likely to die. Hence the risk in an individual poisoning case can be estimated from the known lethal dose and the amount that the victim has eaten, drunk, or inhaled.

In most countries chemicals have to be assessed in relation to the risk they pose to humans and to the environment. Drugs, both human and veterinary, food additives and contaminants, industrial chemicals, pesticides, household chemicals, and cosmetics are all evaluated.[1] There are regulations controlling the use of these chemicals that indicate the specific information required, as it will vary.

Risk is a mathematical concept and is defined as the likelihood that an adverse effect *will occur* as a result of exposure to a chemical. Conversely, safety may be defined as the practical certainty that adverse effects *will not occur* when the substance is used in the manner and quantity proposed for its use. Risk can be expressed mathematically as follows:

$$\text{Risk} = \text{hazard} \times \text{exposure}$$

Hazard is the *inherent* ability of the substance to cause an adverse effect, that is its toxicity (for example, dioxin and ricin are known, potent poisons). Exposure is the level of the substance in the air, water, or food or the dose of a drug.

Risk assessment is a scientific process whereby the level and nature of the risk is determined. If there is no exposure to a particular chemical then there will be no risk at all, for example potassium cyanide in a sealed container is a hazard but not a risk. Risk can therefore be minimized if exposure is minimized. If the chemical is effectively non-hazardous, that is the amount needed to cause harm is unrealistically high, such as with common salt or saccharin, then even if there is a level of exposure, the risk will be so small as to be virtually non-existent. Therefore both the level of exposure and the nature of the chemical (whether or not it is hazardous) must be known if the risk is to be assessed.

Evaluation of exposure

In the human population the level of exposure to a chemical can, in some cases, be known precisely, for example known doses of drugs are given to patients. For industrial chemicals, and especially chemicals in the environment, however, this is much more difficult and sometimes impossible to know. Well-regulated chemical industries in highly developed countries will monitor workers by taking blood or urine samples from them at intervals. The air in the plant or factory can be monitored and workers may wear personal sampling devices that measure how much of a volatile chemical they are exposed to. This is not practised in all industries nor in less well-developed countries. In such cases, where there is environmental exposure to chemicals, the level of exposure will often have to be estimated, a process that is often fraught with uncertainty. For exposure via water or food, this can be estimated based on information about daily water intake, dietary habits, and the known level of the chemical in the water or food. It may be a reasonable method where a staple food eaten daily or at least regularly is involved, but it may not be so simple with contaminants that tend to occur in certain foods, for example mercury which is sometimes found at higher levels in tuna or swordfish. Estimating intake from questionnaires is also liable to error, as people do not always remember the last time they ate a particular food or how much of it they had. With studies in laboratory animals (but not those in the wild) the exposure level can usually be known precisely.

Hazard

It is occasionally possible to discover the hazardous nature of a particular chemical using information from humans, but it is rarely done and, even if information exists, it is usually too limited or too imprecise. Consequently information about the toxicity of chemicals is normally derived from studies in experimental animals or possibly experiments *in vitro*, that is, experiments carried out in isolated cells, organs, or tissues from the body rather than in the whole animal or human subject (see below).

Characterization of hazard in humans

The techniques used to evaluate the possible hazards of chemical exposure in humans fall within the term 'epidemiology', which involves observing human populations and collecting data about the incidence of disease in the population. This is then related to common factors in the population.

One of the first to use this technique was an English physician, John Snow, in the nineteenth century. He was studying the cholera outbreaks in London which had killed large numbers of people but whose cause was unknown. Snow felt sure that it was associated with the victims' living conditions: poverty, overcrowding, and poor sanitation. His observations led him to suspect that the disease was associated with drinking water and the contamination of the supply with sewage. In an outbreak of the disease in a particular area of London in 1848 he plotted the cases on a map and found that they were clustered around and served by a particular water supply, the Broad Street pump. He also noticed that workers in a brewery in the same area and those living in a workhouse had a low incidence of the disease, as they had their own source of drinking water and those in the brewery also drank beer. When he removed the handle of the water pump, the number of cholera cases decreased, implicating the water as the source. His suspicion that contaminated water was to blame was further confirmed when he found that cases in other areas were also associated with drinking water contaminated with sewage.

Epidemiology can therefore show an association between disease and another factor such as location, but it does not give the exact cause. Snow did not discover the cause of cholera, which Robert Koch, a German physician, did later in the century. Koch showed that cholera was caused by bacteria, using experimentation. He produced a set of criteria, known as 'Koch's postulates', describing the necessary experimental procedure. His postulates can be roughly adapted to suit the situation with regard to disease caused by chemicals (see box).

Koch's postulates revisited

Although they were originally designed to apply to the study of bacterial infection, such as that causing cholera, these principles are useful criteria in considering the possibility that a chemical may have caused a disease. They have been modified for that purpose here.

1. The agent (chemical) responsible for the disease must be present in every case. Each person affected must have been exposed to the chemical.
2. The chemical (or a known breakdown product) must be detected in the patient and be shown to be present in each person affected.
3. The pure chemical must be able to cause the disease in a healthy individual (for ethical reasons this would normally have to be an experimental animal).
4. If a metabolic breakdown product is responsible for the effect, then the same product must be shown to be present in affected individuals.

Naturally it can be very difficult to satisfy even some of these criteria in humans exposed to environmental chemicals.

To be certain that a chemical causes a particular disease in a population of humans is clearly no easy task. Simply showing that the chemical is present in the environment in which victims live is not sufficient. Both a means of exposure and a level of exposure must also be shown. One of the problems with using human data and the epidemiological approach is that many human diseases have more than one cause. Without a demonstration that chemical exposure has occurred in each case of the disease the cause cannot be ascribed to the chemical. This may be difficult but the second of Koch's postulates is even more difficult to satisfy because the chemical (or important product) may no longer be present in the body of the victims and, even if it were, collecting samples from hundreds of human subjects and analysing them is no mean task.

Apart from drugs, chemicals cannot ethically or legally be given to humans without safety evaluation, and hence, to confirm the third and fourth postulates, animal experimentation is usually required. For some chemicals information on the toxicity in animals may already exist.

Often it isn't possible to satisfy all the criteria in a study of human populations exposed to chemicals, but other criteria or guidelines have been proposed, for example those set out by Bradford Hill in his article 'The environment and disease: association or causation?',[2] published in 1966:

1. *Consistency and unbiasedness of findings*: confirmation of an effect by different investigators in different populations.
2. *Strength of association*: the frequency with which the factor is found in the disease and the frequency with which it occurs in the absence of the disease.
3. *Temporal sequence*: exposure to the factor should occur before onset of the disease and there should be some relation between exposure to the agent and frequency of the disease.
4. *Dose–response relationship*: a quantitative relationship between the extent of exposure to the factor and the frequency or severity of the disease must be demonstrated.
5. *Specificity*: the factor should be isolated from other factors and shown to cause changes in the incidence of the disease.
6. *Coherence*: the evidence should fit related facts that are known.
7. *Biological plausibility*: the proposed association should fit with biological knowledge.
8. *Analogous situations*: can the situation be compared with others?
9. *Experimental evidence*: does removal of the agent lead to improvement of the disease or symptoms?

Using these guidelines and taking as an example the industrial chemical vinyl chloride (see pp. 168–71), we can see how the evidence can be considered (the weight of evidence), and causation concluded:

1. Different investigators in different countries reported the same type of liver cancer in their patients who were all exposed to vinyl chloride.

2. The particular kind of liver cancer, haemangiosarcoma, was of very low frequency in the general population and the frequency increased after industrial use of vinyl chloride increased.
3. There was a relationship between the time of exposure to the chemical and the appearance of the liver cancer some years later in all the subjects.
4. There was a relationship between the extent of exposure and the development of the liver tumours.
5. The haemangiosarcoma was seen only in industrial workers who were exposed to vinyl chloride.
6. Vinyl chloride was known to cause effects in the liver.
7. Chemicals are known to be able to cause cancer.
8. Other chemicals that affect the liver also cause cancer (for example, aflatoxin).
9. Protection of workers and consequent reduction of exposure led to a reduction in cases of the disease. Vinyl chloride can be shown to cause the same effect in experimental animals and now a mechanism by which this occurs has been formulated.

There is little doubt from the evidence that vinyl chloride is responsible for the effects, and that its use should be regulated and workers protected from exposure, an outcome which duly transpired.

Other diseases that show a strong relationship between exposure to a chemical and an effect are thalidomide-induced malformations in babies and asbestos-related cancer (mesothelioma) (see pp. 56–9, 178–82). In all these cases, the disease or adverse effect was extremely rare, which helped in the detection and determation of the cause.

Epidemiological studies can be carried out in several different ways: retrospectively, where a population that has already been exposed to a chemical is studied; or prospectively, where the population exposed is studied from the beginning of the exposure period. *Retrospective studies* are also described as case-control studies, where the incidence of the disease in the exposed group is compared with that in the control group. Most studies of accidental exposure are retrospective. *Prospective studies* are also described as cohort studies, where the group exposed is followed over time and the appearance of disease is evaluated in comparison with a control group. Studies of possible adverse effects of drugs usually come into this category. There are also *cross-sectional studies*, which evaluate the prevalence of a disease in an exposed group, and *ecological studies*, which compare the incidence of a disease in a particular geographical area with potential exposure to a chemical with that in another area with no exposure.

The design of the studies is very important, and control subjects should be of the same sex as the exposed subjects and as similar as possible in age and lifestyle. Those who set up and interpret epidemiological studies must be aware of sources of bias and any confounding factors, for example smoking or alcohol consumption by members of the population studied, especially when evaluating the incidence of certain diseases (for example, cancer or liver disease).

With prospective studies, for example those used to evaluate adverse drug reactions, the control group (who may have the disease for which the drug is prescribed) will not receive the drug but instead will be given a 'dummy' pill, known as a **placebo**. Ideally the study should be double-blind, that is, where the volunteer or the patient does not know whether they are taking the drug and nor does the research worker. The placebo will be an inert substance, or a substance without the effect of the drug under investigation, but patients or volunteers who have taken it may record that they have suffered adverse effects. Or they may claim that the placebo has had the desired effect, for example, alleviating the symptoms of their disease. This is known as 'the placebo effect' and indicates that some effects can arise from our expectations rather than be produced by a physical cause. The information that can be gleaned from an epidemiological study is whether there is a risk from exposure to a chemical, and this can be quantified as the odds ratio, relative risk, and absolute excess risk (see box).

The results of epidemiological investigations are analysed to show whether the incidence of a particular disease is really greater (that is, statistically significant) in the exposed group. However, with retrospective studies in particular, which may be less well controlled and for which information on exposure to a chemical may be lacking, simply showing a significantly greater incidence of a disease in a group assumed to be exposed is not evidence of a causal association. Bradford-Hill's guidelines and Koch's postulates have to be borne in mind. Thus biological and scientific plausibility should be shown, as *apparent* associations may be statistically significant but have no causal relationship, that is, they may be coincidental rather than causal.

The problem with epidemiological studies, especially the retrospective type, is that there is rarely any exposure data. Thus, while there may be evidence of a chemical being present in the environment, there is often no proof that the individuals with a particular disease were exposed to it. If

Measurements of risk

Odds ratio: this is the ratio of risk of disease in an exposed group compared to a control group. It can be calculated as A × B / C × D = odds ratio, where:

A is the number of cases of disease in the exposed population;
B is the number of unexposed controls without disease;
C is the number of exposed subjects without disease; and
D is the number of unexposed controls with disease.

For example, if there are 5 cases in 100,000 of the control subjects and 10 cases in 100,000 of the exposed population:

$$10 \times 99{,}995 \ / \ 99{,}990 \times 5 = 2$$

There is an odds ratio of 2.

Relative risk: this is the ratio of the occurrence of the disease in the exposed to the unexposed population. This is calculated as A/B, where:

A is the number of cases of disease in the total exposed group per unit of population; and
B is the number of cases of disease in the total non-exposed group per unit of population.

Using the above example, the relative risk is

$$10/5 = 2.$$

Absolute excess risk: this subtracts the number of cases of the disease in the unexposed population from the number of cases in the exposed population, that is A – D. Using the above example, the absolute excess risk would be

$$10 - 5 = 5 \text{ per } 100{,}000.$$

the weight of evidence is great (for example, if many of Bradford-Hill's suggested criteria are met), as with vinyl chloride, the chemical can be said to be the cause with reasonable certainty. The problem of causality is particularly apparent in studies that look at the incidence of a disease in a particular geographical location and try to relate the disease to a possible source of exposure in that location. For example, there have been studies that revealed a slight excess number of cases of childhood leukaemia in a village near a nuclear power plant and reprocessing facility. While these were statistically significant, the number of cases involved was only a handful. Leukaemia is a disease that has many causes and proving causation in this case was hence very difficult. Although some of the criteria might have been satisfied, the association was not strong enough and there was no clear, plausible mechanism by which the children were exposed. Thus statistics alone are not enough and cannot reveal the cause of a disease. There could be many different causes of leukaemia in a particular population; that radiation is one cause of leukaemia does not make it the cause in a particular area. There are also natural sources of radiation, such as radon gas, which can seep into homes from underlying rocks. The proximity of the excess cases of leukaemia to a nuclear power station could be coincidental rather than causal.

Many of the diseases that can be caused by chemicals can also be due to other causes. Cancer is not a single disease but many different diseases, and generally chemicals cause a particular type of cancer rather than a mixture of types. Even so there is usually a background incidence in the population as a whole, and demonstrating an increased incidence is sometimes statistically difficult. Similarly, other disorders, like malformations in babies or liver disease, can have many causes including infections.

Another problem with a retrospective study is that the latent period between exposure and the disease (especially when it is cancer) may be long, and so it can be very difficult to make associations between cause and effect. Furthermore, unlike highly controlled animal studies, human exposure to environmental and industrial chemicals is often chronic and may be intermittent, either being completely random or occurring only during the working day. This makes estimation of true exposures more difficult. Accidental exposure to chemicals, both acute and chronic is, for obvious reasons, normally analysed retrospectively, hence any chance to measure exposure levels is usually lost.

Poorly conceived or badly carried out epidemiological studies can yield data that falsely reassures the public or unnecessarily alarms them, and be impossible to interpret. Unfortunately the general public often demands such studies if they suspect that they are being poisoned by a chemical (for example, at Love Canal in the USA). They fail to understand that, even if a higher than expected incidence of disease is shown in the population studied, and even if some of those afflicted have been exposed, it does not prove that a particular chemical was to blame.

It does not necessarily follow that no action is taken when the exact cause of a disease has not been shown. In the nineteenth century Dr Snow was not able to discover the cause of the cholera outbreaks in London, but with others he managed to improve the quality of the water supply for the population most at risk. The result was a decrease in the incidence of cholera. When the causality and association are tenuous, however, large amounts of money may be spent for no gain (for example, at Love Canal; see pp. 125–6). While epidemiology is useful, in most risk assessment it is used only in addition to animal studies because of the limitations of inadequate exposure information and the lack of appropriate control populations.

Finally another problem associated with studying human populations (or wild animal populations in the same way) is that exposure of the population via water, food, or air is often not to only one chemical but to a mixture. This mixture may be different in different localities and could therefore have a variety of effects on the response of a person to the chemical under investigation. This greatly complicates the situation and is currently a subject of concern and research interest. One particular area of study in relation to this is the effects of oestrogenic compounds, so-called 'gender benders'. It has been suggested that, while each of the chemicals alone is not sufficiently potent to have any effects, when several are combined they may have significant effects. Although this can be shown in the laboratory, there is so far little evidence in the effects of environmental chemicals on humans. It has, however, been shown to be significant in the case of drugs, that is, one drug can affect another. Exposure to an environmental or industrial chemical may also affect the toxicity of a drug.

Characterization of hazard in animals

Because of the ethical and legal restrictions on the exposure of humans to chemicals, and the limitations and uncertainties inherent in using information from available human exposure data, animal experiments are used to characterize the hazard, to determine the toxicity, for most chemicals. This necessity places many toxicologists (scientists who study and assess the risks from chemicals) in a very difficult position, between a rock and a hard place: the public expects drugs to be safe and insists on knowing the adverse effects of chemicals to which they may be exposed, and at the same time many people demand an end to experiments on animals.

The use of animals in biomedical research is now regulated in many countries. The laws in the UK are the strictest in the world: all scientists who use animals in experiments must hold a personal licence, and the procedure and experiments have to be approved in advance by a government inspector from the Home Office and covered by a project licence. All scientists are expected to observe the three Rs:

1. *reduction*: attempt to use fewer animals in experiments and question the necessity of every experiment that uses animals;
2. *refinement*: devise better methods where possible to reduce the use of animals and their discomfort;
3. *replacement*: use alternatives, for example cells *in vitro*, wherever possible.

With some chemicals, especially drugs, it may be possible, for example, to reduce the number of animal experiments as similar chemicals may have been previously tested or the same chemical may have been used for another purpose. Better designed experiments, or the use of more sensitive methods to reduce errors, can lead to fewer animals being needed. The use of alternatives, such as isolated organs, tissues, cells or parts of cells, and human tissue, is growing. For example, the metabolic breakdown of a novel drug can now be studied at least partly in isolated cells in which human enzymes are expressed. The action of a drug on a specific receptor can also be studied *in vitro*, which is increasingly being done in the pharmaceutical industry.

These examples, however, have defined targets and endpoints. When toxicity is being assessed it is more difficult as, at least on current information, known receptors are rarely involved in toxic effects and there are many potential targets for chemicals within cells and within the body.

The first stage for most safety evaluation is the acute toxicity test, in which a number of experimental animals (usually rats or mice) are given various single doses of the chemical and any toxic or adverse effects observed are recorded. After about seven days the animals are humanely killed with an anaesthetic and each animal undergoes a post-mortem, in which all organs and tissues are removed, processed, and examined by a pathologist using a microscope. The result of the experiment is to show the effect, if any, of different doses of the chemical on the animals. The objective of the test is not to kill the animals. (The infamous LD50 test, in which the dose lethal to half the animals was determined, is in most cases and countries no longer carried out or required by regulations.)

Depending on the type of chemical being tested, and its use, testing the toxicity of a single dose may be all that is necessary, while other chemicals, like drugs, require more extensive testing, for example repeated dosing for longer periods of time and studies in more than one species of animal. Repeated dosing usually involves putting the chemical in the animal's food. Blood samples are taken at regular intervals and various parameters measured that will indicate if there are any adverse effects to the animal. The level of the chemical in the blood can also be measured (knowing the exposure level of a chemical is essential). More specialized studies are also necessary for drugs, such as evaluation of the effects of the chemical on the reproductive process and on the embryo and newborn animal, and determining whether it can eventually cause cancer. The result of the safety evaluation should be that any adverse effects can be related to the dose and the level of the chemical in the blood.

The information necessary for the assessment of risk will depend on the nature of the chemical. A chemical that is likely to be released into the environment has to be tested on a variety of animals, for example earthworms and honey-bees, as well as on plants. A drug would be tested on mammals, usually rats and then dogs, and, as soon as the required information about safety and dosage has been acquired, it is given to a few human volunteers in what is called Phase 1 clinical trials. Information on the toxicity and any ability to cause mutations is required before the drug can be given to humans (see also pp. 86–8).

The studies in human volunteers will show whether the drug is effective and whether it causes any adverse effects. If it passes this hurdle, and after further experiments in animals, the drug is given to a small number of patients who are monitored to see whether it is effective and if it has any adverse effects (Phase 2). Knowledge of any effects in animals allows the patients to be monitored more critically when they are given the drug. These studies are now carried out as early as possible in humans, and hence fewer animals are used in the whole testing process.

After the Phase 2 clinical trials a larger number of patients are given the drug (Phase 3) and, if it is deemed effective and there are no significant adverse effects, it is granted a licence by the regulatory authority, for example the Food and Drug Administration (FDA) in the USA or the Medicines and Healthcare Products Regulatory Agency (MHRA) in the UK, after which it is marketed. Even after this, the drug continues to be monitored, by general practitioners and other doctors who report any adverse effects in patients. This is the Phase 4 clinical trial. It is sometimes only during this phase that adverse effects emerge. A proportion of drugs (at least 2.6 per cent over the past 30 years) have been withdrawn because of unacceptable adverse effects after marketing, especially due to adverse effects on the liver. This is because the number of patients exposed to a drug on the market is far higher (usually in the millions) than in the earlier phases of the clinical trials. If, for example, a drug were to cause an adverse effect with an incidence of 1 or 2 in a million, a Phase 3 clinical trial involving perhaps 1,000 or even 10,000 patients would be unlikely to reveal the effect.

The evaluation of the effects of drugs in humans is relatively straightforward and, as we have already seen, epidemiological studies can be reasonably effective, for example prospective, cohort studies with known doses being given. Evaluating the effects of other chemicals is more difficult, however, as exposure has to be estimated in most cases, and will probably be unknown. In well-regulated and responsible chemical companies this is usually done by measuring the level of the chemical in blood or urine samples collected from workers at intervals or by the use of air sampling.

Characterization of hazard *in vitro*

What studies can be carried out without the use of animals, that is *in vitro*? These studies sometimes use isolated cells, usually from animals like rats, but cells from humans, if available, may be used. These primary cells are obtained from a piece of an organ removed from a donor after a fatal accident or during a transplant operation. These cells cannot be kept easily, even when deep frozen, and they change when cultured in the laboratory. Moreover, they do not normally divide and grow for long in culture. Thus they may be usable only for a day or two. Alternatively, 'cell lines' can be used. These are usually cells from a cancer removed from a human or animal. They can be grown in culture and are immortal, that is they can be kept indefinitely and will keep dividing. They are commonly used because they can be bought from suppliers, can be stored deep frozen, and are always available. The other types of cells used for evaluating the toxicity of chemicals *in vitro* are bacteria, which are used mainly to evaluate the ability of a chemical to cause a mutation in DNA.

Primary cells are typically cells from the liver, as these are the most easily obtained and the majority are the same type of cell, whereas those from most other organs are a mixture of types of cells. Primary liver cells are the most likely to be representative of the organ in the whole animal. These cells however, will indicate the effects (if any) that the chemical has only on the liver. They do not necessarily predict the effects on other organs or tissues. Furthermore, research has shown that the way in which isolated cells in the laboratory respond to chemicals is not always the same as the way they respond in the animal or human. They are often less sensitive, sometimes requiring a concentration of chemical many times higher than that which causes an effect in the live whole animal. For example, when isolated human liver cells are exposed to paracetamol, toxic effects are observed only if a concentration of paracetamol is used that is at least ten times higher than the level that causes liver damage in humans after an overdose. Unlike a cell *in vitro*, a cell in an animal is surrounded by other, perhaps different, cells, and is in contact with blood which contains hormones and other substances that can influence cells. This may account for the difference in response. If primary cells *in vitro* do not always respond in the same way as the organ in the body, cell lines, which are derived from tumours and hence are not normal tissue, will often respond very differently. They can, however, indicate whether a chemical would be toxic to all cells, which is useful if chemicals such as potential drugs are being compared in order to select the least toxic. They can also be useful for studying the effect of anti-cancer drugs.

In vitro tests in bacteria, for example the well-known Ames test, and tests in mammalian cells are used to show if a chemical is able to damage DNA and cause a mutation. In the bacterial tests, extracts from mammalian liver (usually from a rat) have to be added to provide the enzymes that convert chemicals into products that might react with DNA.

Chemicals that are liable to come into contact with our skin are tested in the appropriate cells, as human skin is readily obtained and used *in vitro*. The test, which can indicate whether a chemical is likely to be directly damaging to the skin, has now replaced experiments in living animals for some substances, including cosmetics. Human volunteers are also sometimes used to gain information about the effect of chemicals on the skin.

While *in vitro* experiments offer a partial solution to the safety evaluation of chemicals, they cannot completely replace animal experiments for this purpose. Many chemicals are poisonous because they affect organs or systems as well as, or rather than, individual cells. The human body, like that of any mammal, is a highly complex, integrated system and reacts to chemicals in many different ways, far more than can be predicted from one or two different types of isolated cells. Most of the cells in our bodies have some similarities (for example, they all have a nucleus), but brain cells are very different from muscle cells, otherwise we would not be able to think. Different types of cells have specialized functions which can make them a target for chemicals in specific ways. In particular, their metabolic capabilities are different and they may be unable to detoxify chemicals in the same way as liver cells, for example. This is illustrated by the story on p. 296. There is no doubt that the world would be a more hazardous place if the safety of chemicals were evaluated only *in vitro*.

Due to deficiencies in the systems used, the information gained from *in vitro* tests will always be only part of the picture. An animal is an integrated mass of cells of many different types in different organs and tissues. These are influenced by hormones and other substances in the

CASE NOTES

A cautionary tale

A new drug was being tested at a pharmaceutical company. The first tests were carried out in isolated liver cells *in vitro*, and the drug was found to be not toxic to these cells. The drug was then given to experimental animals, in which it caused the destruction of the adrenal glands (which are found close to the kidneys). When cells from the adrenal gland were used instead of liver cells in *in vitro* tests, they were, as expected, destroyed. This was because the cells from the adrenal gland were not able to detoxify the drug whereas the liver cells were. If the ability of the liver cells to detoxify the drug were blocked, they too became susceptible to the toxic effect of the drug. Liver cells are the cells most often used *in vitro* because they are the easiest to prepare and use. They are also often the most readily available kinds of human cell. The toxicity illustrated by this example could easily occur in humans if isolated cells alone were used to test new chemicals.

blood. Some toxic effects result from disturbance in this integration, other toxicity may be the result of changes in blood flow or the rhythm of the heart. These effects are not possible to replicate *in vitro*.

The results of *in vitro* experiments, however, are important in alerting the toxicologist to potential toxic effects that may be expected in whole animals, or indicating the likely potency of the chemical. They might show that a potential drug is not worth further development, perhaps in comparison with others being developed. For these reasons, the use of *in vitro* tests will increase in the safety evaluation of all chemicals. But *in vitro* tests cannot replace evaluation of the effect of a chemical on the whole animal, where possible adverse effects to each organ and system, and the interactions between them, are studied. As we shall see later in the chapter, this is necessary for risk assessment, where the toxic effect that occurs at the lowest dose level *in vivo* is used.

Characterization of hazard in other types of animals

As well as mammals and mammalian cells, animals other than mammals are sometimes used for the evaluation of chemical hazards, for example water fleas, fish, and flatworms (see box). They are used to assess the effects of environmental exposure to chemicals, and data from such species is required by regulatory agencies. They can also give us useful information about the possible toxic effects for mammals.

A simple worm reveals secrets

A flatworm, known scientifically as *Caenorhabditis elegans*, has been widely studied by biologists and biochemists. It is no mere curiosity, as each worm has a known number of cells (about 900), of which the entire genetic code is known. It may be a simple organism but some of its biochemistry is not that different from ours. Its usefulness lies in its simplicity. Much is now known about how genes in the worm control its cells and its behaviour. Thus the effects of chemicals on individual cells can be studied and related to changes in biochemistry. The genetic code of the worm can be manipulated so that particular metabolic processes are absent or changed. This simple organism may yet change the way we carry out the evaluation of chemical hazard and safety.

Safety evaluation and its benefits

If toxicological information had been available in the 1930s the chemical TOCP would probably not have been used as a solvent for Jamaica ginger for human consumption (see pp. 260–2), nor would diethylene glycol have been used as a vehicle for sulphanilamide (see below).

Before chemicals are made available to the public as drugs, or used in industry or in the home, the safety evaluation of the substances in animals is carried out. As well as being a legal requirement in many cases, these

When lack of knowledge is dangerous

In 1937 the first sulphonamide drug, sulphanilamide, became available for the treatment of bacterial infections. In one preparation, known as an elixir, it was dissolved in diethylene glycol along with flavouring agents and colours. The pharmaceutical company making this formulation knew that glycol was a good solvent for the drug but had little if any information about its toxicity. People died from taking the drug, and these deaths were reported in the press. Fortunately, the drug, which was vital in its ability to cure potentially fatal infections, was not banned. It was discovered that it was the solvent that was causing the problem. Seventy-six people died and at least 100 became ill before the cause was discovered and removed. It is also interesting to note that the lethal dose in the human victims ranged from about 30 g to 280 g (almost tenfold), which illustrates human variability.

safety evaluation studies yield a number of benefits. Doctors in Accident and Emergency departments can treat victims of accidental or intentional poisoning with a particular drug or other chemical more effectively if they have information about the likely effects and the doses at which these occur. For example, the lack of information on the hazards from methylisocyanate in the Bhopal disaster (see p. 174) led to the inappropriate treatment of people who had been exposed. Especially useful is knowledge from animals of the blood level at which toxic effects occur, because often the dose that a victim of poisoning has taken or the level of exposure following inhalation are unknown. The blood level of the chemical can be measured in a patient and, given the relevant information available from previous studies in animals, directly related to the severity of a toxic effect. Any existing data resulting from previous exposure of humans to the chemical or drug would also be used. Furthermore, as a result of information gained during evaluation of the toxicity of a chemical in animals, antidotes have sometimes been devised or improved, for example for paracetamol, cyanide, lewisite, methanol, and ethylene glycol.

Knowledge of the relationship between dose and response (effect), and the threshold for this, is crucial in defining the risk of exposure to a chemical. Safety evaluation is a legal requirement for drugs, food additives, and contaminants in food, and a risk assessment has to be carried out in order to set the limits of exposure.[3] The relationship between the dose and the response (effect) can be established and plotted as a graph. This is called a dose–response curve (see Figure 29 and box), which often shows that there is a dose(s) of the chemical that has no effect and another, higher dose(s) which has the maximum effect. It is a visual representation of the Paracelsus principle that, at some dose, all chemicals are toxic. The corollary to this is that there is a dose(s) at which there is no effect.

The dose–response graph

The relationship between the level of exposure to a chemical and its harmful effects is very important in understanding how poisons work and whether they pose a risk. The typical graph can represent either a proportion of a group of individuals (humans or experimental animals) who have been given different doses of a substance and who show a particular effect (for example, the *appearance* of a slowing of reaction time after drinking alcohol), or the magnitude of the effect in a particular individual (for example, the *extent* by which reaction time is slowed by alcohol). Knowing the shape of this curve allows us to predict the effect of a chemical at a particular dose, for example from this graph we can see that there will be a 50 per cent response to a particular chemical at a dose of about 3.8 g, whereas there will be no adverse effect seen at a dose of 1 g or less. The arrow represents a point called the no observed adverse effect level (NOAEL). The NOAEL is the *threshold*, and this value is used in the assessment of risk from chemicals.

For certain types of toxic effect, notably cancer, some scientists believe that there is no threshold, and that the dose–effect graph would look like Graph A in Figure 29. As some chemicals cause cancer by reacting with the DNA molecule in cells, in theory one molecule of chemical would be enough to cause an effect. This notion probably arose from experiments with isolated cells exposed to radiation, but some regulatory authorities, in the USA for example, base their risk assessment of chemicals on this assumption.

If Graph A is correct it means that there is no dose without effect, that is, there is no safe dose. Is this realistic? Not everyone believes so because, even with very potent poisons, there are barriers that reduce the absorption of the chemical into the body. Even if in theory only one molecule were sufficient to cause a cancer, the chances of only one molecule entering the body, reaching its target, and causing sufficient damage are effectively 0.

Take the example: suppose there is a poisonous chemical, a carcinogen, in a foodstuff. Not all the food will be contaminated, so some individuals will escape exposure. If the food eaten is contaminated, suppose it contains 1 mg of carcinogen but only 1 per cent of it is absorbed (0.01 mg). Of that 1 per cent suppose that only 1 per cent is metabolized to the toxic (carcinogenic) substance (0.0001 mg). Of that toxic substance, suppose that only 1 per cent reaches the target, for example the DNA (0.000001 mg). Of the damage caused, suppose that 99 per cent is repaired, and that only 1 per cent is potentially damaging (effectively 0.00000001 mg). This means that only 100-millionths of the original dose reaches the target and causes damage, and damage to DNA does not necessarily result in a cancer.

This is only an illustration, but it shows that there can be a threshold even for potent chemicals. The problem is that it is technically impossible to show effects at very low doses, so the true shape of the dose effect curve A at the bottom is unknown. Experimental data does exist, however, which supports the idea that thresholds can exist for carcinogens that interact with DNA. Because of the technical difficulties, extrapolation of the curve to 0 from a dose where effects have been detected is therefore necessary. Unfortunately, cancer risks are sometimes calculated from this type of graph (the most conservative model), when it should be regarded as potentially very inaccurate.

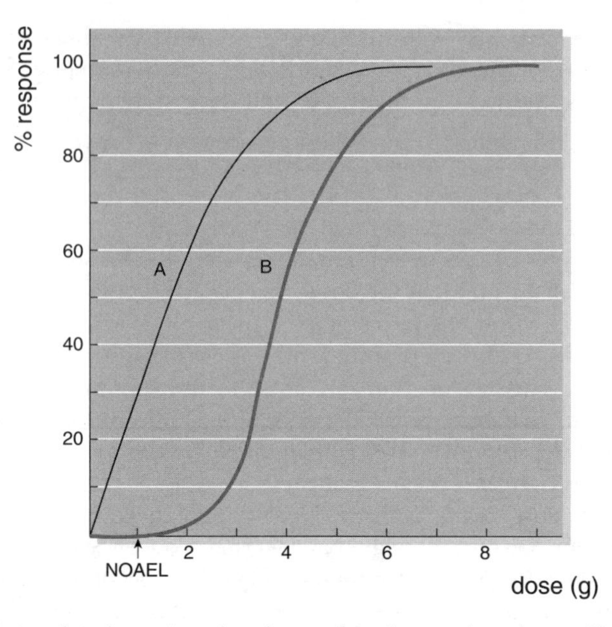

29. The relationship between the dose of a chemical and its effect. Graph B shows the typical relationship with some doses causing no effect (as indicated by the arrow), and some doses causing the maximum effect (100% response). For a full explanation of this figure and graph A see the explanatory box.

The dose–response relationship therefore allows the toxicologist to establish a threshold, the dose at which there is no adverse effect, which is vital for the proper assessment of risk. Information from the dose–response relationship is used to determine the therapeutic index and the margin of safety for drugs, which indicate how safe the drug is. The greater the value the greater the difference between the dose at which there are adverse effects and the therapeutic dose. This is used as part of the risk assessment process.

For chemicals such as food additives, food contaminants, and industrial chemicals the threshold, that is the dose at which toxic effects become apparent, is determined from the dose–response graph and used in the risk assessment process. The threshold value is used, together with safety factors, to determine the acceptable daily intake (ADI) of a food additive, or the tolerable daily intake (TDI) of a food contaminant, or the threshold limit value (TLV in the USA, or maximum exposure limit (MEL) in the UK), for an industrial chemical (see box for calculation). For a drug, information about the dose in animals below which there are no adverse effects will be necessary before human volunteers can be exposed in clinical trials. More extensive safety evaluation is carried out for drugs than for most other chemicals, as described on pp. 86–8, 292–4, and the clinical trials provide direct knowledge of the dose of a particular drug that has no adverse effects in humans. Furthermore, with a drug, certain effects on the patient are desirable, so the dose used must be one that has efficacy but not toxicity.

Terms used in risk assessment

Therapeutic index: the ratio of the dose found to be toxic in 50 per cent of animals (called TD_{50}) to the dose found to be therapeutically effective in 50 per cent of animals (called ED_{50}) given the drug. The larger the number, the safer the drug (that is, ED_{50} is very small or TD_{50} is very large).

Margin of safety: the ratio of the toxic dose in 1 per cent of animals to the therapeutically effective dose in 99 per cent of animals given the drug. The larger the number the safer the drug. This is more discriminating than the therapeutic index.

Acceptable daily intake (ADI): an estimate of the daily intake of a food additive over a lifetime that is considered to be without appreciable health risk. It is calculated as:

ADI = no observed adverse effect level (NOAEL)/safety factor.
The safety factor is usually 100 but may be 1,000 in some cases.

Tolerable daily intake (TDI): an estimate of the daily intake of a substance (a food contaminant) over a lifetime that is considered to be without appreciable health risk. It is calculated in the same way as the ADI.

Threshold limit value (TLV) in the USA; *maximum exposure limit* in the UK: the concentration of a substance in air to which it is believed most workers can be exposed daily without adverse effect. It is calculated in the same way as the ADI.

These values are the estimated amounts of the chemical to which people can be exposed in food, water, and in the workplace without appreciable risk.

Let us take a look at some examples. The first illustrates the problems that may result from a combination of poor science and a cautious and inflexible approach.

The story of saccharin[4]

The tortuous tale of saccharin is a good illustration of the problems that can arise in the risk assessment of chemicals. It also serves to illustrate the process of risk assessment which should be a logical, scientific process, incorporating numerous safety factors. Part of the history of saccharin has already been told on pp. 277–8. We pick up the story in the 1970s, when increased consumption of saccharin by the public and a report showing another sweetener to be carcinogenic prompted further studies to be carried out. At least twelve studies in animals were carried out in the early 1970s and the FDA concluded in 1974 that 'no findings have been reported that caused the safety of saccharin to be questioned'. Then one study, which exposed rats to saccharin over two generations, found bladder tumours, but only in some male animals given very high doses. Another study found similar tumours but at only one, high dose. Because of these findings the US Food and Drugs Administration proposed a ban on the sweetener. There was a public outcry and the American Congress responded by passing a moratorium on the ban to allow time for more research. The further research eventually produced data that was good enough to use in a risk assessment, as described below.

The risk assessment of saccharin

The risk assessment for a food additive or a contaminant requires a body of knowledge usually derived from toxicity or safety evaluation studies like those described above. The data used will be derived from what is deemed to be the 'best' study available, that is the study or studies that best conform to accepted scientific standards. Certain criteria have to be satisfied:

1. a sufficient number of animals must have been used for the results to be statistically significant;
2. several different doses must have been given in order to show a relationship between the dose of the chemical and the response of the animal;
3. the animals should have been well maintained and healthy; and
4. the biochemical analyses and pathology must be properly conducted.

While the assessment may show several effects, from the information gained it should be possible to establish a threshold dose for the most sensitive relevant effect in the animals. From this the no observed adverse effect level (NOAEL) can be estimated and used to determine the acceptable daily intake (ADI).

The safety evaluation study for saccharin, on which the risk assessment was based that allowed the FDA ban to be lifted, was made available in 1983. It established a relationship between the dose of saccharin and the appearance of bladder tumours in male rats (the response). While it was found that saccharin was able to cause tumours in rats, the following points were important considerations in the risk assessment.

1. Saccharin is not metabolized and does not react with DNA. Although it was mutagenic *in vitro*, this was only at high concentrations. It was concluded that saccharin was not a genotoxic carcinogen. Assuming that there was no threshold was, therefore, not appropriate. A threshold model could be applied to the dose–response data from these safety evaluation studies.
2. It was found that the bladder tumours occurred at a dietary level of 3 per cent saccharin, a dose high enough to saturate excretion and cause physiological changes in the animals.
3. Only male rats showed tumours; females and other species did not show this response. The bladder tumours occurred only in rats that had been treated from the beginning of life, but not in rats that were treated only as adults.
4. It was also found that, under the conditions that led to bladder tumours, mineral deposits (microcalculi) were present in the bladder, which is known to cause bladder cancer in rats.
5. There was also information from a number of studies on extensive human exposure which showed no association between the use of saccharin by humans and bladder cancer.

For these reasons the Joint FAO/WHO Expert Committee on Food Additives (JECFA) considered that the tumours in the rat bladder were not relevant to humans. Both JECFA and the European Union Scientific Committee on Food concluded that saccharin was not a hazard and approved it for use. Similarly the FDA withdrew their proposal to ban saccharin in 1991.

In the most recent study at the time adverse effects were detected in animals at a dietary level of 3 per cent and above, and so a NOAEL was set at a level of 1 per cent in the diet. This is equivalent to 500 mg of saccharin per kg of body weight, which in turn is equivalent to a person drinking about 200 cans of diet cola a day! The acceptable daily intake (ADI) is determined by dividing the NOAEL by a safety factor. The safety factor used is usually 100.

$$\text{ADI} = \text{NOAEL} / \text{safety factor}$$

For saccharin:

$$\text{ADI} = 500 \text{ mg per kg bw} / 100$$

or

$$5 \text{ mg per kg body weight}$$

Why is the safety factor 100? This factor takes into account the fact that the safety evaluation is carried out in animals and so there may be differences from humans, who could be more sensitive. This is given a factor of 10. Secondly, the safety factor takes into account the fact that humans vary in their sensitivity: some people are more or less sensitive to chemicals than others. This is also given a factor of 10, and hence the overall factor is 100 (10×10). Retrospective research of available data in humans and animals has confirmed and refined the basis for these factors (each factor can be subdivided). Sometimes, for example when the data is less reliable, larger safety factors may be used (1,000).

By extrapolation from the dose–response information in rats, it has been estimated that a 0.01 per cent dietary level of saccharin, equivalent to about two cans of diet cola a day would lead to a lifetime risk of bladder tumours of 2.5×0.0000000001 (2.5×10^{-10}) or 2.5 in ten thousand million! This does not take into account the fact that (1) the bladder tumours occurred only in male rats treated in a particular way; (2) the doses given were unrealistically high; and (3) there is no human evidence for the effect. Thus even this estimate is erring very much on the side of caution.

The example of saccharin shows how the risk assessment for a food additive is carried out,[5] and the precautions taken by the regulatory authorities, using large safety factors and always erring on the side of caution (the precautionary principle). In this case there was human data available, unusually for a food additive. Today a new food additive would not be used in food for human consumption until it had been evaluated for safety in animals.

If, in the safety evaluation of a new food additive, tumours are found in adult animals given reasonable doses, especially if this occurs in more than one species of animal and if the underlying mechanism is relevant to humans, this would be incorporated into the risk assessment. The chemical would almost certainly not be licensed as a food additive. If the adverse effect caused by a food additive is of a different type and is believed to show a threshold, and provided a NOAEL can be determined from the available data, an ADI can be set, as described above. If a NOAEL cannot be determined, then a larger safety factor will be used. The risks have to be considered in relation to the benefits.

Risk assessment of food contaminants

As we saw in Chapter 5, dioxin (or TCDD) is a chemical contaminant derived from industrial processes that occurs in the environment and which can be found in very small amounts in some kinds of food (for example, oily fish). How do regulatory authorities like the Food and Drug Administration (FDA) in the USA or the Food Standards Agency (FSA) in the UK decide on a value for the tolerable daily intake for this chemical?

There is a considerable amount of information in the scientific literature describing the different toxic effects caused by dioxin in various species of animals, which is reviewed by the relevant government agency. The most scientifically robust papers are selected and discussed by an expert committee. If any of these publications includes a study or studies of effects at different doses, a no observed adverse effect level (NOAEL) can be estimated (from the dose–response curve). If the information is more limited, and there is no indication of a threshold, then the lowest dose that causes a significant toxic effect will be used for setting limits. This is known as the lowest observed adverse effect level (LOAEL). A chemical like dioxin causes a number of different effects, and the agency and its committee take the most sensitive parameter which in the case of dioxin is an effect on the reproductive system. Armed with the no observed adverse effect level (NOAEL) the regulatory agency can set a tolerable daily intake value (TDI) for dioxin in the same way as for saccharin, by dividing by a safety factor, typically 100 (see box, page 301). An additional factor is incorporated if the lowest observed adverse effect level (LOAEL) is used.

If further scientific papers are subsequently published that indicate that there are toxic effects at lower doses than previously thought, or other toxic effects, these values may be used and the TDI will be revised if necessary (as has happened with dioxin). In principle, a similar approach is applied in setting the threshold limit value (TLV) for an industrial chemical.

The upshot is that the standard that is set is very cautious because of the safety factors used, especially if safety factors of 1,000 or higher are used. Moreover, the adverse effects observed in animals usually result from continuous repeated exposure and not single doses, which is clearly different from occasional exposure as a result of eating a foodstuff contaminated with a chemical.

This approach can be used where there is a threshold (or assumed threshold) for the toxic effects, but when cancer is the end-point some regulatory bodies assume that there is no threshold (see box above, page 298, and Figure 29). If there is no threshold for a chemical it means that any dose could be hazardous. This approach is based on the premise that a single molecule is sufficient to cause a toxic effect, because in theory it can reach the DNA in a cell and cause a mutation that can lead to cancer.

Many scientists have been increasingly unhappy with this approach as it has become apparent that there are a number of mechanisms by which chemicals can cause cancer. Even with the chemicals that damage DNA and cause mutations we now know that the damage can be repaired, that cells with damaged DNA are removed by apoptosis, that much of the DNA is redundant, and that a damaged cell must divide in order to become a tumour. Altered cells (that is, cancer cells), moreover, can be removed by the immune system. Furthermore, in order to cause cancer, a chemical has to get inside a cell in the body, and the body has many barriers against chemicals entering it, for example: the membranes of cells form a natural barrier; the blood is able to remove and eliminate chemicals into the urine; and the body has systems to pump chemicals out of cells, and metabolism to detoxify them. The scenario of chemicals entering a body and causing toxic effects can be likened to a few soldiers trying to get into the inner recesses of a heavily guarded fortress who have to go through minefields, scale high walls, cross a moat, scale more walls, to reach the nerve centre, all the while being attacked by the enemy: few if any would make it. Thus the reality is that very small amounts of a chemical are highly unlikely to cause the crucial damage necessary for cancer to develop, even without the body's own mechanisms for repair and removal. Common sense would suggest that in effect a threshold does exist.

So how is a tolerable daily intake (TDI) or a threshold limit value (TLV) determined for a chemical that can cause cancer and for which it is assumed by the regulatory agencies that there is no threshold dose? It requires animals (rats or mice) to be exposed to large (and probably unrealistic) doses of the chemical for their lifetime. Doses need to be high so that the incidence of tumours is (statistically) significantly different from the incidence in the control, untreated animals, say 5 per cent or 10 per cent. This may require a dose of 100 mg per kg body weight, while the level of a food contaminant or industrial chemical in a food may result in a human dose of only nanograms (1 millionth of a mg) per day. Based on the cautious assumption that there is no threshold, the dose–response graph (see Figure 29) is extrapolated to 0. From this extrapolation of the graph, the *theoretical* incidence of cancer at this very low level of exposure can be determined. Alternatively, by using a particular level of cancer incidence, say 1 in a million (which may be considered to be just acceptable), the exposure level giving rise to this incidence can be determined. However, this would be only an estimate, because the shape of the curve at the bottom (at low doses) is unknown and almost certainly shows a threshold. It is practically impossible to determine the incidence at very low doses, as it would require a very large number of animals and much more sensitive methods of detecting tumours than are currently available. Hence significant extrapolation is essential, based on the assumption that the dose–response graph goes to 0 (see Figure 29 and box above, page 298). This is the cautious approach, which does not take into account the factors that in reality mean that a threshold almost certainly does exist, but which it is impossible to discover for the same reasons.

This means that the incidence of tumours would in reality be very much lower than predicted from the graph extrapolated to 0. The extrapolation will give a notional incidence of perhaps one in a million, which means that out of a million humans exposed to a particular level of chemical daily for their lifetime there would be one extra tumour in an individual as a result of the exposure. For the individual, the risk or chance is one in a million, but this is a conservative estimate, based on an extrapolation which overestimates the incidence, so the true risk would probably be many times lower than this.

Half of all chemicals tested in rodents using high doses have been shown to be carcinogens, including natural chemicals. Half the chemicals that occur naturally in plants we eat have been shown to be carcinogens in these rodent tests. This is because the high doses used in the animal tests may cause different or additional effects to lower doses. The tests are crude and insensitive, and the animals are exposed continuously for a long period of time, not occasionally or intermittently. In their extreme caution and an attempt to reduce exposure to impossibly low levels, regulatory authorities could be wasting time and money, resources that could be better spent elsewhere.

With industrial chemicals the situation may be rather different, as some of them may be (and certainly were in the past) encountered at higher levels. The risk assessment using animal studies is more likely to give an estimate of risk closer to reality. For example, the widely used industrial chemical vinyl chloride (see discussion on pp. 168–71) causes rare liver tumours in humans and rodents. Three different models using analysis of several sets of existing animal data gave similar cancer risk estimates to those derived by epidemiological analysis from human subjects exposed to vinyl chloride. Some models based on animal studies still overestimated the actual risk when compared to reported cases of the liver cancer however.

Even when animals are given significant doses of potent known carcinogens, however, not all develop tumours, even inbred mice, due to biological variation in the various factors and stages required for development of a tumour. Humans show much greater variation, and therefore the incidence of cancer could be much lower on the whole, although some individuals could be more sensitive.

How do people view risks?

Throughout our lives we are exposed to many different hazards and to the various risks associated with them. Every time we drive a car, for example, there is a risk of an accident and therefore of injury or death (which can be calculated). Similarly, other means of travel will have different, but known, risks. Other activities, for example smoking, skiing, skydiving, have known risks of death, injury, or disease. We choose whether or not to engage in these activities and take the risk, even though we may not always appreciate the size of the risk. Most people know that a risk of death or disease from an activity is not a certainty, for example not everyone who drives or travels in a car has an accident and suffers injury at some time in their life. When challenged as to why they smoke someone may answer that their Uncle Fred smoked two packs of cigarettes a day and lived to a good age. They know that it *is* possible to avoid the effects many suffer, but they choose to take the risk because they think it won't happen to them.

We can, of course, reduce the risks in our lives by not doing anything with a risk attached, but even staying in bed all day carries a risk, for example, of thromboembolism as a result of inactivity. Life itself is a risky business and most of us learn to accept this. Using chemicals is one of the risks most of us take. If we have a headache or some other minor ailment we take an aspirin and in so doing take a very small but known risk that it may harm us, for it sometimes causes bleeding in the stomach. We can minimize the risk by taking only the necessary dose and by taking it with food, but if we don't accept the risk we should not take the drug. Life and its activities cannot be risk-free: there is no such thing as an absolutely safe drug just as there is no absolutely safe way to travel. We have to decide if the risk is worth taking.

In life absolute safety is impossible. Proving that a chemical, for example a food additive, sweetener, or drug, is absolutely safe, is just not

Table 9 Risks associated with some common voluntary and involuntary actions

RISK	ODDS (death/person/year)
Motorcycling	1 in 50[1]
Smoking (20 cigarettes per day)	1 in 200
Pregnancy (UK)	1 in 4,380
Driving a car (UK)	1 in 5,900[2]
Rock climbing	1 in 7,150[3]
Influenza	1 in 9,000
Drinking (1 bottle of wine per day)	1 in 13,300
Being struck by a car (UK)	1 in 16,600
Taking contraceptive pill	1 in 50,000
Earthquake (California)	1 in 588,000
Lightning	1 in 10,000,000
Release from atomic power station (USA and UK)	1 in 10,000,000

[1] A figure of 1 death per 1,056 registered motorcycles has also been quoted.
[2] Based on deaths per million participants.
[3] Based on deaths per million hours/year spent in sport.

Source: M. D. B. Stephens, J. C. C. Talbot, and P. A. Routledge (eds.), *New Adverse Drug Reactions*, London, Macmillan, 1998, ch. 1, table 5.

possible, because it is impossible to prove a negative, that is, the absence of an effect; it is possible only to prove the presence of an effect. Perhaps because life in the developed world has become much safer and we live much longer, we worry more about the small risks we are exposed to. When life was a constant battle against poverty, life-threatening infections and diseases, and insanitary, unhealthy conditions, people had little time to worry about the risk of taking a drug that could improve their condition.

The way in which we the public perceive risk is not always the same as the way in which experts and governments perceive risk. Individuals perceive risks very differently. Some people engage in climbing without ropes or in skydiving which obviously entail high risk, yet the same individuals may exaggerate other kinds of risks, especially those outside their control. People may ignore or play down the risk of a disease like cancer from excessive cigarette-smoking but worry (unnecessarily) about the risk, for example, from the natural radioactive gas radon, which can seep into houses. Disproportionate anxiety like this is usually more likely between different types of risks, for example between air travel and smoking.

Risks that involve no apparent benefit to the individual are, understandably, less likely to be acceptable. For example, the possibility of traces of a pollutant like dioxin in an incinerator exhaust (where the incinerator may be producing heat or power for a community and removing waste) would be unacceptable to many people.

Sometimes the scientific and/or the official assessment of a risk will be similar to that of the general public, in which case the management and acceptance of the risk will be easier. However, when the public perception of the risk is greater than it really is, politicians may have to take steps (which are really unnecessary) to try to reduce the risk, with the result that huge sums of money may be spent for no real benefit (for example, see pp. 125–6 on Love Canal).

The educational level of the people, and an understanding of the science underlying a risk, will clearly affect the perception of a risk. The media have a crucial role to play in explaining and presenting information, but unfortunately they may sometimes exaggerate risks and thereby increase anxiety.

What is an acceptable risk? This is sometimes given as one in a million, say for the estimate of the risk of a chemical causing cancer. This led someone in the USA (where the population is about 250 million) to point out that permitting the use of a particular chemical with a risk of causing cancer of one in a million would condemn 250 people to die! As we have already seen, this is a mathematical estimate based on a conservative model, whereas a more realistic model would give a very different answer. A more precise detection of adverse effects at low doses in animals or humans would make the estimates more accurate but as yet no method has been developed or validated. In the assessment of the chemical chlordane, which can be found in drinking water, the most conservative estimate (based on the one-hit, not threshold model) indicated that one extra cancer death would result from a level of 0.03 μg/litre, whereas another, less conservative, and possibly more realistic model (the probit model) yielded a figure of 50 μg/litre, that is a level more than 1,000 times higher. Risk assessment is not a precise science. Much of science is not as precise or as certain as many people think.

What are the implications for this imprecision and is it acceptable? The inbuilt safety factors, the conservative assumptions, and the way in which chemicals are tested mean that the risk will always be very much lower than stated.

Furthermore, the benefits must also be considered. The case of contamination of food with chemicals like dioxin and PCBs is a good example of the importance of including risk–benefit considerations in any assessment of a toxic hazard. Dioxin and PCBs are fat-soluble substances and are consequently found in food with a high fat or oil content. It was recently reported in the journal *Science* that higher levels of dioxin, PCBs, and other chemicals were found in farmed salmon (dioxin levels were about ten times the levels in wild salmon). Although the levels were still well below government limits, the message from some sources, predictably, was that people should stop eating such fish regularly, perhaps just once per month, because it could be dangerous. The model on which this conclusion was based is a conservative one using a linear correlation between cancer and exposure, which therefore maximizes the estimated risk. This interpretation of the data was criticized by some toxicologists. Furthermore, there are significant benefits to eating fish and a Finnish scientist, J. T. Tuomisto (one of the world authorities on dioxin), and his colleagues have demonstrated the importance of considering this in a balanced, risk–benefit assessment. They calculated that in the European Union, with a population of 387 million, a restriction on the eating of farmed salmon could lead to the prevention of forty extra deaths from cancer. At the same time, not eating the farmed salmon, which has proven beneficial effects, for example helping to prevent cardiovascular disease, could lead to 5,200 extra deaths.[6] This is the kind of information the public needs to have in order to make an informed judgement. Similarly the detection of dioxin in breast milk at levels higher than were set by regulators has to be viewed in relation to the benefits of breast-feeding, which outweigh the risks, to the newborn child.

What is acceptable to a person often depends on the benefit: if the benefit of taking a risk is to themselves they will accept a higher risk than if it is to an organization they work for, or to the government, or to society. Very often, though not always, people do have the choice of avoiding the risk if they perceive it to be unacceptable, for example some people buy 'organic' food as they perceive it to be more healthy and to have less risk of exposure to chemicals. Organic food won't change their exposure to chemicals, however, as the great majority (at least 99 per cent) of chemicals are there naturally, including chemicals that are able to cause cancer. They could even be exposing themselves to more poisonous chemicals generated, for example, by micro-organisms (as well as to hazardous bacteria) by opting for organic food, but there is no consensus or convincing evidence for or against this.

A few people will not accept the very small risk associated with the chlorination of drinking water due to the production of chemicals in trace amounts that may be carcinogens such as chloroform. They therefore buy and drink bottled water. The chlorination of water, however, is of huge benefit in terms of public health, reducing the risks of bacterial infections from drinking water. The benefits far outweigh the risks in this case, that is, the risks (from waterborne diseases) from not chlorinating water are much higher than the risk of cancer from chlorination products.

Unfortunately, in our increasingly litigious society, risks are becoming less and less acceptable even when the benefits are obvious. For example, the drug Opren which caused serious adverse, and in some cases fatal, effects in the liver in a small number of patients was withdrawn, despite the fact that changing the way the drug was administered could have removed the problem, and that the drug was extremely effective at treating arthritis and safer than the alternatives in many cases. Patients with chronic arthritis were dismayed at the loss of such an effective drug. More recently the drug tolcapone, which was used in the treatment of Parkinson's disease, was withdrawn due to serious liver damage in one in 25,000 patients. Should the patients in whom the drug had had a dramatic effect be denied it because of the risk?

Perhaps the answer lies in greater choice, in allowing the patient or consumer to make the choice by giving them all the information in an intelligible form, though the information may sometimes be complex and difficult to understand. In most instances a patient who doesn't like the side effects of a drug can often choose to stop taking the drug. It does seem bizarre, however, that people smoke, drink alcohol, and drive cars without a moment's thought when these activities carry a much greater risk of harming us than most of the chemicals we are exposed to: 200,000 smokers and 100 drivers out of every million will die as a result of smoking or driving, and these figures are derived from verifiable human records.

So what can we do about chemicals and the risks created by using them? We could stop developing new drugs or using chemicals, but this would have great economic and human consequences, limit our progress, and make our lives generally less pleasant and healthy. Most people would consider this to be a retrograde step, but society has to make these choices and people need to be informed in order to do so. I believe we have to continue using chemicals but we should be vigilant, not complacent, and treat chemicals with respect, and we should also use common sense and good science in regulating their use. We should certainly not succumb to what Alice Ottoboni has called the 'poison paranoia'.[7]

Our use of man-made chemicals is increasing, and the presence of these chemicals in our environment as well as their risk assessment creates a number of uncertainties. Despite this, the incidence of most cancers and most diseases is not increasing, but decreasing. Health problems and illnesses that are increasing—for example, lung cancer, liver cirrhosis, diabetes, and obesity—are mostly due to lifestyle (smoking, drinking, overeating, poor diet and lack of exercise). We can expect a longer lifespan than at any other time in human history, and our lives are healthier. This would be unlikely if the synthetic chemicals we regularly come into contact with were really as hazardous as some would have us believe.

In conclusion, this is the poison paradox: chemicals have useful and beneficial effects but under different conditions the same chemicals can be harmful. Some chemicals can cure our ailments, improve our mood, or make our food taste better; others can be used to make valuable products such as brightly coloured paints, parts of computers, or cars, and many kinds of plastic or flame-resistant clothes. There is a multitude of ways in which chemicals enrich our society, yet at the same time, if they are misused or used without respect, they can be hazardous. When a person takes or is given a dose of a drug that is too high, or is exposed to a concentration of a chemical that is too high (say in an industrial accident), adverse consequences inevitably follow. As Paracelsus commented centuries ago: 'All substances are poisons; there is none that is not a poison. The right dose differentiates a poison from a remedy.'

Endnotes

1. Poisons: Old Art, New Science

1. *The Popular Oxford Dictionary of Current English*, 6th edn., Oxford, OUP, 1979.
2. *Chambers English Dictionary*, 7th edn., Cambridge, W. & R. Chambers/CUP, 1988.
3. A. M. Ottoboni, *The Dose Makes the Poison: A Plain-Language Guide to Toxicology*, 2nd edn., New York, Van Nostrand Reinhold, 1997.
4. A. Furst, 'Profiles in toxicology: Moses Maimonides', *Toxicological Sciences*, 59 (2001), 196–7.
5. E. Efron, *The Apocalyptics: How Environmental Politics Controls What We Know about Cancer*, New York, Simon & Schuster, 1984.
6. J. Brody, 'Managing planet earth: personal health; risks and realities: in a world of hazards, worries are often misplaced', *New York Times* (20 Aug. 2002).

2. Bodily Functions: What Chemicals Do to Us and What We Do to Them

1. E. J. Calabrese and L. A. Baldwin, 'Toxicology rethinks its central belief: hormesis demands a reappraisal of the way risks are assessed', *Nature*, 421 (13 Feb. 2003), 691–2.
2. S. E. Rau *et al.*, 'Grapefruit juice–terfenadine single dose interaction: magnitude, mechanism, and relevance', *Clinical Pharmacology and Therapeutics*, 61 (1997), 401–9.

3. Keep Taking the Medicine: There are No Safe Drugs, Only Safe Ways of Using Them

1. L. F. Prescott, *Paracetamol: A Critical Bibliographic Review*, London, Taylor & Francis, 1996.
2. J. J. Fenton, *Toxicology: A Case-Oriented Approach*, Boca Raton, Fla., CRC Press, 2002, 347–9.
3. J. M. Tredger *et al.*, 'Metabolic basis for high paracetamol dosage without hepatic injury: a case study', *Human and Experimental Toxicology*, 14 (1995), 8–12.
4. 'Heroin', *BBC News* [website] (8 Feb. 2003) ⟨http://news.bbc.co.uk/1/low/health/medical_notes/85691.stm⟩

5. B. Morris, 'Heroin death girl's parents set for body to be exhumed', *Sunday Herald* (4 May 2003) ⟨http: / / www.sundayherald.com. / print33546⟩
6. U. D. McCann *et al.*, 'Positron emission tomographic evidence of toxic effect of MDMA ("Ecstasy") on brain serotonin neurons in human beings', *Lancet*, 352 (1998), 1437.
7. T. Stone and G. Darlington, *Pills, Potions, Poisons: How Drugs Work*, Oxford, Oxford University Press, 2000, 404–6.
8. K. Jansen and A. R. W. Forrest, 'Toxic effect of MDMA on brain serotonin neurons', *Lancet*, 353: 9160 (10 Apr. 1999), Letter to the Editor.
9. G. A. Ricaurte *et al.*, 'Retraction', *Science*, 301 (2003), 1479.
10. 'FDA warns consumers to discontinue use of botanical products that contain aristolochic acid', *FDA* [website] (11 Apr. 2001) ⟨http: / / vm.cfsan.fda.gov / ~dms / addsbot.html⟩; J. Kelly, 'Chinese herb leads to kidney failure, cancer', *WebMD Medical News Archive* [website] (7 June 2000) ⟨http: / / my.webmd. com / content / article / 36 / 1728_58270⟩
11. J. L. Nortier *et al.*, 'Urothelial carcinoma associated with the use of a Chinese herb (Aristolochia fangchi)', *New England Journal of Medicine*, 342 (2000), 1686–92.
12. 'Herbs: not always a remedy', *CBS News* [website] (2000) ⟨http: / / www.cbsnews.com / stories / 2000 / 06 / 08 / national / main203943.shtml⟩
13. 'Drug Safety and Quality; Centre for Drug Evaluation and Research, Report to the Nation, FDA [website] (2002) ⟨http: / / www.fda.gov / cder / reports / rtn. / 2002 / rtn2002-3.htm⟩

4. Blood, Sweat, and Tears: Pesticides

1. L. W. Smith, 'Key challenges for toxicologists in the 21st century', *Trends in Pharmacological Sciences*, 22 (2001), 281–5.
2. N. F. R. Snyder and V. J. Meretsky, 'California condors and DDE: a re-evaluation', *Ibis*, 145: 1 (2003), 136.
3. Smith, 'Key challenges for toxicologists in the 21st century', 281–5.
4. M. Jacobs and B. Dinham (eds.), *Silent Invaders, Pesticides, Livelihoods and Women's Health*, London, Zed Books, 2003, 111.
5. W. N. Aldridge, *Mechanisms and Concepts in Toxicology*, London, Taylor & Francis, 1996.
6. Fenton, *Toxicology*, 174–6.

5. First the Cats Died: Environmental Contaminants

1. J. Woodall, *The Surgeon's Mate or Military & Domestic Surgery*, 1639.
2. Fenton, *Toxicology*, 317.
3. 'Minamata disease: the history and the measures', *Ministry of the Environment* [Japan] [website] (2002) ⟨http: / / www.env.go.jp / en / topic / minamata2002 / index.html⟩

4. C. J. Polson, M. A. Green, and M. R. Lee, *Clinical Toxicology*, 3rd edn., London, Pitman, 1983, 423.
5. F. Pierce, 'Danger in every drop', *New Scientist* (12 Feb. 2000), 16–17.
6. 'Arsenic in drinking water', Fact Sheet no. 210, *World Health Organization* [website] (2001) ⟨http://www.who.int/mediacentre/factsheets/fs210/en/⟩
7. G. Reggiani, 'Medical problems raised by the TCDD contamination in Seveso, Italy', *Archives of Toxicology*, 40 (1978), 161–88.
8. G. P. Daston, J. C. Cook, and R. J. Kavlock, 'Forum: Uncertainties for endocrine disruptors: our view on progress', *Toxicological Sciences*, 74 (2003), 245–52; T. Danstra *et al.*, *Global Assessment of the State of the Science of Endocrine Disruptors*, WHO publication no. WHO/PCSEDC/02.2, Geneva, WHO, 2002.
9. Polson, Green, and Lee, *Clinical Toxicology*, 468.
10. J. M. Bruce and W. J. Dilling, *Materia Medica and Therapeutics*, 12th edn., London, Cassell & Co., 1921, 88–92.
11. R. L. Canfield *et al.*, 'Intellectual impairment in children with blood lead concentrations below 10 μg per deciliter', *New England Journal of Medicine*, 348 (2003), 1517–26; B. P. Lanphear *et al.*, 'Primary prevention of childhood lead exposure: a randomized trial of dust control', *Pediatric*, 103 (1999), 772–7.
12. S. Milloy, 'Get the lead hysteria out', *JunkScience.com* [website] (16 Mar. 2001) ⟨http://www.junkscience.com/foxnews/fn031601.htm⟩
13. P. Altmann *et al.*, 'Disturbance of cerebral function in people exposed to drinking water contaminated with aluminium sulphate: retrospective study of the Camelford water incident', *British Medical Journal*, 319 (1999), 807–11; V. Murray *et al.*, 'Cerebral dysfunction in Camelford: inappropriate study, inappropriate conclusions', *British Medical Journal* (13 Dec. 1999), Letters; T. I. Lidsky, '(Altmann) Study has several methodological shortcomings', ibid.

6. Natural Born Killers: Poisonous Chemicals Designed by Nature

1. W. Sperl *et al.*, 'Reversible hepatic veno-occlusive disease in an infant after consumption of pyrrolizidine-containing herbal tea', *European Journal of Pediatrics*, 154 (1995), 112–16.
2. V. Stenkamp, M. J. Stewart, and M. Zuckerman, 'Clinical and analytical aspects of pyrrolizidine poisoning caused by South African traditional medicines', *Therapeutic Drug Monitoring*, 22 (2000), 302.
3. J. Mann, *Murder, Magic, and Medicine*, Oxford, Oxford University Press, 1994, 53.
4. J. H. Gaddum, *Pharmacology*, 5th edn., Oxford, Oxford University Press, 1959.
5. R. A. Christison, *Treatise on Poisons*, Edinburgh, A. & C. Black, 1836, quoted in Polson, Green, and Lee, *Clinical Toxicology*, 402.
6. British Pharmacopoeia (1973), quoted in Polson, Green, and Lee, *Clinical Toxicology*, 403.

7. Polson, Green, and Lee, *Clinical Toxicology*, 405–6.
8. R. J. Flanagan and A. L. Jones, *Antidotes*, London, Taylor & Francis, 2001, 97.
9. Mann, *Murder, Magic, and Medicine*, 40.

7. The Mad Hatter and a Bad Case of Acne: Industrial Chemicals

1. E. Hodgson, *A Textbook of Modern Toxicology*, 3rd edn., New York, John Wiley & Sons, 2004.
2. 'Vinyl chloride: hazard summary', *US Environmental Protection Agency* [website] (2003) ⟨http://www.epa.gov/ttn/atw/hlthef/vinylchl.html⟩
3. J. Kielhorn *et al.*, 'Vinyl chloride: still a cause for concern', *Environmental Health Perspectives*, 108 (2000), 579–88.
4. *Guardian* (8 Dec. 2001).
5. D. Mackenzie, 'Opinion: Class act', *New Scientist* (8 Jan. 2000).
6. M. Day, 'Bad medicine deepens Bhopal's misery', *New Scientist*, 152: 2060 (1996), 6.
7. 'Chemical taxi spread Bhopal toxin', *New Scientist*, 132: 1799 (14 Dec. 1991), 18.
8. P. Martin, 'Dust to dust', *Sunday Times Magazine* (16 May 2004).
9. M. L. Newhouse and H. Thomson, 'Mesothelioma of pleura and peritoneum following exposure to asbestos in the London area', *British Journal of Industrial Medicine*, 22 (1965), 261–9.

8. Under the Sink and in the Garden Shed: Household Poisons

1. Polson, Green, and Lee, *Clinical Toxicology*, 206.
2. N. A. Memon and A. R. Davidson, 'Multisystem disorder after exposure to paint stripper (Nitromors)', *British Medical Journal*, 282 (1981), 1033–4.
3. Gaddum, *Pharmacology*, 112.
4. E. Klarreich, 'Wine fights heart foe', *Nature Science Update* [website] (20 Dec. 2001). See ⟨http://www.innovations-report.de/html/berichte/medizin_gesundheit/bericht-6775.html⟩ (accessed 1 Feb. 2005).
5. 'Medical Students' Handbook: Alcohol and Health', 3rd edn., *The Medical Council on Alcohol* [website], published online ⟨http://www.medicouncilalcol.demon.co.uk/handbook/hb_intro.htm⟩

9. Rasputin's Revenge: Chemicals Used to Kill

1. Mann, *Murder, Magic, and Medicine*, 51.
2. A. Beéche, 'The evil monk: the life and times of Gregory Efimovich

Rasputin', *European Royal History* [website] ⟨http://www.eurohistory.com/Rasputin.html⟩

3. Polson, Green, and Lee, *Clinical Toxicology*, 429; K. Watson, *Poisoned Lives: English Poisoners and their Victims*, London, Hambledon and London, 2004, 3–5; C. Wilson and P. Pitman, *Encyclopaedia of Murder*, London, Pan Books, 1961, 92.
4. Bruce and Dilling, *Materia Medica and Therapeutics*, 129.
5. Polson, Green, and Lee, *Clinical Toxicology*, 429; Watson, *Poisoned Lives*; Wilson and Pitman, *Encyclopaedia of Murder*, 445.
6. Polson, Green, and Lee, *Clinical Toxicology*, 412; Watson, *Poisoned Lives*; Wilson and Pitman, *Encyclopaedia of Murder*, 497.
7. Polson, Green, and Lee, *Clinical Toxicology*, 449–52. C. Wilson and D. Seaman, *Encyclopaedia of Modern Murder 1962–1983*, London, Pan Books, 1986, 354–5; J. McCord, 'I was lucky to escape with my life', *Herald Express* [Hemel Hempstead] (1 Feb. 2001), 1; 'The casebook collection: The man behind the Bovingdon Bug (the Graham Young case of poisoning with thallium)', *Watford Observer* (8 Jan. 2002).
8. Polson, Green, and Lee, *Clinical Toxicology*, 449–52.
9. Wilson and Pitman, *Encyclopaedia of Murder*, 37–8.

10. Ginger Jake and Spanish Oil: Toxic Food Constituents and Contaminants

1. P. Grandjean and S. Tarkowski (eds.), *Toxic Oil Syndrome: Mass Food Poisoning in Spain*, Report on a WHO Meeting Madrid, 21–25 Mar. 1983, Copenhagen, World Health Organization, 1984.
2. S. H. Henry, F. X. Bosch, and J. C. Bowers, 'Aflatoxin, hepatitis and worldwide liver cancer risks', *Advances in Biology and Medicine*, 504 (2002), 229–33.
3. C. P. Wild *et al.*, 'Molecular dosimetry of aflatoxin exposure: contribution to understanding multifactorial etiopathogenesis of primary hepatocellular carcinoma with particular reference to hepatitis B virus', *Environmental Health Perspectives*, 99 (1993), 115–22.
4. 'Selected mycotoxins, ochratoxins, trichothecenes, ergot', *Environmental Health Criteria*, 105, Geneva, World Health Organization, 1990.
5. M. K. Matossian, *Poisons of the Past: Molds, Epidemics and History*, New Haven, Conn., Yale University Press, 1989; 'Selected mycotoxins, ochratoxins, trichothecenes, ergot'.
6. Matossian, *Poisons of the Past*.
7. Mann, *Murder, Magic, and Medicine*, 41–2.
8. Ibid. 86–7.
9. Matossian, *Poisons of the Past*; 'Selected mycotoxins, ochratoxins, trichothecenes, ergot'.
10. J. A. Timbrell, *Introduction to Toxicology*, 3rd edn., London, Taylor & Francis, 2001, 149.

11. Mann, *Murder, Magic, and Medicine*, 36–7.
12. B. Max, 'This and that: risk and resurrection', *Trends in Pharmacological Sciences*, 8 (1987), 16–18.
13. D. J. C. Shearman, 'Vitamin A and Sir Douglas Mawson', *British Medical Journal*, 1 (1978), 283–5.
14. R. Uhlig, 'Vitamin pill alert will be hard to swallow', *Daily Telegraph* (8 May 2003); J. Leake, 'Health risk warning over high doses of vitamin pills', *Sunday Times* (4 May 2003).
15. H. Rosling, *Cassava Toxicity and Food Security*, Report for UNICEF, Uppsala, Tryck Kontakt, 1988.
16. S. Wheeler, 'Cancer scare over Belgian chickens', *BBC News* [website] (31 May 1999) ⟨http://news.bbc.co.uk/2/hi/europe/357529.stm⟩
17. B. Hileman, 'Belgium has a problem: dioxin-tainted food', *Chemical and Engineering News*, 77: 24 (14 June 1999), 9.
18. 'Contamination with dioxin of some Belgian foods: 4. Toxicological evaluation of dioxins in Belgian poultry meat and eggs', *Food Safety Authority of Ireland* [website] (2004) ⟨http://www.fsai.ie/surveillance/food/surveillance_food_contamination.asp⟩
19. J. Parascandola, 'Pharmacology and public health: the Jamaica ginger paralysis episode of the 1930's', *Pharmacy in History*, 36: 3 (1994), 123–43, published online ⟨http://www.herbalgram.org/wholefoodsmarket/herbalgram/articleview.asp?a=957⟩
20. Bruce and Dilling, *Materia Medica and Therapeutics*, 191.
21. Parascandola, 'Pharmacology and public health', 123–43.
22. J. P. Morgan and T. C. Tulloss, 'The Jake Walk Blues', *Annals of Internal Medicine*, 85 (1976), 804–8.
23. H. V. Smith and J. M. K. Spalding, 'Outbreak of paralysis in Morocco due to ortho-cresyl phosphate poisoning', *Lancet*, 2 (1959), 1019–21.
24. Grandjean and Tarkowski (eds.), *Toxic Oil Syndrome*.
25. M. G. Ladona *et al.*, 'Pharmacogenetic profile of xenobiotic enzyme metabolism in survivors of the Spanish Toxic Oil Syndrome', *Environmental Health Perspectives*, 109 (2001), 369–75.

11. Butter Yellow and Scheele's Green: Food Additives

1. M. Hanssen and J. Marsden, *E for Additives*, Wellingborough, Thorsons, 1984.
2. L. Insall, 'Food additives and why they are used', in M. Saltmarsh (ed.), *Essential Guide to Food Additives*, Leatherhead, Surrey, LFRA, 2000.
3. J. Emsley and P. Fell, *Was it Something You Ate? Food Intolerance: What Causes It and How to Avoid It*, Oxford, Oxford University Press, 1999, 119–24.

4. Timbrell, *Introduction to Toxicology*, 101–2; Emsley and Fell, *Was it Something You Ate?*
5. Timbrell, *Introduction to Toxicology*, 101–2.
6. D. M. Conning, 'Toxicology of food and food additives', in B. Ballantyne, T. Marrs, and T. L. M. Syversen (eds.), *General and Applied Toxicology*, 2nd edn., Basingstoke, Macmillan, 1999; S. Barrett, 'The Feingold diet: dubious benefits, subtle risks' (2004), published online ⟨http://www.quackwatch.org/01QuackeryRelatedTopics/feingold.html⟩
7. Timbrell, *Introduction to Toxicology*, 101–2.
8. Ibid.
9. Ibid.
10. L. Juhlin, 'Intolerance to food and drug additives', in A. L. Weck and H. Bundgaard (eds.), *Allergic Reactions to Drugs*, Berlin, Springer Verlag, 1983.
11. Timbrell, *Introduction to Toxicology*, 101–2.
12. Ibid.
13. Conning, 'Toxicology of food and food additives'; Emsley and Fell, *Was it Something You Ate?*; Insall, 'Food additives and why they are used'; Barrett, 'The Feingold diet'.
14. Conning, 'Toxicology of food and food additives'.
15. Timbrell, *Introduction to Toxicology*, 102.
16. Calorie Control Council, 'Backgrounder on saccharin', *Saccharin* [website] (2003) ⟨http://www.saccharin.org/backgrounder.html⟩
17. B. L. Oser, 'Highlights in the history of saccharin toxicology', *Food and Chemical Toxicology*, 23: 4/5 (1985), 535–42.
18. D. Benford, *Principles of Risk Assessment of Food and Drinking Water related to Human Health*, Washington, DC, ILSI Press, 2001.
19. A. G. Renwick *et al.*, 'Risk characterization of chemicals in food and diet', *Food and Chemical Toxicology*, 41 (2003), 1211–71.

12. A Risky Business: Assessment of Chemical Hazards and Risks

1. UNEP/IPCS, *Chemical Risk Assessment*, Training Module no. 3, Geneva, World Health Organization, 1999, published online at ⟨http://www.chem.unep.ch/irptc/Publications/riskasse/coverpg.pdf⟩; Timbrell, *Introduction to Toxicology*, 163–79.
2. A. Bradford-Hill, 'The environment and disease: association or causation?', *Proceedings of the Royal Society of Medicine*, 58 (1966), 295–300.
3. UNEP/IPCS, *Chemical Risk Assessment*; Timbrell, *Introduction to Toxicology*, 163–79.
4. 'Saccharin: current status: report of an expert panel', *Food and Chemical Toxicology*, 23: 4/5 (1983), 543–6.

5. A. G. Renwick, 'Acceptable daily intake and the regulation of intense sweeteners', *Food Additives and Contaminants*, 7 (1990), 463–75; A. G. Renwick, 'A data derived safety (uncertainty) factor for the intense sweetener saccharin', *Food Additives and Contaminants*, 10 (1993), 337–50.
6. J. T. Tuomisto *et al.*, 'Risk benefit analysis of eating farmed salmon', *Science*, 305 (2004), 476.
7. Ottoboni, *The Dose Makes the Poison*, 1.

Further Reading

A. Albert, *Xenobiosis: Food, drugs and poisons in the human body*, London, Chapman and Hall, 1987.

B. Ballantyne, T.C. Marrs, T. Syversen, (eds.) *General and Applied Toxicology*, 2nd edn, Oxford, Macmillan Reference Ltd. 2000.

R. Bate, (ed.) *What Risk?* Oxford, Butterworth-Heinemann, 1997.

British National Formulary , BNF 45, Wallingford, Oxon, Pharmaceutical Press, 2004.

M. Burke, *Don't worry, it's organic* Chemistry World. Vol. 1, no 6; 30–35, 2004.

R. Carson, *Silent Spring*, Houghton Mifflin, 1962.

J. Emsley, *The Consumers Good Chemical Guide*, Oxford, WH Freeman, 1994.

J. Emsley, *The Elements of Murder*, Oxford, Oxford University Press, 2005.

J. Emsley, *Nature's Building Blocks*, Oxford, Oxford University Press, 2001.

J. Emsley, *Vanity, Vitality and Virility*, Oxford, Oxford University Press, 2004.

J. Emsley and P. Fell, *Was it Something You Ate? Food Intolerance: What Causes It and How to Avoid It*, Oxford, Oxford University Press, 1999.

E. Efron, *The Apocalyptics: How Environmental Politics Controls What We Know about Cancer*, New York, Simon & Schuster, 1984.

S.M. Hersh, *Chemical and Biological Warfare: the hidden arsenal*, London, Panther, 1970.

International Programme on Chemical Safety, *Hazardous Chemicals in Human and Environmental Health. A resource book for school, college and university students*, Geneva, World Health Organisation 2000.

United Nations Environment Programme/International Programme on Chemical Safety, *Chemical Risk Assessment*, Geneva, World Health Organisation 1999.

J. J. Fenton, *Toxicology: A Case-Oriented Approach*, Boca Raton, Fla., CRC Press, 2002.

R. J. Flanagan and A. L. Jones, *Antidotes*, London, Taylor & Francis, 2001.

M. Hanssen, *E for Additives*, Wellingborough, Thorsons, 1994.

J. Mann, *Murder, Magic, and Medicine*, Oxford, Oxford University Press, 1994.

M. K. Matossian, *Poisons of the Past: Molds, Epidemics and History*, New Haven, Conn., Yale University Press, 1989.

K. Mellanby, *Pesticides and Pollution*, London, Collins Fontana, 1967.

A. M. Ottoboni, *The Dose Makes the Poison: A Plain-Language Guide to Toxicology*, 2nd edn, New York, Van Nostrand Reinhold, 1997.

C. J. Polson, M. A. Green, and M. R. Lee, *Clinical Toxicology*, 3rd edn, London, Pitman, 1983.

D. L. Ray, *Trashing the Planet: How science can help us deal with acid rain, depleting of the ozone, and nuclear waste (among other things)*, Harper Perennial, 1992.

M. Saltmarsh (ed.), *Essential Guide to Food Additives*, Leatherhead, Surrey, LFRA, 2000.

M. D. B. Stephens, J. C. C. Talbot, and P. A. Routledge (eds.), *New Adverse Drug Reactions*, London, Macmillan, 1998.

T. Stone and G. Darlington, *Pills, Potions, Poisons: How Drugs Work*, Oxford, Oxford University Press, 2000.

J. A. Timbrell, *Introduction to Toxicology*, 3rd edn, London, Taylor & Francis, 2001.

C. H. Walker, S.R. Hopkin, R.M. Sibley, D.M. Peakall, *Principles of Ecotoxicology*, 2nd edn, London, Taylor & Francis, 2001.

K. Watson, *Poisoned Lives: English Poisoners and their Victims*, London, Hambledon and London, 2004.

C. Wilson and P. Pitman, *Encyclopaedia of Murder*, London, Pan Books, 1961.

C. Wilson, and D. Seaman, *Encyclopaedia of Modern Murder 1962–1983*, London, Pan Books, 1986.

Glossary

α-tocopherol
Vitamin E

acceptable daily intake (ADI)
An estimate of the daily intake of a food additive over a lifetime that is considered to be without appreciable health risk. It is based on the known toxicity of the additive.

acetylator phenotype
A genetic characteristic in humans that determines how rapidly a particular individual metabolizes certain drugs.

acetylcholine
A chemical released from the ends of nerves as a (neuro)transmitter when these are stimulated.

acetylcholinesterase
The enzyme that breaks down acetylcholine.

acute (toxicity)
The toxic effect of a single (acute) dose of a chemical. It is a relatively immediate effect.

adrenaline (epinephrine)
A hormone release in the body in response to a stimulus such as fear.

alcohol dehydrogenase
An enzyme that breaks down alcohols.

aldehyde dehydrogenase
An enzyme that breaks down aldehydes.

alkaloids
Natural chemicals containing nitrogen that are often found in plants, for example nicotine and morphine.

amino acids
Natural chemicals which are the building blocks of proteins. Some are obtained from the diet, others are made in our bodies.

anaemia
Disease in which the level of red blood cells is low.

anaphylactic reaction/shock
A potentially severe, possibly fatal, immune reaction.

antibodies
Proteins produced by the immune system that recognize substances as foreign to the body and remove them.

anticholinergic
Drugs or chemicals that inhibit nerves, using acetylcholine as a neurotransmitter, or block its effects.

antidote
A substance that specifically blocks the effect(s) of a poison, removes it, or stops it being activated.

antigen
A substance (for example, a protein) that is recognized by the body's immune system as foreign.

anti-inflammatory
A drug that reduces inflammation.

antioxidants
Substances that remove reactive chemicals such as reactive forms of oxygen or inhibit their ability to cause damage to cells.

antipyretic
A drug that reduces body temperature in fever.

antivenins (antivenom)
Specific antidotes against venoms produced by animals.

apoptosis
Programmed cell death; the removal of single damaged cells.

arrythmia(s)
Abnormal rhythm(s) of the heart.

arsenic
A chemical element found naturally in rocks and water in various forms, for example arsenite or arsenate. The word 'arsenic' is often used to mean arsenic trioxide.

arsenic trioxide
A form of arsenic commonly used in poisoning.

asbestos
A fibrous mineral found underground that exists in many forms, some of which can cause cancer, while others do not.

ascorbate
Vitamin C.

atom
The smallest unit of a chemical element that still maintains the characteristics. Groups of atoms form molecules.

azo dye
A group of synthetic chemicals that contain an azo group and are coloured.

benzene
An organic chemical used as a solvent in industry. It occurs in petrol.

beta carotene (carotene)
A substance related to vitamin A and found in plants like carrots.

bioaccumulation
The process of accumulation of a chemical in a living organism as a result of persistence and solubility in fat.

biodegradable
A chemical that can be broken down by living organisms or in the environment.

biomagnification
The process by which there is an increase in the concentration of a chemical in the species at higher levels in the food chain.

British Pharmacopoeia
A detailed reference book on drugs used by pharmacists.
cadmium
A toxic metal and chemical element found naturally in various forms in rocks. It is used in industry and in batteries.
capillaries
Very small blood vessels.
carboxyhaemoglobin
The product formed when carbon monoxide combines with haemoglobin in blood.
carcinogen
A chemical that causes cancer.
cell
The basic unit (building block) of animals and plants.
chloracne
A severe form of acne caused by chlorinated hydrocarbons such as dioxin.
chlorine
An element, a highly reactive, irritant gas.
chloroform
A volatile chlorinated solvent that may be produced by the breakdown of other chlorinated compounds. It was one of the first anaesthetics.
chronic (toxicity)
The toxic effect of repeated (chronic) dosing with or exposure to a chemical. It is usually a delayed effect.
cross-reactivity
Where hypersensitivity to one chemical (allergy) may lead to hypersensitivity to a another.
cyanide
A chemical grouping (CN) that is toxic. Cyanide may exist as an ion, as in potassium or sodium cyanide (KCN, NaCN), or attached to another molecule, as found in plants.
DDE
A commonly used abbreviation for dichloro-diphenyl-dichloroethylene, a breakdown product of DDT.
DDT
A commonly used abbreviation for dichloro-diphenyl-trichloroethane, an organochlorine insecticide.
detoxication
The breakdown of chemicals in the body (biotransformation) by enzymes. (Detoxification refers to the removal of a toxic chemical from the body.)
digitalis
A drug derived from the foxglove plant. It contains digoxin which causes the heart to beat more strongly.
dioxins
A group of related chemicals containing chlorine atoms. The word 'dioxin' is usually used with reference to TCDD (2,3,7,8-tetrachlorodibenzodioxin), the most toxic of the dioxins.

DNA
Deoxyribonucleic acid, a molecule (a double helix) that forms the basis of our chromosomes, the essential part of the heredity mechanism.

DNA adducts
Fragments of chemicals attached to DNA.

dopamine
A neurotransmitter chemical found in the brain.

double-blind (placebo-controlled) trial
A study to look at the effects of a chemical, such as a new drug, in which neither the patient nor research workers know whether the patient is taking the drug (or a placebo).

ecstasy (E)
Methylene dioxy-metamphetamine, abbreviated to MDMA. It is an amphetamine derivative.

emetic
A drug that induces vomiting which can be used for the treatment of poisoning.

endocrine disruptor
A (natural or synthetic) chemical that causes changes in endocrine (hormonal) function, leading to adverse effects in an animal or its offspring.

enzymes
Biological catalysts which consist of protein molecules.

epidemiology
The study of diseases in human populations, for example those caused by chemicals.

esters
Chemicals that often smell fruity (for example, like pear drops) and may be used as flavouring agents for food.

ethylene glycol
The major constituent of antifreeze. It is an alcohol.

ethynyloestradiol
Synthetic oestrogen, used in the contraceptive pill.

fatty acids
A type of natural organic acid found in fats.

free radicals
Products of chemical reactions that are normally very reactive and damaging if formed inside the body.

fungicide
A chemical substance (synthetic or natural) that destroys a fungus. It is a type of pesticide.

genotoxic
A chemical that is toxic to the genetic system, damaging DNA or chromosomes.

glutathione
An important natural substance, found especially in the liver, that has protective properties. It is an antioxidant.

gut
The stomach and small and large intestine.

half-life
The time taken for the concentration of a chemical in the body or environment to decrease by half.

hallucinogenic(ic)
A chemical that causes hallucinations, for example LSD.

hazard
Hazard is the *inherent* ability of a chemical to cause an adverse effect.

herbicide
A chemical (natural or synthetic) that destroys plants. It is a type of pesticide.

histology
The examination of sections of organs or body tissue under a microscope to look for damage or abnormality.

immune response
The response of the body's immune system to challenge by a foreign organism or chemical.

induce/induction (of enzymes)
The phenomenon where exposure of an animal to a chemical increases the amount of an enzyme, thereby increasing the potential for metabolism.

insecticide
A chemical (natural or synthetic) that destroys insects. It is a type of pesticide.

intravenous injection
Injection of a drug into a vein.

in vitro
Experiments using isolated parts of a living system such as cells or parts of cells.

ion
A positively or negatively charged atom or molecule.

isomers
Forms of a chemical that are identical in composition but with different arrangements in space.

jaundice
The accumulation of bile in the blood and tissues which makes the patient look yellow. It is due to liver failure.

lead
A toxic metal and one of the chemical elements, found naturally in rocks in various forms.

liver cirrhosis
Progressive and degenerative disease of the liver.

mercury
A toxic liquid metal and a chemical element, found naturally in rocks as the metal and in various other forms.

metal ions
Positively charged forms of metals, for example iron (Fe^{++}).

molecule
A group of atoms, the smallest unit of a chemical compound that retains its characteristics.

monoglycerides and diglycerides
Types of fats.

mucosa(e)
Soft, internal tissues, for example those lining the mouth.

mutation
A change in DNA or chromosome(s) that can be inherited.

mycotoxin
A toxin derived from a fungus.

necrosis
Death of an area of tissue that may be caused by chemical exposure. Necrosis can cause an organ to fail.

neurotoxin
A substance that is toxic to nerves.

nitrite
An inorganic chemical group (found as a salt, for example, sodium nitrite), used as a preservative and found naturally in water. It can be toxic at high levels.

no observed adverse effect level (NOAEL)
The threshold for a toxic effect. It is determined from the graph of the relationship between dose and effect.

oestrogen
Female hormone.

oestrogenic
Substances that act like the natural hormones oestrogens.

oncogenes
Cancer genes.

organochlorine compound
Organic chemical containing chlorine atoms. It tends to be persistent and not biodegradable.

oxidative stress
The production of reactive oxygen derivatives and other oxidizing chemicals in the body, which is potentially hazardous for cells and structures.

oxidized
The addition of oxygen to a molecule, the loss of hydrogen, or certain other types of chemical change.

oxyhaemoglobin
The form of haemoglobin that contains oxygen and transports it around the body.

pancreatitis
Inflammation of the pancreas, which can be fatal.

paraquat
Weedkiller. In the UK it is marketed as Weedol for home use and as Grammoxone for agricultural use.

parathion
An organophosphate insecticide.

PBBs and PCBs
Polybrominated and polychlorinated biphenyls. Chemicals that have various industrial uses, for example as fire retardants and electrical insulators. They are persistent if released into the environment.

peripheral neuropathy
Damage to peripheral nerves such as those in the arms and legs and leading to the fingers and toes.

pH
The level of acidity or alkalinity, pH 1 being very acid and pH 14 very alkaline.

phthalates
Plasticizers, that is, chemicals added to plastic to make them more flexible.

physico-chemical characteristics
Characteristics of chemicals, for example, their state (gas or liquid), solubility, acidity.

placebo
An inert substance used in a drug trial as a control.

polyneuritis
Irritation/inflammation of the nerves, leading to tingling in the hands and feet.

post-mortem
Literally, 'after death'. The term also refers to an examination of a body after death.

potentiation
The phenomenon whereby one chemical increases the toxic or other effects of another.

ppb
Parts per billion (see Units of Measurement).

ppm
Parts per million (see Units of Measurement).

promoter
A substance that does not initiate the process but is essential for the development of carcinogenesis.

prophylactic(ally)
Treatment with a drug to prevent a disease or ailment.

prostaglandins
Substances in our bodies that are mediators produced in response to damage to cells and which tell the body that there is damage. They are involved with pain and inflammation.

protein
One of the main building blocks of the body. Proteins are relatively large molecules.

pulmonary oedema
Accumulation of fluid in the lungs.

receptors
Structures (usually proteins) in the body, sometimes inside cells, that bind chemicals (such as drugs) and initiate an effect.

risk
The likelihood of a chemical causing an adverse effect. For a chemical, risk is defined as hazard × exposure.

RNA
A complex chemical structure (molecule) similar to DNA that is involved in the translation of information from DNA into protein.

Rodenticide
Pesticide which targets rodents such as rats and mice.

Salmonella
A group of bacteria some of which are responsible for diseases like food poisoning.

slow acetylators
Individuals who metabolize certain chemicals slowly.

solvent
Any liquid that dissolves solid substances, for example, water, alcohol, petrol, white spirit, and benzene.

sulphur
A chemical element that is important in living systems.

sulphydryl group
A chemical group containing sulphur, that is present in many biologically important molecules. The group can be oxidized and so damaged.

teratogen
A substance that causes birth defects (teratogenesis) if given to a pregnant animal.

testosterone
Male hormone.

thiols
Chemical groups containing sulphur (chemical notation SH) that are vital in proteins and structures. They are targets for toxic chemicals.

tolerable daily intake (TDI)
An estimate of the daily intake of a substance (a food contaminant) over a lifetime that is considered to be without appreciable health risk. It is calculated in the same way as the ADI.

tolerance
The phenomenon where repeated exposure to a chemical results in a reduction in its effect.

triglycerides
A type of fat.

tri-orthocresyl phosphate (TOCP)
Very toxic organophosphate that is used as a solvent.

urticaria
Skin rash that can be caused by an immune reaction.

vinyl chloride
The basic building block of pvc plastic. It is also known as vinyl chloride monomer.

w/v
Weight per volume (see Units of Measurement).
w/w
Weight per weight (see Units of Measurement).

Units of Measurement

Weight

ng	nanogram	1 hundred millionth of a gram (0.000000001 g or 1×10^{-9} g)
µg	microgram	1 millionth of a gram (0.000001 g or 1×10^{-6} g)
mg	milligram	1 thousandth of a gram (0.001 g or 1×10^{-3} g)
g	gram	1 thousandth of a kilogram (0.001 kg or 1×10^{-3} kg) (28 g = approx. 1 oz)
kg	kilogram	1 thousand grams (1,000 g or 1×10^{3} g) (1 kg = approx. 2.2 lb)

Volume

ml	millilitre	1 thousandth of a litre (0.001 l or 1×10^{-3} l)
l	litre	1 thousand millilitres (or 1,000 cubic centimetres) (1000 ml or 1×10^{3} ml). (1 l = approx. 1.75 imperial pints or 1.056 US quart)

Concentration

Concentration may be expressed as weight per volume or, alternatively, as ppm or ppb.

mg/l	1 mg dissolved in 1 l of water or other solvent
µg/ml	1 µg dissolved in 1 ml of water or other solvent
ppm	parts per million, e.g. 1 mg/kg (1 mg/1,000,000 mg) 1 mg/l or 1 µg/ml = 1 part per million.
ppb	parts per billion, e.g. 1 µg/kg (1 µg/1,000,000 mg) 1 ng/g or 1 µg/l = 1 part per billion
w/v	weight per volume, e.g. 20 g/100 ml = 20 % w/v
w/w	weight per weight, e.g. 20 g/100 g = 20 % w/w

Length

µm	micron	1 millionth of a metre (0.000001 m or 1×10^{-6} m)

INDEX